Lecture Notes in Electrical Engineering

Volume 609

The book series *Lecture Notes in Electrical Engineering* (LNEE) publishes the latest developments in Electrical Engineering - quickly, informally and in high quality. While original research reported in proceedings and monographs has traditionally formed the core of LNEE, we also encourage authors to submit books devoted to supporting student education and professional training in the various fields and applications areas of electrical engineering. The series cover classical and emerging topics concerning:

- Communication Engineering, Information Theory and Networks
- Electronics Engineering and Microelectronics
- Signal, Image and Speech Processing
- Wireless and Mobile Communication
- Circuits and Systems
- Energy Systems, Power Electronics and Electrical Machines
- Electro-optical Engineering
- Instrumentation Engineering
- Avionics Engineering
- Control Systems
- Internet-of-Things and Cybersecurity
- Biomedical Devices, MEMS and NEMS

For general information about this book series, comments or suggestions, please contact leontina.dicecco@springer.com.

To submit a proposal or request further information, please contact the Publishing Editor in your country:

China

Jasmine Dou, Associate Editor (jasmine.dou@springer.com)

India

Aninda Bose, Senior Editor (aninda.bose@springer.com)

Japan

Takeyuki Yonezawa, Editorial Director (takeyuki.yonezawa@springer.com)

South Korea

Smith (Ahram) Chae, Editor (smith.chae@springer.com)

Southeast Asia

Ramesh Nath Premnath, Editor (ramesh.premnath@springer.com)

USA, Canada:

Michael Luby, Senior Editor (michael.luby@springer.com)

All other Countries:

Leontina Di Cecco, Senior Editor (leontina.dicecco@springer.com)

**** Indexing: The books of this series are submitted to ISI Proceedings, EI-Compendex, SCOPUS, MetaPress, Web of Science and Springerlink ****

More information about this series at http://www.springer.com/series/7818

S. N. Singh · R. K. Pandey ·
Bijaya Ketan Panigrahi ·
D. P. Kothari
Editors

Advances in Power and Control Engineering

Proceedings of GUCON 2019

Springer

Editors
S. N. Singh
Madan Mohan Malaviya University
of Technology
Gorakhpur, Uttar Pradesh, India

Bijaya Ketan Panigrahi
Department of Electrical Engineering
Indian Institute of Technology Delhi
New Delhi, Delhi, India

R. K. Pandey
National Power Training Institute
Faridabad, Haryana, India

D. P. Kothari
Indian Institute of Technology Delhi
New Delhi, Delhi, India

ISSN 1876-1100 ISSN 1876-1119 (electronic)
Lecture Notes in Electrical Engineering
ISBN 978-981-15-0312-2 ISBN 978-981-15-0313-9 (eBook)
https://doi.org/10.1007/978-981-15-0313-9

This Springer imprint is published by the registered company Springer Nature Singapore Pte Ltd.
The registered company address is: 152 Beach Road, #21-01/04 Gateway East, Singapore 189721, Singapore

GUCON 2019 Organization

Chief Patrons

Mr. Suneel Galgotia, Chancellor, Galgotias University, India
Mr. Dhruv Galgotia, CEO, Galgotias University, India

Patron

Prof. Renu Luthra, Vice-Chancellor, Galgotias University, India

General Chairs

Mr. Deepak Mathur, Director-Elect, IEEE Region 10 (Asia and Pacific)

Conference Chair and Chairman, Oversight Committee

Prof. Rabindra Nath Shaw, Galgotias University, India

Conference Secretary

Prof. Priyabrata Adhikary, NHEC, India
Prof. Saravanan D., Galgotias University, India

Technical Chairs

Prof. Yen-Wei Chen, Ritsumeikan University, Japan
Prof. Maria Virvou, HOD, Department of Informatics, University of Piraeus, Greece

Publication Chairs

Prof. George A. Tsihrintzis, University of Piraeus, Greece
Prof. Valentina E. Balas, University of Arad, Romania
Prof. Dilip Kumar Sharma, GLA University, Mathura

Honorary Chairs

Prof. Vincenzo Piuri, University of Milan, Italy
Prof. Georges Zissis, President, IEEE IAS
Prof. Lakhmi C. Jain, University of Technology Sydney
Dr. Tamas Ruzsanyi, Ganz-Skoda Electric Ltd., Hungary

Honorary Co-chairs

Prof. C. Boccaletti, Sapienza University of Rome, Italy
Prof. Mukhopadhyay, Ex-VC, Lingaya's University, India
Dr. Nishad Mendis, Det Norske Veritas, Australia
Dr. Akshay Kumar, Concordia University, Canada

Springer/GUCON Liaison

Dr. Aninda Bose, Senior Editor, Springer Nature

International Advisory Board

Prof. Valentina Balas, University of Arad, Romania
Prof. N. R. Pal, President, IEEE CIS
Prof. George T., University of Piraeus, Greece
Prof. Yen-Wei Chen, Ritsumeikan University, Japan
Prof. Milan Simic, RMIT University, Australia

Prof. M. Paprzycki, Polish Academy of Sciences
Prof. Maria Virvou, University of Piraeus, Greece
Prof. Vincenzo Piuri, University of Milan, Italy
Prof. D. P. Kothari, Ex-Director, IIT Delhi, India
Prof. S. N. Singh, VC, MMMUT, Gorakhpur, India
Prof. Bijaya Ketan Panigrahi, IIT Delhi, India
Prof. R. K. Pandey, DG NPTI, India

Technical Program Committee and Reviewers

Dr. A. R. Abhyankar
Dr. Aditi Sharan
Dr. Ajay Mittal
Dr. Sudhir Kumar Sharma
Dr. Ajai Jain
Dr. Alok Kushwaha
Dr. Amit Agarwal
Dr. Amalendu Patnaik
Dr. Anil K. Ahlawat
Dr. Anil K. Singh
Dr. Anuradha
Dr. Arun Kumar Verma
Dr. Aseem Chandel
Dr. Asheesh K. Singh
Dr. Ashutosh Dixit
Dr. Asif Ekbal
Dr. B. Dushmanta Kumar Patro
Dr. Baij Nath Kaushik
Dr. Bhaskar Biswas
Dr. Bharat Singh Rajpurohit
Dr. C. Patvardhan
Dr. C. Rama Krishna
Dr. C. K. Nagpal
Dr. Chandra Sekaran
Dr. Chiranjeev Kumar
Dr. Chittaranjan Hota
Dr. D. Bhagwan Das
Dr. D. A. Mehta
Dr. D. S. Kushwaha
Dr. D. S. Yadav
Dr. Desh Deepak Sharma
Dr. Dhram Singh

Dr. Dimple J. Gupta
Dr. Diwakar Bhardwaj
Dr. Girish Patnaik
Dr. Jai Govind Singh
Dr. Joy Deep Mitra
Dr. K. V. Arya
Dr. Kiran Kumar Pattanaik
Dr. Kishor K. Bhoyar
Dr. Komal Kumar Bhatia
Dr. Lalit Kumar Awasthi
Dr. M. K. Dutta
Dr. M. P. Singh
Dr. Madhavi Sinha
Dr. Manisha Sharma
Dr. Mohd. Rihan
Dr. Mayank Pandey
Dr. Munesh C. Trivedi
Dr. N. Badal, Knit
Dr. Nanhay Singh
Dr. Narendra Kohli
Dr. Naresh Chauhan
Dr. Naveen Kumar
Dr. Neelam Duhan
Dr. Neeraj Tyagi
Dr. O. P. Verma
Dr. Pooja Jain
Dr. Pooja Pathak
Dr. Prabhat Ranjan
Dr. Prabhakar Tiwari
Dr. Prabin Panigrahi
Dr. Pragya Dwivedi
Dr. Pradeep Sharma
Dr. Pramod Kumar
Dr. Pramod Kumar Singh
Dr. Punam Bedi
Dr. R. K. Singh
Dr. R. S. Yadav
Dr. R. S. Rao
Dr. Rahul Rishi
Dr. Rajesh Prasad
Dr. Reena Dadhich
Dr. Ruchika Malhotra
Dr. S. P. Tripathi
Dr. Sapna Gambhir
Dr. Suneeta Agarwal

Dr. Sujoy Das
Dr. Sukomal Pal
Dr. Sunil Kumar Khatri
Dr. Tanveer Siddiqui
Dr. Tarun Shrimali
Dr. Vasudha Bhatnagar
Dr. Vishal Bhatnagar
Dr. Yashpal Singh
Prof. Herbert H. C. Lu
Dr. Senthilrajan Agni
Dr. Abhineet Anand
Dr. Anurag Baghel
Dr. Balamurugan Balusamy
Dr. Priti Bansal
Dr. Sonia Bansal
Dr. Annappa Basava
Dr. Rohit Beniwal
Dr. Vandana Bhasin
Dr. Rodrigo Bortoletto
Dr. John Moses Cyril
Dr. Pinaki Chakraborty
Dr. Sansar Chauhan
Dr. Rahul Chaurasiya
Dr. Surya Deo Choudhary
Dr. Anurag Dixit
Dr. Nripendra Narayan Das
Dr. Indrani Das
Dr. Aparna Datt
Dr. Parneeta Dhaliwal
Dr. Chandrakant Divate
Dr. Rajesh Dubey
Dr. Arman Faridi
Dr. Ankush Ghosh
Dr. Utkarsh Goel
Dr. Pallavi Goel
Dr. Amit Goel
Dr. Priyanka Goyal
Dr. Deepak Gupta
Dr. Suneet Gupta
Dr. Raza Haidri
Dr. Syed Shabih Hasan
Dr. Manas Hati
Dr. Brijesh Iyer
Dr. Manisha Jailia
Dr. Prashant Johri

Dr. Jegathesh Amalraj Joseph
Dr. Sandeep K. Singh
Dr. Vinay Kumar
Dr. Amita Kapoor
Dr. Sandhya Katiyar
Dr. Anvesha Katti
Dr. Ruqaiya Khanam
Dr. Aanchal Khatri
Dr. Shrawan Kumar
Dr. Devendra Kumar
Dr. Avneesh Kumar
Dr. Arun Kumar
Dr. Sanjeev Kumar
Dr. Vipin Kumar
Dr. Sanjay Kumar
Dr. Bhavnesh Kumar
Dr. Sandeep Kumar
Dr. Neetesh Kumar
Dr. Mohanraj M.
Dr. Ramakrishnan Malaichamy
Dr. Manas Kumar Mishra
Dr. Baibaswata Mohapatra
Dr. Thiyagarajan Muthunatesan
Dr. Rashid Mahmood
Dr. Yogendra Meena
Dr. Gitanjali Mehta
Dr. A. K. Mishra
Dr. Keshav Niranjan
Dr. Manoj Panda
Dr. Sanjeev Pippal
Dr. V. A. Sankar Ponnapalli
Dr. Shiv Prakash
Dr. Sheetla Prasad
Dr. Mohammed Abdul Qadeer
Dr. R. Gunasundari Ranganathan
Dr. Ranjeet Kumar Ranjan
Dr. Rohit Raja
Dr. Bharti Rana
Dr. Mukesh Rawat
Dr. Navaid Zafar Rizvi
Dr. Pravinth Raja S.
Dr. Anil Kumar Sagar
Dr. Rajeev Sharma
Dr. Birendra Kumar Sharma
Dr. Shrddha Sagar

Dr. Jyoti Sahni
Dr. Mohd. Saifuzzaman
Dr. Kavita Saini
Dr. Kamalesh Sethuramalingam
Dr. Priestly Shan
Dr. Gavaskar Shanmugam
Dr. Dilip Kumar Sharma
Dr. R. P. Sharma
Dr. Mayank Sharma
Dr. Sudhir Sharma
Dr. Lokesh Kumar Sharma
Dr. Vishnusharma
Dr. Jitendra Singh
Dr. Girish Singh
Dr. Karan Singh
Dr. Harikesh Singh
Dr. Prashant Singh
Dr. Neetay Singh
Dr. Ajay Shanker Singh
Dr. Arun Solanki
Dr. Subhranil Som
Dr. Ritesh Srivastava
Dr. Vijayalakshmi Subramanian
Dr. Hardeo Kumar Thakur
Dr. Pradeep Tomar
Dr. Shashi Kant Verma
Dr. Sohan Kumar Yadav
Dr. Vinod Yadav
Dr. Dileep Yadav
Dr. Chandra Yadav
Dr. Emre Yay
Dr. Aasim Zafar
Dr. Usha Chauhan
Dr. Chetna Dabas
Dr. Sanjoy Das
Dr. Sumithra Gavaskar
Dr. Vimal Kumar

Preface

The book constitutes selected high-quality papers presented in the International Conference on Computing, Power, and Communication Technologies 2019 (GUCON 2019) organized by Galgotias University, India, in September 2019. It discusses the issues in electrical, computer, and electronics engineering and technologies. The selected papers are organized into three sections—power and energy, control system and power electronics, and drives and renewable energy. In-depth discussions on various issues under topics provide an interesting compilation for researchers, engineers, and students.

We are thankful to all the authors who have submitted papers for keeping the quality of the GUCON 2019 at high levels. We would like to acknowledge all the authors for their contributions and the reviewers. We have received invaluable help from the members of the International Program Committee and the chairs responsible for different aspects of the workshop. We also appreciate the role of special session organizers. Thanks to all of them, we had been able to collect many papers on interesting topics, and during the conference, we had very interesting presentations and stimulating discussions.

Our special thanks go to Leopoldo Angrisani (Editor in Chief, Springer, Lecture Notes in Electrical Engineering Series) for the opportunity to organize this guest-edited volume.

We are grateful to Springer, especially to Aninda Bose (Senior Editor, Hard Sciences Publishing), for the excellent collaboration, patience, and help during the evolvement of this volume.

We hope that the volume will provide useful information to professors, researchers, and graduated students in the areas of power, control, and renewable energy and applications, and all will find this collection of papers inspiring, informative, and useful. We also hope to see you at a future GUCON event.

Gorakhpur, India — S. N. Singh
Faridabad, India — R. K. Pandey
New Delhi, India — Bijaya Ketan Panigrahi
New Delhi, India — D. P. Kothari

Contents

About the Editors

S. N. Singh is currently the Vice-Chancellor of Madan Mohan Malaviya University of Technology, Gorakhpur, India, and a Professor at the Department of Electrical Engineering, Indian Institute of Technology Kanpur, India. He is also the recipient of the German Humboldt Fellowship and the Danish Otto-Monsted Fellowship. In 2013, Professor Singh became the first Asian to receive the IEEE Educational Activity Board Meritorious Achievement Award in Continuing Education. His research interests include power system restructuring, FACTS, power system optimization and control, wind power, and security analysis. Professor Singh has published more than 440 papers in international/national journals/conferences, and written two books.

R. K. Pandey is a Professor at the Department of Electrical Engineering, Indian Institute of Technology (BHU), Varanasi, and is the Director General, National Power Training Institute (NPTI). He has over 36 years' experience in the fields of high-voltage direct current (HVDC) transmission technology and flexible AC transmission systems (FACTS) device control, intelligent power control along with the operation of power systems in open access. He has published more than 150 peer-reviewed papers in both national and international journals and at leading conferences.

Bijaya Ketan Panigrahi is a Professor of Electrical and Electronics at the Engineering Department, Indian Institute of Technology, Delhi, India. He received his Ph.D. degree from Sambalpur University. He is Chief Editor of the International Journal of Power and Energy Conversion. His interests include power quality, FACTS devices, power system protection, and AI applications in power systems.

D. P. Kothari is an educationist and a Professor who has held leadership positions at various engineering institutions in India, including IIT Delhi, Visvesvaraya National Institute of Technology, Nagpur, and VIT University, Vellore.

In recognition of his contributions to engineering education, he was made an IEEE Fellow. He has published and presented over 842 papers in national and international journals and conferences. He has authored and co-authored about fifty books. He is the recipient of numerous lifetime achievement and other awards.

An Investigation on Multi-junction Solar Cell for Maximum Power Point Tracking Using P&O and ANN Techniques

Prachi Rani and Omveer Singh

Abstract Renewable energy resources are becoming a vital part for the energy production in today's world. Generation of energy by using solar equipment is increasing day by day. A multi-junction solar cell is the one which consists of multiple junctions taking into consideration the tunnel junction in the solar cell so that efficiency of the solar cell is improved. Temperature and solar irradiance are consequential factors in order to determine the ability of the solar cell. A noticeable change in any one factor shows a clear change in the values in the Voltage-Current (V-I) and Power-Voltage (P-V) curves of the solar cell. The conversion efficiency can be improved by some MPPT techniques that are applied on the solar cell. Perturb & observe (P&O) and Artificial Neural Network (ANN) techniques are implemented and compared here to attain a better technique for the extraction of solar energy. The plots and graphs here depict the miscellaneous characteristics of the multi-junction solar cell including effect of strategies applied on it in MATLAB/Simulink.

Keywords Multi-junction · Tandem cell · Perturb & observe · Artificial neural network · MATLAB/Simulink

1 Introduction

A source of energy can continue only until it is depleted. Well due to this fact, the conventional sources that are depleting will no longer be the source of generating energy. This fatal problem is affecting almost every section contributing in the economy of any country. Existence of renewable energy resources in ample amount is a boon for generating unlimited energy. The trail of these renewable energy resources includes many resources yet much has been said and done on solar energy [1].

For bringing forth solar energy into play, solar cells are required which later became a vital topic of research. These researches lead us to multiple types of solar

P. Rani (✉) · O. Singh
Department of Electrical Engineering, Gautam Buddha University, Greater Noida 201312, India
e-mail: 05prachi05@gmail.com

O. Singh
e-mail: omveers@gmail.com

S. N. Singh et al. (eds.), *Advances in Power and Control Engineering*, Lecture Notes in Electrical Engineering 609, https://doi.org/10.1007/978-981-15-0313-9_1

cells, but the only motto was to maximize the efficiency of the solar cell. Multi-junction solar cell (MJSC) is one of these solar cells that have high conversion efficiency as compared to some of the solar cells [2]. These types of Photovoltaic (PV) cells have space and terrestrial applications in which many researches are done every now and then. Tandem solar cells have been studied since 1960. Earlier days, researchers encouraged R&D of tandem cells based on their computer analysis [3].

MJSCs consist of different semiconductor material that forms diversified p–n junction integrated to cell stack which is connected together with the help of tunnel junctions and window layer. The conversion efficiency of the whole solar cell is increased as each semiconductor material absorbs wide range of wavelength of light which extracts more sunlight. At present, InGaP/GaAs/Ge solar cell which is a MJSC efficiently absorbs the broad spectrum of sunlight and is used in space [3–8]. MJSCs are the solar cells in which maximum power point is greater than the single-junction silicon solar cell.

Higher resistance towards irradiance, increased conversion efficiency and lower temperature coefficient are some of the properties of MJSC. The efficiency of multi-junction solar cell is far more than that of silicon solar cell. These solar cells can be used in several other industrial, domestic and agriculture sectors. Implementation of solar cells in these section leads to more economic aspect. The operation of these devices is more effective as there is luminescent coupling that occurs in the multi-junction devices [9, 10].

As there are different semiconductor materials, the bandgap energy of each sub-cell is different depending upon the material. The top layer of the MJSC uses the largest band gap, and the lower layers reduce the band gap [5]. The bandgap energy of every material and the open-circuit voltage vary with respect to temperature. These solar cells convert sunlight to DC electricity which is then connected to inverter that converts into AC source through which it can be used in further applications in day-to-day life. Such connections form a PV system that can be certainly used in household chores as well as in street lights.

In this study, triple-junction solar cell is implemented, and then, its V–I and P–V characteristics are studied at standard as well as different radiations. Afterwards, P&O and ANN based MPPT techniques are exercised on the MJSC. This helps in enhancing the conversion efficiency at economical level.

2 Methodologies

When the laboratory experimental procedures were done, some challenges were faced in combining semiconducting material for the stack. These combinations are based on some basic criteria like lattice matching, bandgap energy matching and current matching.

Lattice constant of the material of the solar cell should match because even a small mismatch in lattice will lead to lattice dislocation. The bandgap energy matching deals with the fact that the energy levels should be in such a limit that absorption of the spectrum of sunlight is done in an efficient way. Lastly, current matching is the most important factor as the materials are stacked in a series form so it is required that the current of each sub-cell is equal. These criterions lead to higher conversion efficiency and also coherent application of the multi-junction solar cell. Keeping all the factors in mind, a triple-junction solar cell consisting of a GaInP solar cell as the top layer, InGaAs in the middle layer and Ge layer as the bottom layer [6], is formed and implemented.

Standard Test Condition (STC), i.e. 1000 W/m^2 irradiance and 25 °C temperature, is the condition under which the simulation is performed.

2.1 Perturb & Observe Technique

Out of all the techniques, P&O is the most primary and prominent technique. The algorithm of P&O method is to measure the instantaneous current and instantaneous voltage of the output power. Then, a comparison of output power in current time point is made to previous time point's power output [3].

In this approach, yield voltage of sun-based board is perturbed and yield control is always recognized. The sun-powered board current and voltage are evaluated, and therefore, similar power is figured. Essential advance size is given by the external circuit. Perturbation in voltage is resolved, and many experiments are compared and observed as per control scheme. In case that the negative change is recognized concerning the change in voltage, the duty ratio is diminished; it is pretended by the nonlinear P–V characteristics that MPP is navigated and following inverse way begins. Just as the fixed step is given by the outer circuit, it significantly influences the rate of convergence and ripple content; however, the tracking speed enormously diminishes.

2.2 Artificial Neural Network

ANN is one of the best-utilized strategies to accomplish MPPT for PV frameworks of nonlinear ecological circumstances. It is a soft computing technique which helps in improving the conversion efficiency of the given MJSC. ANN technique works similar to that of a human brain.

This technique is used for exploring the MPP of the solar cell by the help of feed-forward network. These networks are characterized by acyclic graph which describes the topological structure of network [10]. Operating temperature and irradiance are served as inputs to the feed-forward network including the weights and the target output, and the network is processed just the same way as brain of human functions to obtain the most efficient output. This output is then served to pulse width modulation sub-system which produces the pulse passed on to the DC converter. The final outcomes of the whole set-up lead to more accurate result.

3 Mathematical Structure of the System

The basic model of MJSC in the software MATLAB/Simulink is implemented when the basic relation between current density and voltage is obtained. Therefore, the current density of a single cell of the MJSC can be represented by the following equation:

$$J = J_{\text{ph}} - J_{\text{diode}} - J_{\text{pr}} \tag{1}$$

where J_{ph} is the photo current density which is equal to the short-circuit current density at reference temperature and irradiance. J_{diode} is the diode current density, and J_{pr} is the shunt current density.

The diode current density is formulated as follows:

$$J_{\text{diode}} = J_{0i}\left(e^{\frac{q(V+J\times A\times R_S)}{n_i\times kB\times T}} - 1\right) \tag{2}$$

where J_{0i} is the diode's reverse saturation current density, q is charge constant, A is the cell area, n_i is a constant, kB is Boltzmann constant and T is the operating temperature.

J_{0i} which characterizes diode saturation current density is given by:

$$J_{0i} = k_i \times T^{\frac{3+\Delta i}{2}} \times e^{\frac{-E_g}{(n_i\times kB\times T)}} \tag{3}$$

where k is a constant, J_{pr} is the shunt current density which is conventionally neglected as the shunt resistance is very large. E_g is the bandgap energy in eV whose value depends on operating temperature since its energy is inversely proportional to operating temperature [8].

Open-circuit voltage per cell of the multi-junction solar cell is given by the following equation:

$$V_{\text{oc}} = \frac{n_i \times kB \times T}{q}\ln\left(\frac{J_{\text{sci}}}{J_{0i}} + 1\right) \tag{4}$$

The sum of all three open-circuit voltages is the total open-circuit voltage of the multi-junction solar cell.

Figure 1 shows the equivalent circuit of multi-junction solar cell which is further implemented in the MATLAB/Simulink software for study and analysis keeping the physical conditions and limitations in mind.

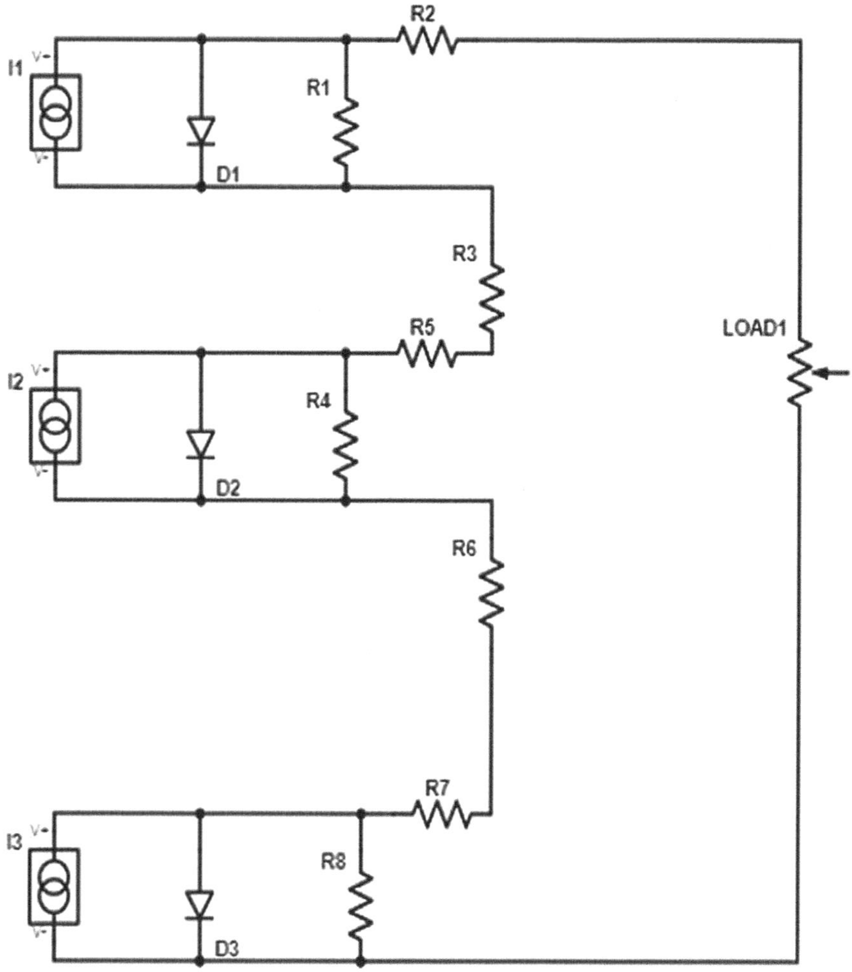

Fig. 1 Equivalent circuit diagram of MJSC

4 Way of Implementation

A MATLAB/Simulink of multi-junction solar cell is created on the basis of above theories and equations. Triple-junction solar cell is simulated, and then, various results are examined. The block diagram of single cell of MJSC is drawn in Fig. 2.

Current density and voltage and also power and voltage characteristics can be easily obtained by the simulation model of the solar cell. These curves give the value of short-circuit current, open-circuit voltage, maximum voltage and maximum power. The simulated results show how each stacked cell can perform for PV energy generation. Note that sampling curves were created at 1-sun radiation and fixed

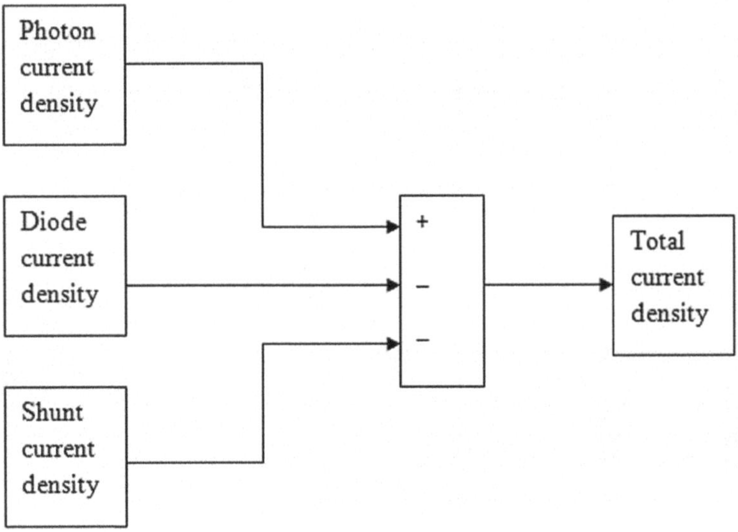

Fig. 2 Block diagram of single cell of MJSC

ambient temperature same as reference [7]. Implemented model of MJSC is presented in the Fig. 3.

Perturb & observe is a consistent procedure; it is recasting intermittently up until the MPP occurs. The Simulink model of the system is given as in Fig. 4. The simulation model of MPPT controller, i.e. a sub-system for fundamental simulation defined in the light of the ANN, appears in the given Fig. 5.

5 Results and Analysis

Figure 6 gives the V–I curve of solar cell and its sub-cell at 1000 W/m^2 radiation and temperature of 25 °C. Due to the values of the parameters in the above formulae, the following graphs are attained with Ge having the lowest values then InGaAs and after that InGaP. The total of all the sub-cells is equal to MJSC graph.

Following graphs depict the various characteristics of triple-junction solar cell.

Figure 7 shows the V–I curve of multi-junction solar cell at different radiations. The following graph depicts that as the radiation decreases, similarly the value of current density in the graph decreases. The values of the irradiance vary from 300 to 1000 W/m^2.

In the Fig. 8, plot gives the P–V curve of solar cell at 1000 W/m^2 radiations and temperature of 25 °C. The peak of this curve gives the maximum power point at which the voltage attained is maximum voltage.

Figure 9 shows the P–V curve of multi-junction solar cell at different radiations.

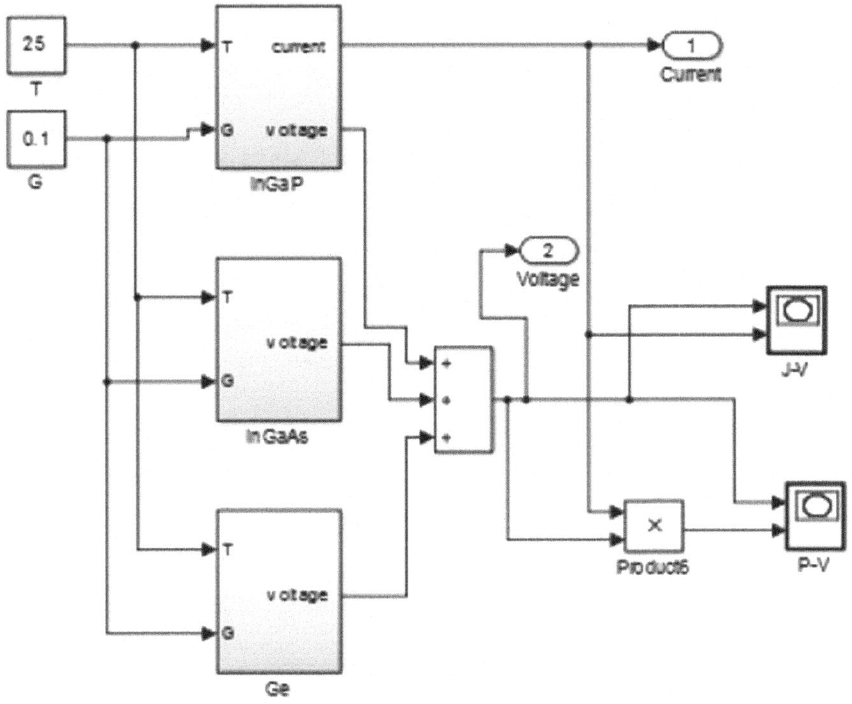

Fig. 3 Implemented model of MJSC

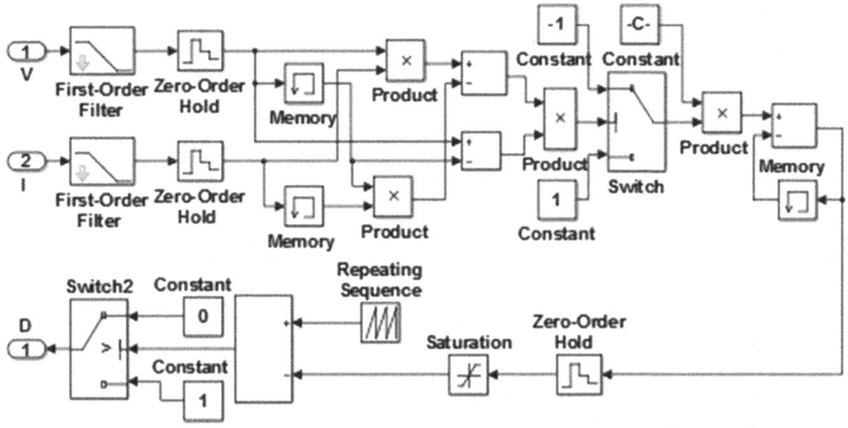

Fig. 4 MATLAB/Simulink model using P&O technique

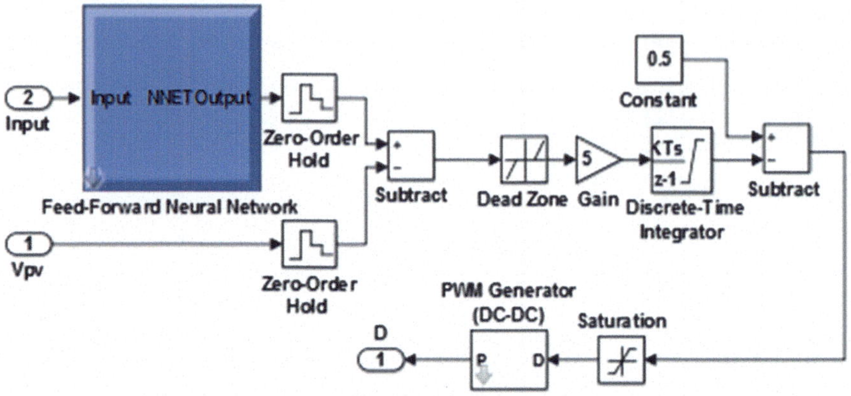

Fig. 5 MATLAB/Simulink model using ANN technique

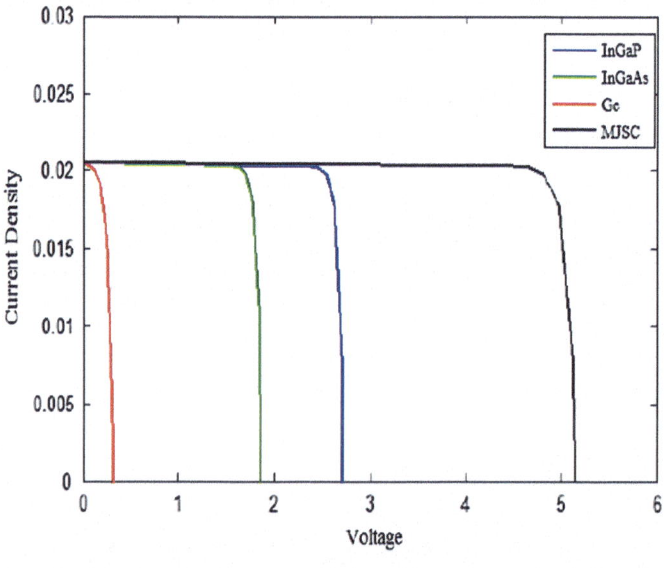

Fig. 6 V–I characteristics of MJSC and its sub-cell

It can be inferred from the given below graph that as the radiation changes the P–V graph also changes accordingly.

The response of output power after implementing P&O strategy is shown in Fig. 10. It is noticed that P&O is an effective control method to overcome the non-linear characteristics of PV cell and improve the efficiency of PV power generation system but still causes some oscillations and the output produced by the system.

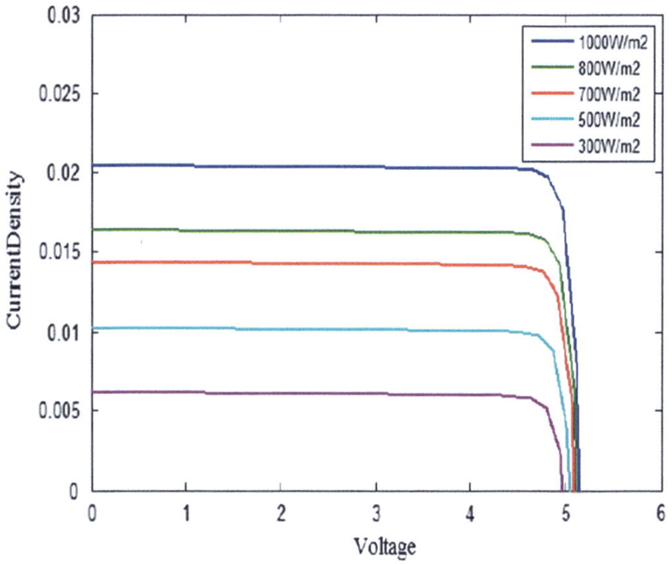

Fig. 7 V–I characteristics of MJSC at different radiations

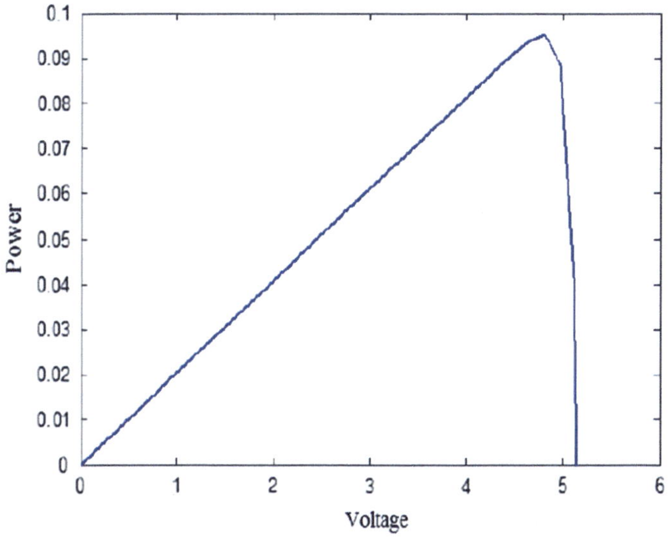

Fig. 8 P–V characteristics of MJSC

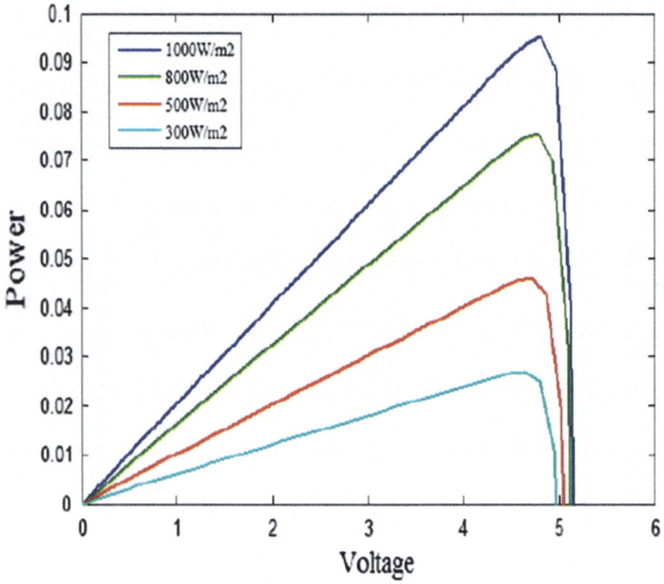

Fig. 9 P–V characteristics of MJSC at different radiations

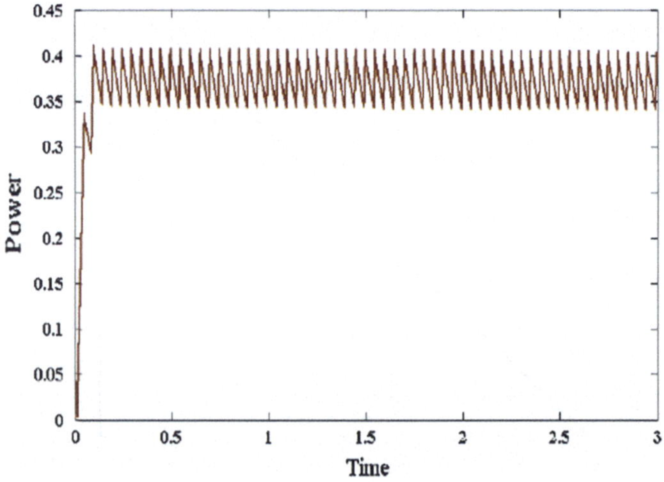

Fig. 10 Power vs Time graph after implementing P&O MPPT technique

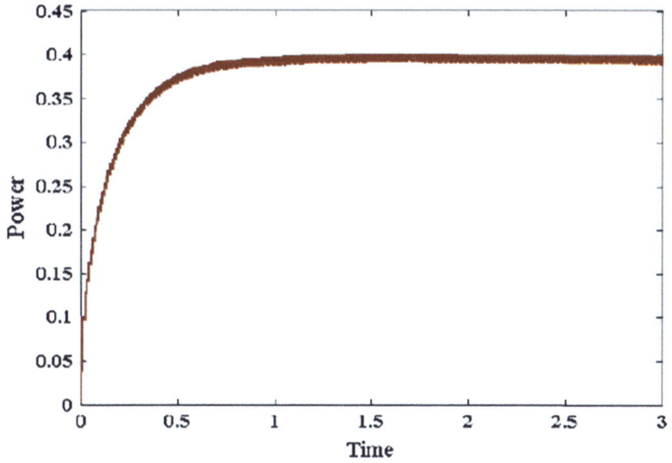

Fig. 11 Power vs Time graph after implementing ANN MPPT technique

The response of output power after implementing ANN is shown in Fig. 11. Through simulation results, it is observed that the system completes the MPPT successfully, and it has fewer fluctuations as compared to the other strategy. The plot given in Fig. 12 depicts the comparison of the two MPPT techniques, i.e. P&O and ANN. This clearly shows that ANN technique is the most appropriate technique as it has less oscillation and is more efficient than the P&O technique. ANN is faster, accurate, efficient and accessible as compared to the conventional strategies.

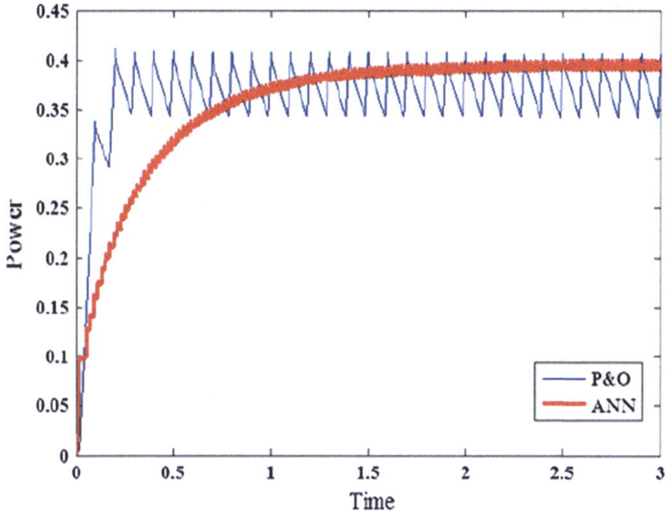

Fig. 12 Power vs Time graph depicting comparison of P&O and ANN techniques

6 Conclusion

Stacking of InGaP as top layer, InGaAs as middle layer and Ge as bottom layer demonstrate the most optimal performance as a MJSC. The overall voltage of this solar cell depends on all the three sub-cells, whereas the current is dependent on the minimized current in all the three sub-cells. Further, the efficiency of solar cell can be improved by implementing various MPPT techniques. These techniques are useful in improving conversion efficiency. P&O is the most preliminary and conventional technique with more oscillations, whereas the ANN technique is a soft computing technique with less oscillations and is more efficient.

References

1. Gupta S, Singh O, Urooj S (2017) A review on single and multi-junction solar cell with MPPT techniques. In: 3rd IEEE international conference on nanotechnology for instrumentation and measurement, GBU, India, 16–17 Nov 2017
2. Luceno-Sanchez JA, Diez-Pascual AM, Capilla RP (2019) Materials for photovoltaics: state of art and recent developments. Int J Mol Sci 2(976)
3. Yamaguchi M, Takamoto T, Araki K, Daukes NE (2005) Multi-junction III–V solar cells: current status and future potential. Solar Energy 79:78–85
4. Yamaguchi M, Nishimura K, Sasaki T, Suzuki H, Arafune K, Kojima N, Ohsita Y, Okada Y, Yamamoto A, Takamoto T, Araki K (2008) Novel materials for high-efficiency III-V multi-junction solar cells. Sol Energy 82:173–180
5. Wang Z, Das N, Helwig A, Ahfock T (2017) Modeling of multi-junction solar cells for maximum power point tracking to improve the conversion efficiency. In: Australasian universities power engineering conference (AUPEC)
6. Babar M, Rizvi AA, Al-Ammar EA, Malik NH (2014) Analytical model of multi-junction solar cell. Arab J Sci Eng 39:547–555
7. Das N, Wongsodihardjo H, Islam S (2015) Modeling of multi-junction photovoltaic cell using MATLAB/Simulink to improve the conversion efficiency. Renew Energy 74:917–924
8. Hussain AB, Abdalla AS, SidiMukhtar A, Elamin M, Alammari R, Iqbal A (2016) Modeling and simulation of single and triple junction solar cell using MATLAB/SIMULINK. Int J Ambient Energy
9. Hinzer K, Beattie MN, Wilkins MM, Valdivia CE (2018) Modeling techniques for multi-junction solar cells. In: International conference on numerical simulation of optoelectronic devices (NUSOD)
10. Galphade R (2017) Electrical characterization of a photovoltaic module through artificial neural network: a review. Int J Electr Compon Energy Convers 3:14–20

Generalized Decoupling and Robust Control of Unstable MIMO System

Sumit Kumar Pandey, Jayati Dey and Subrata Banerjee

Abstract This paper presents the decoupling and control of the unstable MIMO systems. To design the decoupler for an unstable MIMO system, a state feedback method is adopted to stabilize the unstable MIMO system by placing the poles at the left half of s plane. After that, a decoupler is designed using generalized decoupling method for the stabilized transfer matrix of the MIMO systems. At last, the PID controller is designed for the decoupled MIMO system using Kharitonov's theorem. A numerical example is considered to test the proposed method in MATLAB/Simulink. The simulation results obtained are found to be satisfactory for the proposed method.

Keywords Decoupling · MIMO · Unstable · PID

1 Introduction

The importance of decoupling for multi-inputs multi-outputs (MIMO) systems gained significant attention in recent years. Decoupled systems are defined as a class of MIMO systems in which each output is affected by one input only. It simplifies the implementation of a controller to control the system with respect to the coupled systems. The researchers have shown significant interest to control the unstable MIMO systems in recent years [1–3]. In [1], the authors develop a method to control the unstable nonlinear MIMO systems by MIMO fractional order (FO) controllers. The authors propose a linear quadratic 2-DOF controller structure to control the unstable MIMO system in [2]. A graphical user interface is also developed by the authors to access important data quickly. A multi-model adaptive control technique is developed and used in delay compensation schemes MIMO plants in [3]. The

S. K. Pandey (✉) · J. Dey · S. Banerjee
Department of Electrical Engineering, National Institute of Technology, Durgapur 713209, India
e-mail: skpdmk@gmail.com

J. Dey
e-mail: deybiswasjayati@gmail.com

S. Banerjee
e-mail: bansub2004@yahoo.com

S. N. Singh et al. (eds.), *Advances in Power and Control Engineering*, Lecture Notes in Electrical Engineering 609, https://doi.org/10.1007/978-981-15-0313-9_2

delay problem is analyzed and solved as an identification problem. The importance of the developed method is verified on a second-order delayed unstable process. The authors designed the controller to control the velocity, heave, and attitude by designing the yaw controller for an autonomous model chosen as helicopter [4]. A state space is adopted to stabilize the helicopter by using pole placement design. The state feedback approach is mainly adopted to stabilize the unstable MIMO systems for the design of the decoupling control scheme [5–8]. In [6], a special state variable feedback method is introduced to provide an easier method for decoupling a MIMO system. The developed method is verified on a model of a physical system. A generalized decoupling method is proposed by the authors [9] to decouple the MIMO system. It is observed during the literature review that the Gershgorin bands are used as a powerful tool for decoupling of the MIMO systems [10, 11]. The present work also aimed to design the robust PID controller for the decoupled system due to its simple structure and easy implementation. The Kharitonov theorem played an important role to obtain the robust range of the PID controller gains [12]. The presented work is said to be unique because the proposed approach not only decouples and controls the MIMO stable system but implemented to the MIMO unstable system. The rest of the paper is categorized as below. Next section describes the decoupling scheme of the unstable MIMO system followed by the robust PID controller design technique. A numerical example is taken in Sect. 4 to test the proposed method and the last section concludes the paper.

2 Generalized Decoupling of Unstable System

To implement the decoupling technique, first objective is to stabilize the unstable system. The open-loop unstable MIMO system is stabilized by the state feedback method [5] as shown in Fig. 1. The feedback gain matrix is calculated by pole placement or pole assignment technique.

The pole placement method is almost similar to the root locus method in which the poles of closed-loop systems are placed to the desired locations. It is considered

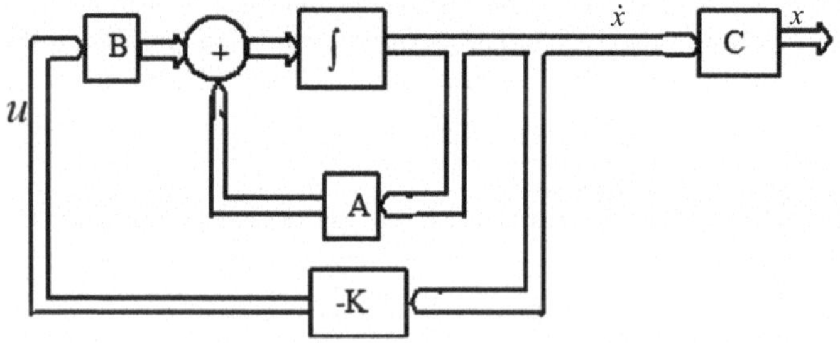

Fig. 1 Stabilization of unstable MIMO system

that the all the state variables of the system are measurable and available for the feedback. The transfer matrix of MIMO system $G(s)$ can be written in state-space form [5] as below.

$$\left.\begin{aligned} \dot{x} &= Ax + Bu \\ y &= cx \end{aligned}\right\} \tag{1}$$

where $u =$ Control signal considered as,

$$u = -Kx \tag{2}$$

$K =$ State feedback gain matrix. Substituting Eq. (2) in Eq. (1), one can find that,

$$\dot{x} = (A - BK)x(t) \tag{3}$$

The solution of the equation yields,

$$x(t) = e^{(A-BK)t}x(0) \tag{4}$$

If a matrix K is suitably selected, the matrix $[A - BK]$ becomes a stable matrix. The state feedback gain matrix is obtained in MATLAB by command "place" for the MIMO system. If $G(s)$ is considered as the $p \times p$ transfer matrix of the unstable MIMO system, then the stabilized system transfer matrix is represented as $\hat{G}(s)$ and the decoupler is designed for $\hat{G}(s)$ as shown in Fig. 2.

To eliminate the effect of undesirable cross-couplings, one must have output y_i, $i = 1, 2, 3, \ldots, p$ as a function of input v_i only. The objective of decoupling scheme shown in Fig. 2 is to obtain a decoupler $D(s)$ for the stabilized plant $\hat{G}(s)$ in order that the input–output relationship should contain only diagonal elements and the values of the other than the diagonal elements should reach to zero. In consequence of the above statement, decoupler $D(s)$ is obtained in order that the original MIMO system $\hat{G}(s)$ should be transformed to $p \times p$ number of diagonal MIMO systems as below,

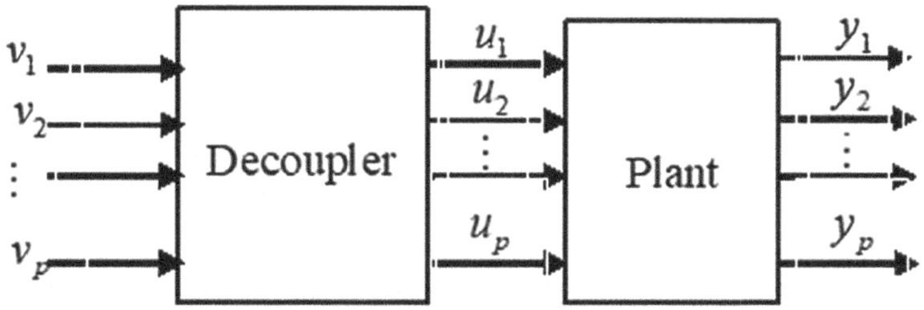

Fig. 2 Decoupling of the unstable MIMO system

$$\hat{G}(s)D(s) = [G_R(s)] \tag{5}$$

where $D(s)$ = decoupling matrix of $\hat{G}(s)$ $\hat{G}^{-1}(s)$ = inverse matrix of $\hat{G}(s)$ and $G_R(s)$ is Diagonal matrix of $\hat{G}(s)$. Therefore, $G_R(s)$ is the $p \times p$ transfer matrix stated as,

$$G_R(s) = \begin{bmatrix} \hat{g}_{11}(s) & 0 & 0 & 0 \\ 0 & \hat{g}_{22}(s) & 0 & 0 \\ \vdots & \vdots & \ddots & \vdots \\ 0 & 0 & 0 & \hat{g}_{pp}(s) \end{bmatrix} \tag{6}$$

One can obtain $D(s)$ as below,

$$D(s) = \left[\hat{G}^{-1}(s)\right]_{s=0} [G_R(s)] \tag{7}$$

The decoupled system $G_D(s)$ is obtained as,

$$G_D(s) = \left[\hat{G}(s)\right] [D(s)_{s=0}] \tag{8}$$

3 PID Controller Design

The PID controller by using Kharitonov's theorem is designed for the decoupled MIMO system as shown in Fig. 3.

PID controller is represented in the transfer function form is as,

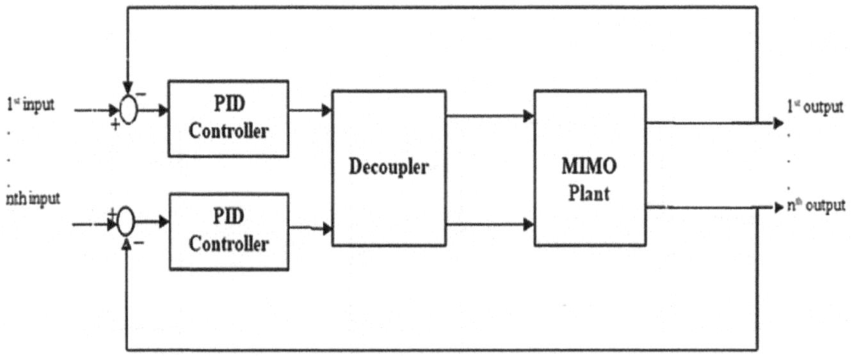

Fig. 3 Design of PID controller for the MIMO system

$$C(s) = K_{\mathrm{p}} + \frac{K_i}{s} + K_{\mathrm{d}}s \tag{9}$$

The loop transfer function with the controller and the plant is written as,

$$G_{\mathrm{c}}(s) = G_{\mathrm{D}}(s) \cdot C(s) \tag{10}$$

Substituting $s = j\omega$, one can find,

$$G_{\mathrm{D}}(j\omega) \cdot C(j\omega) = \alpha e^{j\beta} \tag{11}$$

where $|G_{\mathrm{D}}(j\omega) \cdot C(j\omega)| = \alpha$ l and $\angle G_{\mathrm{D}}(j\omega) \cdot C(j\omega) = \beta$.

The above equation can be represented as [12] below

$$1 + Ae^{j\gamma}(G_{\mathrm{D}}(j\omega) \cdot C(j\omega)) = 0 \tag{12}$$

where $A = \frac{1}{|G_{\mathrm{D}}(j\omega) \cdot C(j\omega)|}$, and $\gamma = 180 + \beta$.

The characteristics equation can be written as,

$$P(j\omega) = 1 + Ae^{j\gamma} G_{\mathrm{D}}(j\omega) \cdot C(j\omega) \tag{13}$$

The above equation is divided into two parts real and imaginary to obtain the three unknown controller parameters values. Since there are three unknowns and two equations, one controller parameter value is suitably assumed in the first and the rest two are determined by solving the two equations.

Considering $P(s)$ of n degree real polynomial as

$$\delta(s) = \delta_0 + \delta_1 s + \delta_2 s^2 + \cdots + \delta_n s^n \tag{14}$$

Here, the coefficient is in between,

$$\delta_0 \in [x_0,\ y_0],\ \delta_1 \in [x_1,\ y_1],\ \ldots,\ \delta_n \in [x_n,\ y_n]$$

By the Kharitonov theorem, set $P(s)$ is said to be Hurwitz stable if the following polynomial is Hurwitz stable.

$$\begin{aligned} K_1(s) &= x_0 + x_1 s + y_2 s^2 + y_3 s^3 + x_4 s^4 + \cdots, \\ K_2(s) &= x_0 + y_1 s + y_2 s^2 + x_3 s^3 + x_4 s^4 + \cdots, \\ K_3(s) &= y_0 + x_1 s + x_2 s^2 + y_3 s^3 + y_4 s^4 + \cdots, \\ K_4(s) &= y_0 + y_1 s + x_2 s^2 + x_3 s^3 + y_4 s^4 + \cdots, \end{aligned} \tag{15}$$

4 Numerical Example

4.1 Design of Decoupler

The unstable MIMO system is considered as described by Eq. (16).

$$G(s) = \begin{bmatrix} \frac{1}{s-1} & \frac{1}{s+5} \\ \frac{1}{s+2} & \frac{1}{s+1} \end{bmatrix} \tag{16}$$

A state feedback control method is employed to stabilize the unstable MIMO system by suitably placing the poles at the left half side of s plane as described in the previous section. The stabilized transfer matrix is obtained as below,

$$\hat{G}(s) = \begin{bmatrix} \frac{s^3+9s^2+24s+20}{s^4+10s^3+33s^2+44s+20} & \frac{s^3+4s^2+5s+2}{s^4+10s^3+33s^2+44s+20} \\ \frac{s^3+6s^2+3s-10}{s^4+10s^3+33s^2+44s+20} & \frac{s^3+8s^2+17s+10}{s^4+10s^3+33s^2+44s+20} \end{bmatrix} \tag{17}$$

At steady state, $\hat{G}(s)$ can be written as,

$$\hat{G}(0) = \begin{bmatrix} 1 & 0.1 \\ -0.5 & 0.5 \end{bmatrix} \tag{18}$$

The inverse of $\hat{G}(0)$ is obtained as,

$$\hat{G}^{-1}(0) = \begin{bmatrix} 0.9091 & -0.1818 \\ 0.9091 & 1.8182 \end{bmatrix} \tag{19}$$

The decoupling matrix $D(s)$ is evaluated by following the Eq. (7) is as below,

$$D(s) = \begin{bmatrix} \frac{0.91s^3+8.18s^2+21.82s+18.18}{s^4+10s^3+33s^2+44s+20} & \frac{0.18s^3-1.45s^2-3.09s-1.81}{s^4+10s^3+33s^2+44s+20} \\ \frac{0.91s^3+8.18s^2+21.82s+18.18}{s^4+10s^3+33s^2+44s+20} & \frac{1.81s^3+14.54s^2+30.91s+18.18}{s^4+10s^3+33s^2+44s+20} \end{bmatrix} \tag{20}$$

At steady state, decoupler is written as,

$$D(0) = \begin{bmatrix} 0.9 & -0.09 \\ 0.9 & 0.9 \end{bmatrix} \tag{21}$$

The decoupled system obtained by following the Eq. (8) is as below,

$$G_D(s) = \begin{bmatrix} \frac{1.8s^3+16.2s^2+43.2s+36}{s^4+10s^3+33s^2+44s+20} & \frac{0.81s^3+2.79s^2+2.34s+1.62}{s^4+10s^3+33s^2+44s+20} \\ \frac{1.8s^3+12.6s^2+20.23s}{s^4+10s^3+33s^2+44s+20} & \frac{0.81s^3+6.66s^2+15.03s+9.9}{s^4+10s^3+33s^2+44s+20} \end{bmatrix} \tag{22}$$

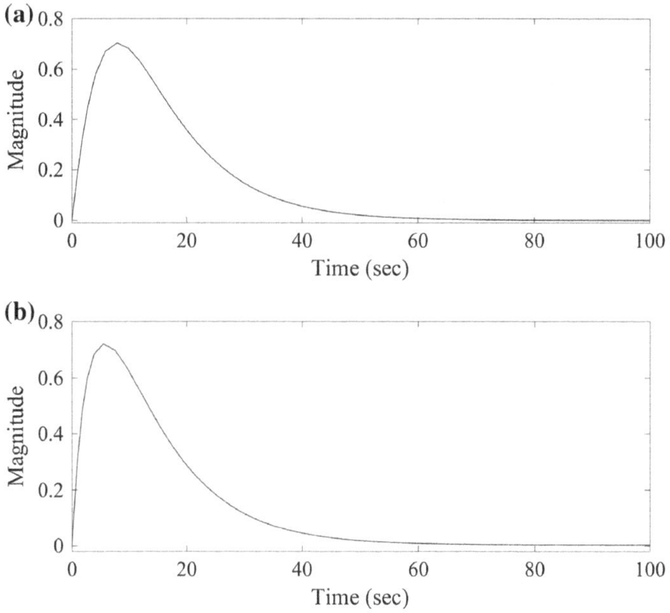

Fig. 4 **a** Coupling effect on second output due to step response in first input. **b** Coupling effect on first output due to step response in second input

The designed decoupler is connected to the system as shown in Fig. 2. To test the effect of the decoupler, a step input is applied to one input and no response is applied to other input at the same time. The results obtained are shown in Fig. 4 which shows that the coupling effects are completely eliminated in the resulting decoupled system. Figure 4a shows the coupling effect on second output when step response is applied to first input while Fig. 4b shows the coupling effect on first output when step response is applied to second input.

4.2 Design of Robust PID Controller

Two PID controllers are designed for $G_{\mathrm{D11}}(s)$ and $G_{\mathrm{D22}}(s)$ of the decoupled MIMO system written by Eq. (18).

$$\begin{aligned} G_{\mathrm{c11}}(s) &= G_{\mathrm{D11}}(s)\cdot C_1(s) \\ &= \left(\frac{1.8s^3+16.2s^2+43.2s+36}{s^4+10s^3+33s^2+44s+20}\right)\cdot\left(\frac{K_{\mathrm{p1}}s+K_{i1}+K_{\mathrm{d1}}s^2}{s}\right) \end{aligned} \tag{23}$$

The characteristics equation is determined as,

Table 1 Range of controller parameter for $G_{D11}(s)$

Controller parameter	K_{p1}	K_{i1}	K_{d1}
Lower value	0.48	8.6	1.5
Upper value	0.56	9.1	2

Table 2 Range of controller parameter for $G_{D22}(s)$

Controller parameter	K_{p2}	K_{i2}	K_{d2}
Lower value	0.02	1	1.6
Upper value	1.56	1.5	3.6

$$\begin{aligned} &s^5[1 + 1.8K_d] + s^4[10 + 1.8K_p + 16.2K_d] + s^3[33 + 1.8K_i + 16.2K_p + 43.2K_d] \\ &+ s^2[44 + 16.2K_i + 43.2K_p + 36K_d] + s[20 + 43.2K_i + 36K_p] + 36K_i = 0 \end{aligned} \tag{24}$$

Substituting $s = j\omega$, the equation is written as below.

$$\begin{aligned} &j\omega^5[1 + 1.8K_d] + \omega^4[10 + 1.8K_p + 16.2K_d] - j\omega^3[33 + 1.8K_i + 16.2K_p + 43.2K_d] \\ &- j\omega^2[44 + 16.2K_i + 43.2K_p + 36K_d] + j\omega[20 + 43.2K_i + 36K_p] + 36K_i = 0 \end{aligned} \tag{25}$$

The above equation is divided into two parts: real and imaginary, as below.

$$\omega^4[10 + 1.8K_p + 16.2K_d] - \omega^2[44 + 16.2K_i + 43.2K_p + 36K_d] + 36K_i = 0 \tag{26}$$

$$\begin{aligned} &\omega^5[1 + 1.8K_d] - \omega^3[33 + 1.8K_i + 16.2K_p + 43.2K_d] \\ &+ \omega[20 + 43.2K_i + 36K_p] = 0 \end{aligned} \tag{27}$$

The controller gain ranges are determined by solving Eqs. (21) and (22) assuming one controller parameter value at first as depicted in Table 1. Similar procedure is obtained to find the range of the controller parameters for $G_{D22}(s)$ and presented in Table 2.

The above-obtained range of the controller parameters satisfy the Routh–Hurwitz stability criteria of the interval polynomials shown by Eq. (15). Therefore, the obtained ranges of the controller parameters are said to be robust controller parameter.

4.2.1 Time-Domain Analysis

The PID controller gains values are suitably chosen from the obtained range of gains based on time response performance. Step response is applied to both the inputs at the same time. Figure 5a shows the step response of the first output and Fig. 5b shows the step response of second output.

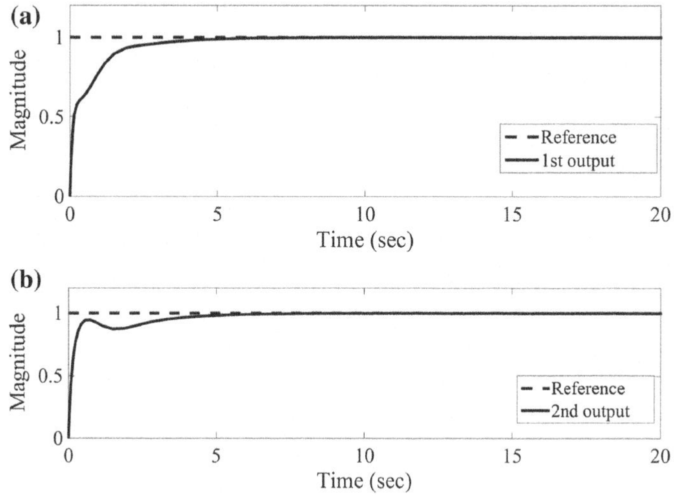

Fig. 5 **a** Step response of first output. **b** Step response of second output

4.2.2 Robustness Analysis

The robustness characteristics of the MIMO system are analyzed by the simultaneous gain variation. A simultaneous multiplicative gain of the form of diag. $[1-\sigma,\ 1+\sigma]$ and diag. $[1-\sigma,\ 1+\sigma]$ is applied and the range σ is obtained as $[-0.45, 0.38]$ for which system is stable as shown in Figs. 6 and 7.

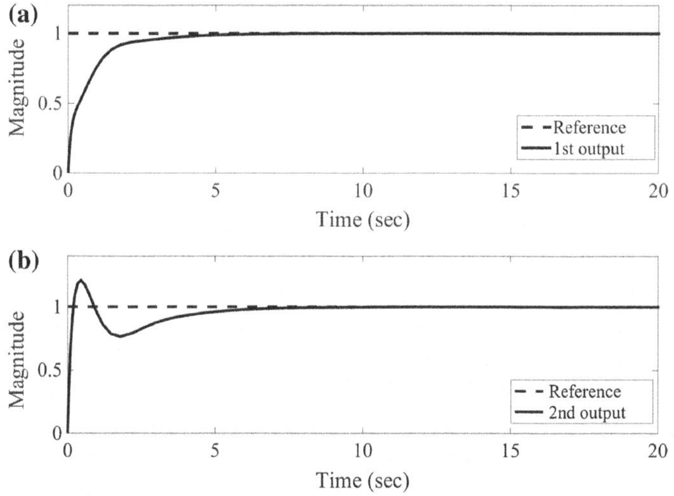

Fig. 6 **a** Step response of first output at $\sigma = 0.38$. **b** Step response of second output at $\sigma = 0.38$

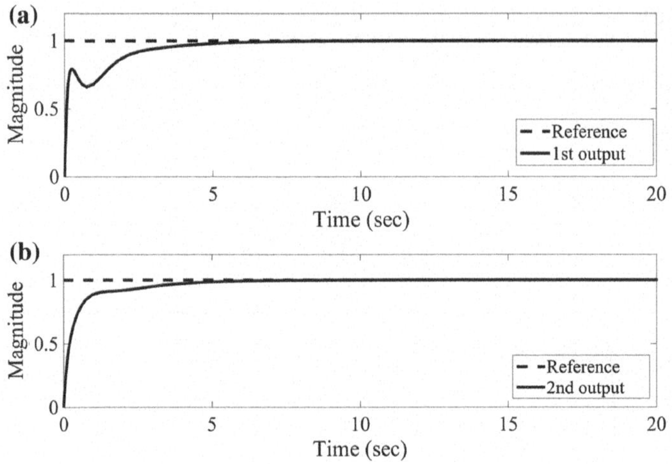

Fig. 7 **a** Step response of first output at $\sigma = -0.4$. **b** Step response of second output at $\sigma = -0.4$

5 Conclusion

In this work, a decoupler is designed for the unstable MIMO systems by using the generalized decoupling technique. At first, the unstable system is stabilized by state feedback method then the decoupler is designed for this stabilized system. At last, robust PID controller based on Kharitonov's theorem is designed for the decoupled MIMO system. The proposed approach is verified through a numerical example in simulation and obtained responses are satisfactory. The robustness feature of the designed controller is tested through the simultaneous gain variations.

References

1. Moreno AR (2016) An approach to design MIMO FO controllers for unstable nonlinear plants. IEEE J Autom Sinica 3(3):338–344
2. Spacek L, Vojtesek J, Bobal V (2017) Educational model of unstable MIMO system. In: International conference on engineering, technology and innovation
3. Herrera J, Ibeas A, Sen M, Alcántara S, Alonso Quesada S (2013) Identification and control of delayed unstable and integrative LTI MIMO systems using pattern search methods. Adv Differ Equ 331–343
4. Wahab AA, Mamat R, Shamsudin SS (2009) The effectiveness of pole placement method in control system design for an autonomous helicopter model in hovering flight. Int J Integr Eng 33–46
5. Oggata K (2010) Modern control engineering. Prentice Hall
6. Ghartemani MK, Mobed M (2008) A state feedback strategy for decoupling a class of multi-variable systems. In: International conference on systems
7. Angelico BA, Barbosa FS, Toriumi FY (2017) State feedback decoupling control of a control moment gyroscope. J Control Autom Electr Syst 28:26–35

8. Gilbert EG (1969) The decoupling of multivariable systems by state feedback. SIAM J Control 7(1)
9. Pandey SK, Dey J, Banerjee S (2018) Design of robust proportional–integral–derivative controller for generalized decoupled twin rotor multi-input-multi-output system with actuator non-linearity. J Syst Control Eng 232(8):971–982
10. Rajabioun R, Mamizadeh A (2017) Gershgorin bands minimization for decoupling MIMO systems and decentralized PID design using COA. In: 13th international conference on technical and physical problems of electrical engineering, Turkey
11. Dasgupta S, Sadhu S, Ghoshal TK (2014) Internal model based V-norm decoupling control for four tank system. In: International conference on control, instrumentation, energy and communication (CIEC)
12. Pandey SK, Dey J, Banerjee S (2016) Design and real-time implementation of robust PID controller for Twin Rotor MIMO System (TRMS) based on Kharitonov's theorem. In: IEEE 1st international conference on power electronics, intelligent control and energy systems, India

Economic Approach to Design of a Level 2 Residential Electric Vehicle Supply Equipment

Samir M. Shariff, Mohammad Saad Alam, Shadman Faraz, Mohd Aiman Khan, Ahmad Abbas and Mohammad Amir

Abstract Increase in atmospheric adulteration due to exponential growth in urbanization, industrialization, and number of on-road vehicles have necessitated the need for emission-free transportation. The major challenges in the proliferation of electric vehicles are the lack of international standards. This paper presents simulation and hardware implementation of a low-cost residential EVSE based on SAE J1772 standard. This is accomplished by using Arduino microcontroller for detection and communication with an electric vehicle. The main difference of the proposed system to existing EVSE is its low cost and amperage adjustment for better load control. The resultant system is capable of detecting various states and provides communication as specified by the CCS standard. The proposed EVSE includes necessary protective features like ground fault circuit interrupter (GFCI), continuous monitoring of temperature, humidity, and potential earth for protection against short circuit. Simulation of the pilot circuit is done using Labcenter Electronics: Proteus software and simultaneously a hardware model was developed and tested. Results indicate system to be economical and compliant with future improvement in energy storage devices and smart grid.

S. M. Shariff
Department of Electrical Engineering, Taibah University, Medina, Saudi Arabia
e-mail: samshariff@yahoo.com

M. S. Alam · S. Faraz · M. A. Khan (✉) · A. Abbas · M. Amir
Department of Electrical Engineering, Aligarh Muslim University, Aligarh, India
e-mail: maimankhan@zhcet.ac.in

M. S. Alam
e-mail: saad.alam@zhcet.ac.in

S. Faraz
e-mail: farazshadman@gmail.com

A. Abbas
e-mail: ahmadabbasamu91@gmail.com

M. Amir
e-mail: mdamir.caret@gmail.com

S. N. Singh et al. (eds.), *Advances in Power and Control Engineering*, Lecture Notes in Electrical Engineering 609, https://doi.org/10.1007/978-981-15-0313-9_3

Keywords Electric vehicle supply equipment (EVSE) · J1772TM Std. · Electric vehicle charging (EVC) · Signaling circuit (SC) · Combined charging system (CCS)

1 Introduction

With rapid development in energy storing devices and the need for emission-free transportation demand for electric vehicles have increased rapidly and according to projections, it will continue to grow [1, 2]. The electric powertrain not only provides a safe means of transportation but also helps in the reduction of greenhouse gas emitting fossil fuels [3]. Along with the reduction in carbon footprint, an EV has various other benefits also—the energy storing capability of plug-in EVs make the bidirectional flow of power from EV to a power grid (V2G) and from EV batteries to home (V2H) using bidirectional EVSEs possible [4–8]. Thus, an EV can be used to stabilize the power grid, act as emergency backup power, and support large-scale renewable energy integration as presented in [9]. Owning to these numerous advantages EV offers, they are dubbed as the future of transportation. The only factors impeding the proliferation of EVs are lack of implementation of international standards, high charging time, and lack of infrastructure as presented in [10–12].

Both battery electric vehicles (BEVs) and plug-in hybrid electric vehicles (PHEVs) are charged through the electric power system via a charging station [13]. The success of EV technology depends on the number of charging stations accessible to EV owners [14]. With an increase in sales of global electric car stock, global EVSE outlets also surpassed 2 million in 2016. Even with increased sales, electric cars still outnumber public charging stations by more than six to one, indicating that most drivers rely primarily on private charging stations as shown in Fig. 1. An increasing

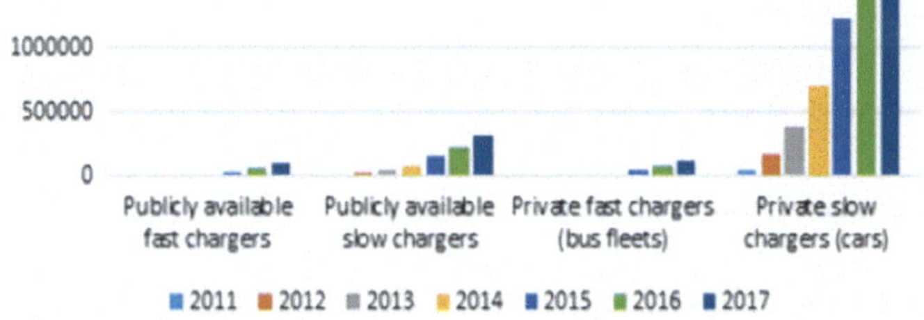

Fig. 1 EVSE outlets availability [15]

Table 1 EV charging standards

EV system standard	Country
SAE	USA
IEC	Europe Union
JEVS	Japan
Bharat EV charger	India
GB	China

number of EVs increasingly impact the power system, e.g., peak power demand, load profile, and voltage magnitude deviations.

Since commercial chargers are still in the primary stages of deployment, residential charging level 2 stations are mainly employed by EV owners to charge their vehicles. For this operation, the residential owner buys the EVSE and pays for installation, a total cost of $1,500–$2,000. The low acceptance of publicly available chargers can be attributed to their high costs of deployment and competition between consumers for the required parking spots [16]. Accelerating the deployment of the charging system and their infrastructure will invigorate economic development, augment consumer welfare, create new employment prospects, and contribute to a climate-safe future [17].

A variety of electric vehicle standards have been published by different countries due to lack of international standards. A list of different standards and their origin is provided in Table 1. With no specific standard to follow, Electric Vehicles from different manufactures [18] are incompatible with each other which severely limits the options of various alternatives like battery swapping, energy transfer and V2H chargers and is thus hindering the penetration of EVs in the market.

This paper proposes the design of an economic level 2 residential electric vehicle supply equipment (EVSE) based on SAE J1772 standard due to its wide acceptability, high demand, and ability to incorporate future development in energy storage devices [19]. This charger has to provide flexibility adjustment for consumers to choose between the costs of charging or charging time. The design focuses on the proper safety of consumer against shocks and over voltages and provides various safety equipment [20, 21]. To increase faster EV adoption, public perception is also needed to address the safety, cost, and driving range of electric vehicles. Therefore, the main contribution of this paper is to develop a flexible charger for dynamic control of the EV battery charging process, according to user preference. Nevertheless, the other important contribution is the cost-effectiveness of the charger without any compromise with the safety of the user [22].

The rest of this manuscript is organized as follows. Section 2 discusses the designing of the charger, the methodology adopted, charging, and the protection scheme of the charger. This section also gives a brief discussion about the communication scheme of the charger. The simulation of the designed charger and the results of the simulation are presented in Sect. 3. The final hardware implementation of the charger is carried out in Sect. 4. Section 5 presents the results obtained. Finally, Sect. 6 presents the concluding remarks.

2 Proposed Topology of EVSE

2.1 Control Circuit of an EVSE Charging Scheme

EVSE based on J1772 can be divided into four main parts: EVSE controller (a), EVSE connector (b), vehicle inlet (c), and onboard vehicle controller (d). Figure 2 shows a basic J1772 schematic [23]. The model adopted in this paper focuses on the safety of the user against shocks by not providing power if EV is not correctly matched. Further risk of electric shock is avoided by monitoring the availability of the ground to prevent earth leakage and also continuously monitors the connection between EV and the charger.

Communication between EVSE and EV charger is done by a modulated 1 kHz square wave which oscillates between −12 V to +3/6/9 V to indicate various states.

The user sets a desired charging current and the controller performs the necessary steps to modify the duty ratio and thus changing the current that can be drawn from the charger. (The controller determines the current to be drawn by charger by determining the duty ratio). The onboard charger handles the changes in resistance which are detected by the EVSE controller. Rectification of AC to DC for charging of batteries and protection against overcharge is handled by the onboard charger. An energy meter is provided for measurement of current, power supplied, peak power, and kilowatt per hour [24, 25]. All this information is provided to the user by an LCD keypad shield, a user interface for changing the charging current is provided.

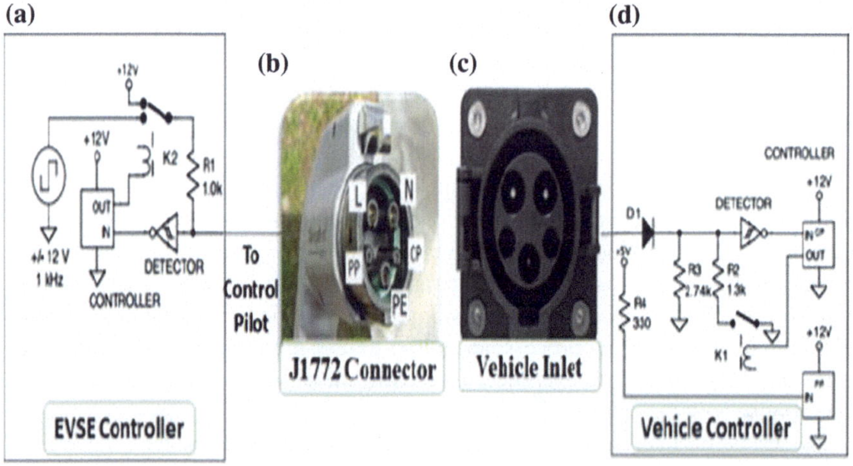

Fig. 2 Signaling circuit schematic for the J1772 standard

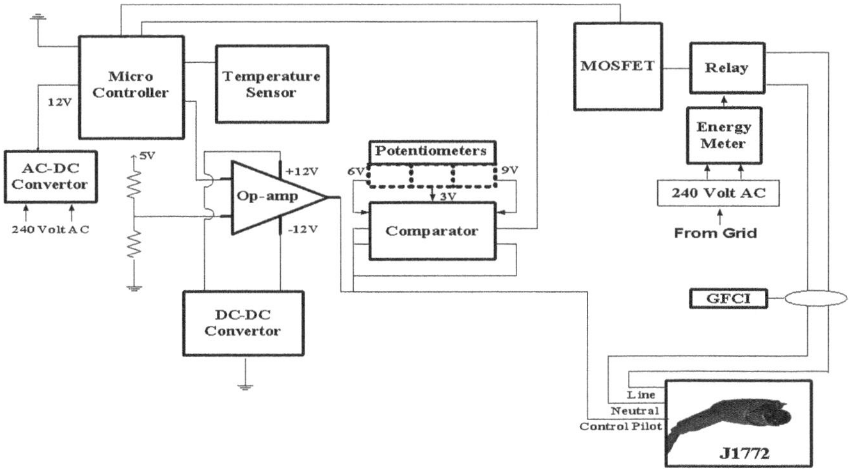

Fig. 3 Proposed architecture of residential EVSE

2.2 Methodology Adopted

In this paper, the general circuit design of the EVSE to comply with the working of an off-board charger is first simulated in Proteus environment and the results of the simulation are obtained. Simultaneously, a hardware model satisfying the simulation results is made and tested. The general block diagram of the level 2 charger is shown in Fig. 2.

A single-phase 230–12 V power supply unit is used to power the circuit. Temperature and humidity sensor provided in Fig. 3 is part of a protection circuit to maintain the operation under specified temperature and humidity. The microcontroller makes use of a comparator to measure the voltage difference between the control pilot and the proximity earth pin of the J1772 connector. Using the value of this voltage difference, the microcontroller identifies which state is the EV in and takes action accordingly. An op-amp is used to convert 5 V PWM provided by the microcontroller to ±12 V PWM for high-level communication between electric vehicle and charger. A normal open semiconductor relay is used to provide power to a vehicle which is tripped by MOSFET. A GFCI is provided for further protection against short circuits.

2.3 Designing of the EV Charger

The off-board charger consists of a microcontroller unit which controls the overall charging of EV whose algorithm is shown in Fig. 4.

The J1772 control pilot is at +12 V at normal standby mode, on being connected to the EV the +12 V changes into a 1 kHz. +9 V to −12 V square wave, indicating

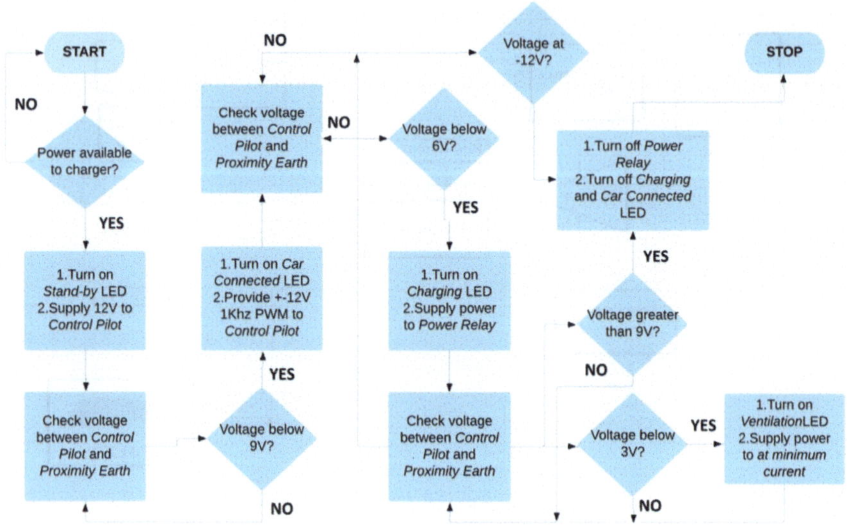

Fig. 4 Flow chart of charging algorithm

the connection of an EVSE [26]. The voltage present on the control pilot defines the current state of EV. The EV adds resistance between control pilot to proximity earth to vary the voltage. The EVSE reads the voltage difference and changes in state accordingly. The pilot characterizes different stages by a drop in voltage and EVSE controller takes the required action as shown in Table 2.

The pulse-width modulation is the utility for low-level communication between the EV and the EVSE. The PWM signal is applied to the circuit of control pilot and PE. Three different kinds of information can be transmitted:

- Low duty cycle 3–7% = digital communication required
- High duty cycle 8–97% = defines maximum current available

Table 2 SAE J1772 charging states

State	Pilot voltage	Frequency	EV resistance	Description
A	+12 V	1000 Hz	N/A	No coupler engagement
B	+9 to −12 V	1000 Hz	2.74 k	Coupler engagement detected Vehicle not yet ready EVSE supply energy: Off
C	+6 to −12 V	1000 Hz	882	Vehicle ready EVSE supply energy: ON
D	+3 to −12 V	1000 Hz	246	Vehicle ready EVSE supply energy: ON Vent required
E	0	N/A	N/A	Short of CP to PE (connection lost) Unlock plug after max. 30 ms
F	−12 V	N/A	N/A	EVSE not available

- Invalid duty cycle >97% = Charging not possible.

Once the PEV initiates charging (States C or D), the amperage is then set by the duty cycle, D, of the 1 kHz square wave oscillator according to the following formulas:

For 6–51 A:
Amp = Duty Cycle × 0.6
Duty Cycle = Amps/(0.6 Amps)
For 51–80 A:
Amp = (Duty Cycle × 64)25
Duty Cycle = Amps/(0.6 Amps) + 64.

2.4 *Protection of Charging Scheme*

The EVSE addresses safety and protection [27] by incorporating a temperature and humidity sensor which signals the microcontroller to stop charging when temperature and/or humidity reaches beyond a certain specified limit as shown in Fig. 5. These limits can differ from region to region and define the maximum temperature to which a user can charge a vehicle without any risk. This function severely limits the function of electric vehicles during rainy or extremely hot temperatures over ICEVs. With charging infrastructure still in development stage, preference should be given to the

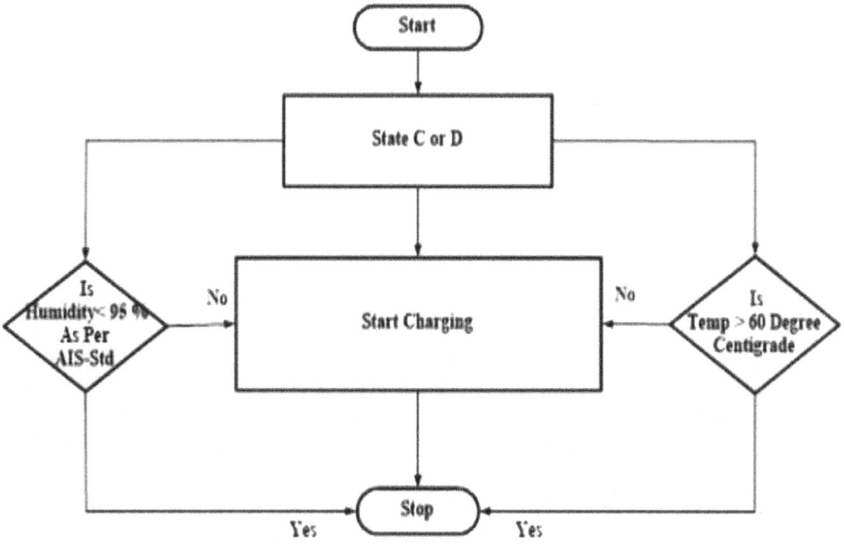

Fig. 5 Flowchart of protection against temperature and humidity based on CCS Std. [24]

protection of consumers and various schemes like battery swapping can be provided to circumvent this problem.

To protect electronic components, charging current can also be reduced through temperature sensor to continue operation under safe limits. This feature is also preferred in residential chargers to protect the user in case of sudden increase in temperature or humidity.

3 Simulation of Signaling Circuit

The simulation of the charger is prepared using Labcenter Electronics: Proteus software. The simulation model prepared is as shown in Fig. 6.

The model consists of an Arduino: Nano microcontroller as the main processing unit, which provides the necessary control signals and allows for monitoring of signal on control pilot. The voltage drop between the control pilot and proximity earth is detected by a quad voltage comparator (LM339), which sends the signals to the microcontroller. The microcontroller provides a +12 V signal indicating the standby mode. Once the charger connects simulated by inserting a resistance between the control pilot and the proximity earth, a 1 kHz 5 V PWM signal is provided by the microcontroller which is converted into a +9 V to −12 V signal by an operational amplifier in comparator mode as shown by Fig. 7.

State C or the charging state is simulated by inserting another resistance between the control pilot and proximity earth which lower its voltage to +6 V. Once state C is reached, Arduino sends a signal which forward biases the transistor (2N7000)

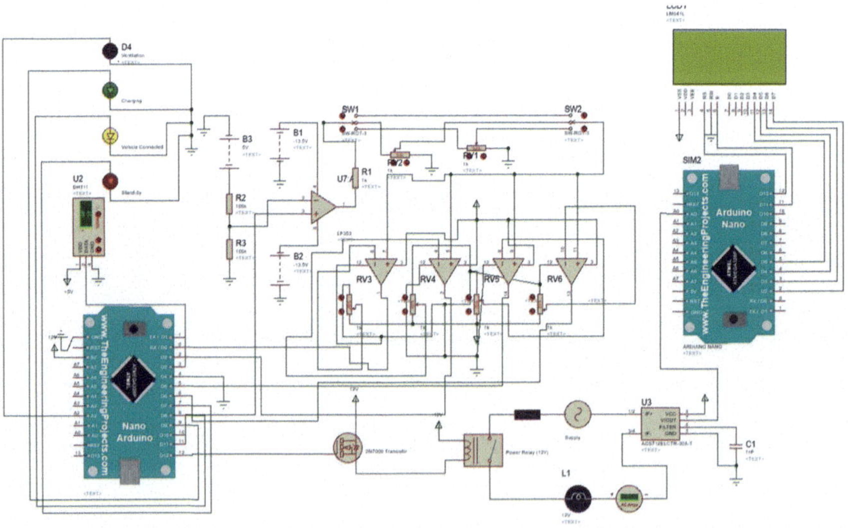

Fig. 6 Proteus simulation of signaling circuit

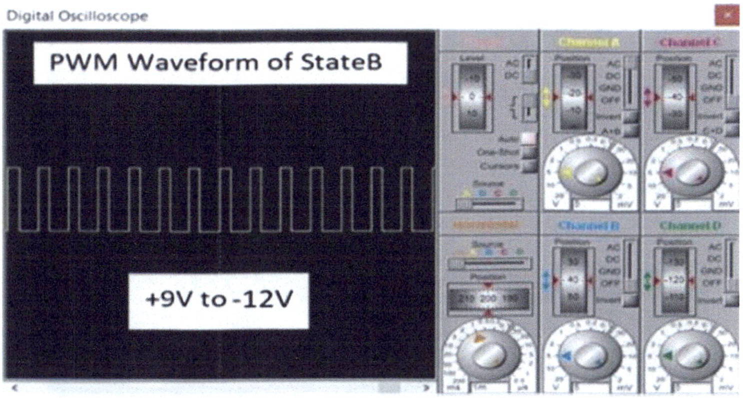

Fig. 7 PWM waveform obtained from State B

which trips the relay and initiate charging. A working model of an energy meter using ACS712 current sensor is also developed. A three-way switch with variable resistance is provided to simulate voltage drop. A lamp is provided to indicate charging of electric vehicle. The energy meter provides a user with an interface to measure the amount of power consumed in a charging cycle. A user interface in the form of a switch is also provided for the customer to change his charging current [28, 29]. The PWM waveform gained from state B (vehicle connected) as shown in Fig. 7.

4 Hardware Implementation

The apparatus that were used to perform the experiment is as follows: oscilloscope (1), power supply unit (2), circuit connection (3), and relay (4) as shown in Fig. 8.

The voltage change is provided with 2.74 k ohms resistor (State B). The hardware model of the charger is made such that not only the basic working of the charger is satisfied but the effective safety and protection features of the charger are also incorporated. Figures 9, 10, 11, and 12 show the various states detected by the EVSE controller. The controller trips the relay in State C and alerts the user when ventilation is required in State D.

5 Result and Discussion

The results obtained from Proteus simulation and consecutive hardware implementation show that the proposed EVSE controller can detect all six states as required for proper operation of J1772 standard. Simultaneously a hardware model of EVSE was developed as shown in Fig. 13. The proposed EVSE ensures the safety of con-

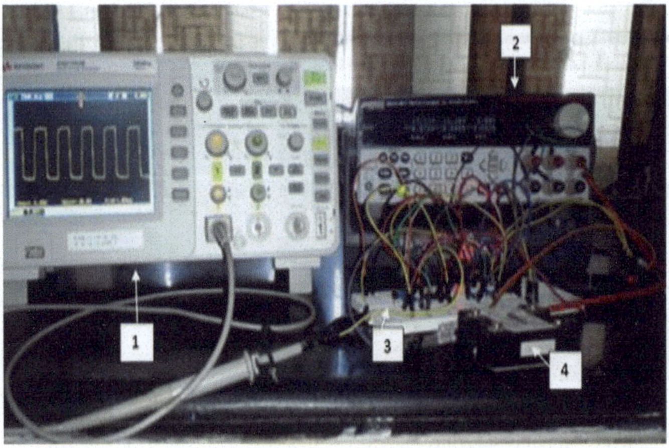

Fig. 8 Hardware setup of EVSE controller

Fig. 9 State A (12 V DC)

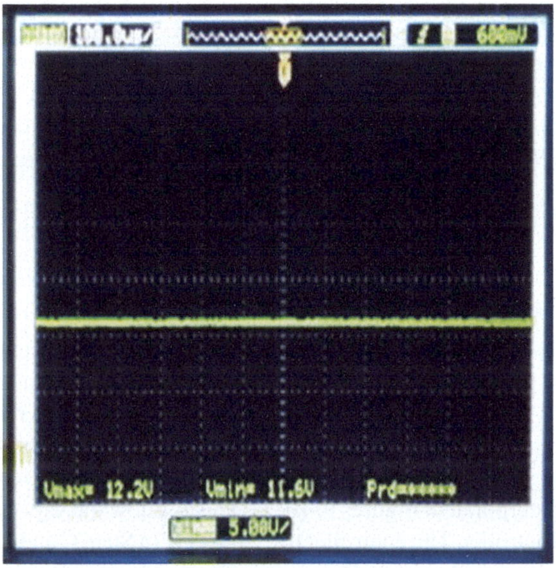

sumer by varying amperage in case of high temperatures and high humidity to avoid insulation failures. Further risk of shocks is reduced by continuous monitoring of the availability of ground. A ground fault circuit interrupter (GFCI) circuit with the differential current limit of 5ma maximum for protection against short circuit is also provided. The EVSE allows to charge at variable currents. Figures 14, 15, 16, and 17 show different charging currents produced by varying duty ratio.

Fig. 10 State B (+9 to −12 V PWM)

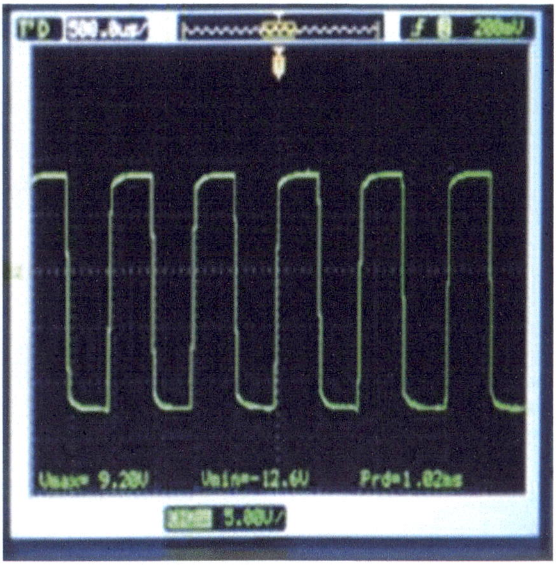

Fig. 11 State C (+6 to −12 V PWM)

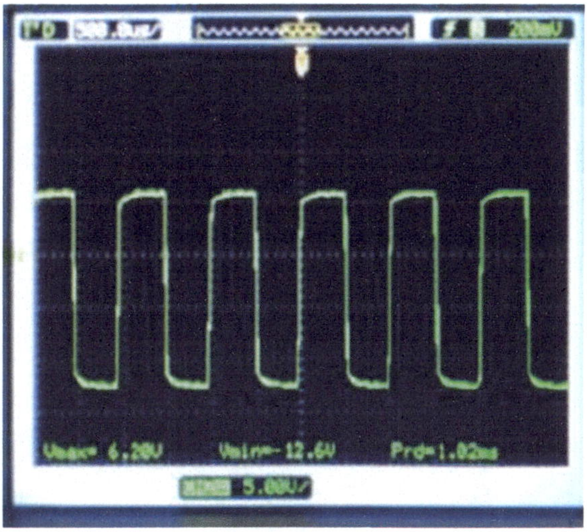

6 Conclusions and Future Scope

The EVSE proposed in the paper can detect the voltage changes and can charge a residential electric vehicle at user-defined variable charging current. The charger features many safety features and is comparatively cheaper than chargers available in the market. Although many governments provide different incentives for residential chargers, only customers with private parking can avail those offers. There are also

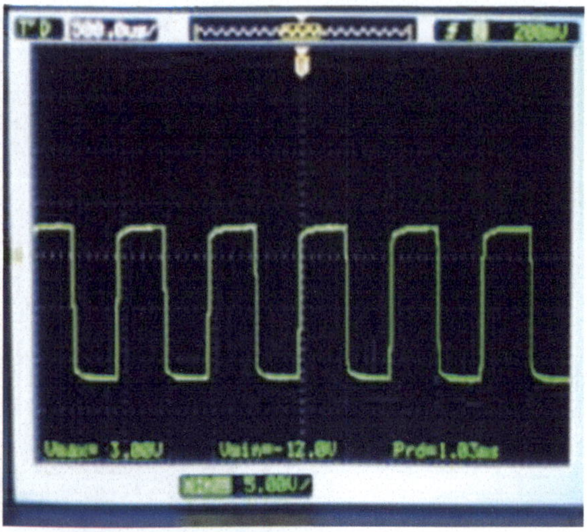

Fig. 12 State D (+3 to − 12 V PWM)

Fig. 13 EVSE hardware implementation

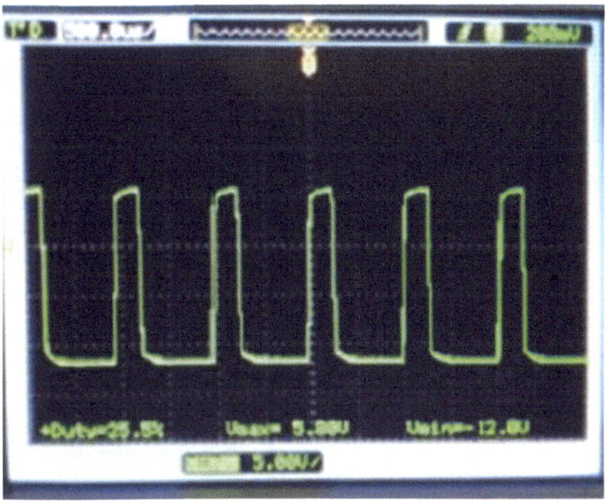

Fig. 14 State C with 25% duty cycle (15 A)

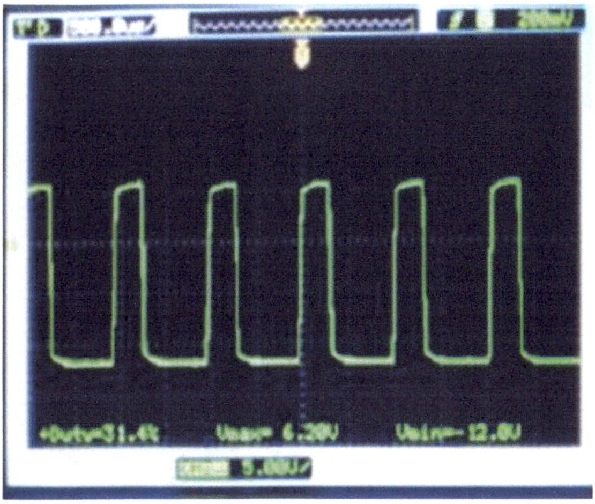

Fig. 15 State C with 30% duty cycle (18 A)

various workarounds to bypass J1772 protection but they offer little protection and can cause damage to consumer and charging equipment. The charger satisfactorily complies with the J17772 standards along with various safety features and has energy metering and current changing capability. Although the charger claims completeness many challenges still remain and the real problems can only be observed with the widespread adoption of electric vehicles.

The proposed charger can vary current up to 30 A and also provides room for further development in battery chargers to increase the amperage up to 80 A. The residential charger can be combined with smart meters to provide two-way communication to facilitate current adjustment for better load management, which will reduce stress on the grid. Time of day tariff can also be implemented to improve flexibility

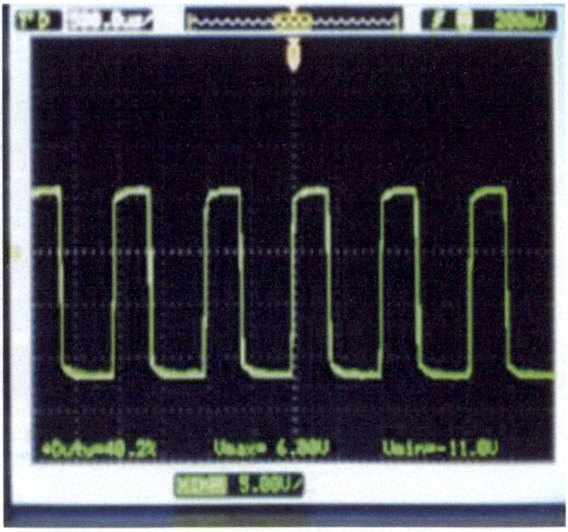

Fig. 16 State C with 40% duty cycle (24 A)

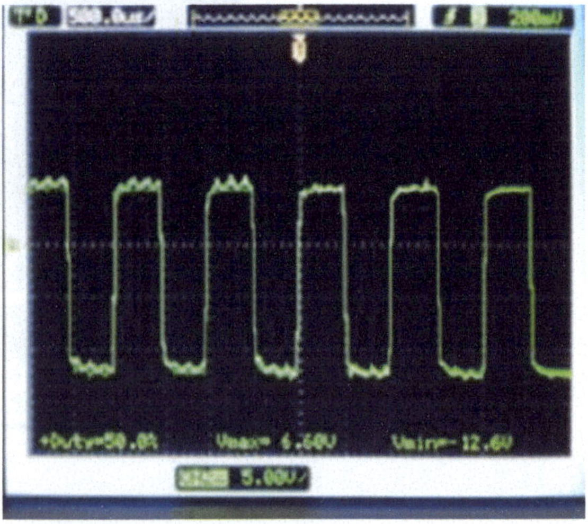

Fig. 17 State C with 50% duty cycle (30 A)

and provide the user with data that can help in making an informed decision when using the charger.

References

1. Ahmad F, Alam MS, Asaad M (2017) Developments in xEVs charging infrastructure and energy management system for smart microgrids including xEVs. Sustain Cities Soc 35:552–564

2. Electric Vehicle Transportation Center (EVTC) (2017) Prediction of electric vehicle penetration. EVTC report number: FSEC-CR-2069-17 [Online]. Available: http://fsec.ucf.edu/en/publications/pdf/FSECCR-2069-17.pdf
3. US Department of Energy Multi-Year Program Plan 2011–2015: Vehicle Technologies Program (2010) [Online]. Available on: https://www1.eere.energy.gov/vehiclesandfuels/pdfs/program/vt_mypp_2011-2015.pdf
4. Botsford C (2012) The economics of non-residential level 2 EVSE charging infrastructure
5. Kempton W, Tomic J (2005) Vehicle-to-grid power fundamentals: calculating capacity and net revenue. ELSEVIER J Power Sources 144(1):269–279
6. Yilmaz M, Krein PT (2013) Review of the impact of vehicle-to-grid technologies on distribution systems and utility interfaces. IEEE Trans Power Electron 28(12):5673–5689
7. Puentes MP, Ramos G (2013) Level 2 bidirectional charger for small electric vehicles: topologies comparison. In: 2013 power electronics and power quality applications, PEPQA 201
8. Kempton W, Tomic J (2005) Vehicle-to-grid power fundamentals: calculating cand net revenue. J Power Sources 144:268–279
9. Kempton W, Tomic J (2015) Vehicle-to-grid power implementation: from stabilizing the grid to supporting large-scale renewable energy. ELSEVIER J Power Sources 144:280–294
10. Khan S, Ahmad A, Ahmad F, Shafaati Shemami M, Saad Alam M, Khateeb S (2018) A comprehensive review on solar powered electric vehicle charging system. Smart Sci 6(1):54–79
11. Ahmad A, Alam MS, Chabaan R (2018) A comprehensive review of wireless charging technologies for electric vehicles. IEEE Trans Transp Electrification 4(1):38–63
12. Ruiz MA, Abdallah FA, Gagnaire M, Lascaux Y (2013) TeleWatt: an innovative electric vehicle charging infrastructure over public lighting systems. In: Proceedings of international conference on connected vehicles and expo, ICCVE 2013, pp 741–746
13. Krein PT (2013) Review of battery charger topologies, charging power levels, and infrastructure for plug-in electric and hybrid vehicles. IEEE Trans Power Electron 28(5). Available on https://doi.org/10.1109/IEVC.2012.6183208
14. Khan W, Ahmad A, Ahmad F, Alam MS (2018) A comprehensive review of fast charging infrastructure for electric vehicles. J Smart Sci 6(3):256–270
15. International Energy Agency (IEA) (2018) Global EV Outlook 2018 [Online]. Available on: https://www.iea.org/gevo2018/
16. Khan W, Ahmad F, Ahmad A, Alam MS (2018) Feasibility analysis of electric vehicle charging infrastructure deployment in India. Lect Notes Electr Eng 487:193–206
17. Khan S, Alam MS, Asghar MSJ, Khan MA, Abbas A (2018) Recent development in level 2 charging system for xEV : a review. In: International conference on computational and characterization techniques in engineering and sciences (CCTES), pp 83–88
18. Kramer B, Chakraborty S, Kroposki B (2008) A review of plug-in vehicles and vehicle-to-grid capability. In: IEEE IECON industrial electronics conference, pp 2278–2283
19. Moghaddam Z, Ahmad I, Habibi D, Phung QV (2017) Smart charging strategy for electric vehicle charging stations. IEEE Trans Transp Electrification 4(1):76–88
20. Schürmann D, Timpner J, Wolf L (2016) Cooperative charging in residential areas. IEEE Trans Intell Transp Syst 18:1–13. https://doi.org/10.1109/TITS.2016.2590960
21. Knipping E, Duvall M (2007) Environmental assessment of plug-in hybrid electric vehicles, vol 2. In: United States air quality analysis based on AEO-2006 assumptions for 2030 Palo Alto, CA, Jul. 23, 2007, Tech. Rep. 1015326, EPRI
22. Freschi F, Mitolo M, Tommasini R (2015) Electrical safety of electric vehicles. Hydro-Québec, Electric vehicle charging stations
23. Charging Interface Initiative E.V. (CharIN E.V) 29 Dec 2018. Available on https://www.charinev.org/index.php?id=170
24. Chaudhari S, Rathod P, Shaikh A, Vora D, Ahir J (2017) Smart energy meter using Arduino and GSM. In: International conference on trends in electronics and informatics (ICEI)
25. Metering AS, Visalatchi S, Sandeep KK (2017) Smart energy metering and power theft control using arduino and GSM. In: 2017 2nd international conference for convergence in technology (I2CT)

26. Hunter B (2018) Electric vehicle demand response via J1772TM charging signal interruption. In: 2018 IEEE green technologies conference (GreenTech), pp 81–88
27. CCS Standard (2018). Available: https://www.charinev.org/index.php?id=170. Accessed 20 Dec 2018
28. Pinto JG, Vitor M, Henrique G, Bruno E, Monteiro V, Afonso JL (2013) Bidirectional battery charger with grid-to-vehicle, vehicle-to-grid and vehicle-to-home technologies. In: IECON proceedings (industrial electronics conference), pp 5934–5939
29. Shamami MS, Alam MS, Asghar MSJ (2018) Reliable residential back up power control system through home to plug-in electric vehicle (H2V). J Technol Econ Smart Grids Sustain Energy

FPGA Based Speed Control and Back EMF Extraction from Line Voltages Using IIR Digital Filters for BLDCM

Umesh Kumar Soni, Maloth Naresh and Ramesh Kumar Tripathi

Abstract Back EMF detection is important aspect of the sensorless control of BLDC motor. In this paper, the BEMF detection using FPGA based digital low pass filtering of the line to line voltage using Least *P*th-norm IIR filter has been realized. Lower switches of the legs are fed with PWM and upper switches are fundamentally controlled. The HON_LPWM strategy is used for the speed control as well as for the BEMF detection. Various quantities, e.g., percentage current ripple, current magnitude, line to line back EMF, speed, and switching signals were analyzed for the case of varying voltage with fixed duty as well as for varying duty cycles with fixed DC voltage. It has been analyzed that the back EMF detection with this method has some limitations on duty cycle because intermittent up and down peaks in filtered BEMF waveform become high as we increase or decrease duty out of certain band. During the implementation of BEMF based sensorless techniques, these up and down peaks in BEMF waveforms have to be removed. For this, further improved filtering has to be searched so as to assure proper wave shaping with least delays in line to line BEMF zero crossings for sensorless application. Whole system has been implemented using Xilinx based System Generator using Vivado 2014.4 design suite for FPGA based programming in connection with MATLAB/Simulink and WAVECT controller. Simulation results and hardware results have been verified.

Keywords FPGA · BEMF · HON_LPWM · PWM · LPF · Low pass filter

---

U. K. Soni (✉) · M. Naresh · R. K. Tripathi
MNNIT, Allahabad, India
e-mail: ree1456@mnnit.ac.in

M. Naresh
e-mail: ree0114@mnnit.ac.in

R. K. Tripathi
e-mail: rktripathi@mnnit.ac.in

S. N. Singh et al. (eds.), *Advances in Power and Control Engineering*, Lecture Notes in Electrical Engineering 609, https://doi.org/10.1007/978-981-15-0313-9_4

1 Introduction

The association of electrical equipment and medical imaging manufacturers, i.e., NEMA design B defines minimum nominal efficiency of 78% for 1–4 HP and 92.4% for more than 125 HP DC brushed motors. The brush drop increases with speed and current [1]. Due to presence of electronic commutation through the sensors or sensorless schemes, the permanent magnet brushless DC (PMBLDC) motor has added advantage over the brushed DC motor [2] which results in the improvement of efficiency. In BLDC drive system, the trapezoidal back EMF permanent magnet motor is fed from six-switch three-leg CSI or VSI triggered from sensor signals. Special logic designs for commutation are needed. These logic signals may be obtained from a sensor fitted with motor or with sensorless techniques. With the recent development in power switching devices, microcontrollers, and digital technology, the use of BLDC has increased day by day [3]. BLDC motor has the advantage of long lifetime, faster response, and capability of high speed operation comparatively to other motors [3]. Permanent magnet BLDC motor is widely used in computers, electric vehicles, aerospace, robotics, military, automotive, industrial, and domestic purpose due to its inherent ability of high torques, compactness, and high efficiency [4]. Higher power density and higher speed ranges are available with BLDC motors comparatively to others with respect to same steady state values of input power. The efficiency up to 94.6% has been recorded for a four-pole BLDC motor [5]. BLDC can be operated efficiently at higher speed ranges if the motor construction is appropriately modified. A BLDC machine of 1KW with special mechanical and electromagnetic construction and a special converter control scheme giving excessive reduction in various losses, proposed by Zwyssig et al. [6, 7], has achieved ultrahigh speed of 500,000 rpm intended for the use in microgas turbine generator and turbo compressor systems.

Hall sensors are used where low cost and low resolutions are needed [8]. For higher resolution of angle measurement, optical encoders are used [8]. Apart from these various other sensors, e.g., magnetic sensors are also used. However, hall sensors are better way to sense the position at low speeds or zero speed. Sensors used for position and speed have number of drawbacks, e.g., increased cost and complexity in control schemes [9]. Vibration and heating of machine and sensor circuitry, during operation on high speed, leads to reduced reliability and failure possibility of sensing system [9]. The device noise and delay [9] together with reduced drive robustness [10] and increased machine size [11] limits the scope of the sensor based control. Hall sensors lose its sensing capability above 120 °C [12]. These limitations of the sensor based system lend us to explore the techniques of sensorless operation of BLDC motor. Due to problems associated with the use of hall sensors or decoders, the sensorless operation is being developed. The milestone work on the position sensorless was first initiated by Frus and Kuo [13], in which they used the "waveform detection" technique for calculating the rotor position of voltage-fed variable-reluctance stepping motors by analyzing the current waveforms [14]. Many researchers have reported their work on sensorless techniques which enable the control of position speed and torque without shaft-mounted position sensors [11, 15, 16]. Generally, the sensorless operation involves some errors, e.g., phase shifter position error, commutation error, error due to wrong threshold which may deteriorate the performance of drives implementing sensorless control scheme [17], and should be taken care of during operation. Due to these problems, sensorless operation has been a major obstacle in the use of BLDC drives for industrial applications. Also, the sensorless techniques cannot be used at very low speed or zero speed because sensorless technique always needs the back EMF obtained by either sensing or estimation which is not possible in standstill. The sensorless control at zero speed is still a challenge.

In the indirect back EMF method, we estimate the rotor position directly by detection of ZCP of the terminal voltage of idle phase with respect to neutral which may be independent or may be connected to negative terminal of DC source. Indirect back EMF method presented in [17] in which the back EMF is obtained from the difference between the voltage floating terminal or off phase (during freewheeling) and DC source ground and does not require the motor neutral point. The virtual neutral point in place of motor neutral was originally proposed by Iizuka et al. [18] and Erdman [19]. Virtual neutral point based method uses the resistive and capacitive network connected at middle of the virtual neutral point and floating terminal for comparison [20]. A similar indirect method was reviewed in [21]. The main problem of indirect back EMF method is that the motor neutral or auxiliary neutral point is needed. Voltage dividers comprised of resistive network and low pass filters for signal processing are connected. Filters wipe out the switching noises in motor neutral but introduce a phase delay in ZCP which depends on speed. So making this method is suitable only for narrow range of operating speed. Small signal to noise ratio in the voltage measurement occurs during low speeds when using indirect back EMF detection method.

In direct back EMF method, two PWM strategies were used in [21, 22]. In "one side chopping mode," one of the switches in the circuit is given PWM and another switch is ON during complete step. In "two side chopping mode," both the switches in closed circuit with two motor phases are fed from same PWM signal. It is further processed with the help of comparators and logic circuits to get back EMF information directly. For further improvement in extraction of back EMF, midpoint of DC link comprised of two registers is used for calculating reference voltage, and this voltage was compared with the midpoint voltages of the resistive network connected between phases and DC ground [21].

The detected back EMF may not be same as the real back EMF but it may be vary depending upon the electromagnetic design of the motor. A method of restoration of real back EMF in case of slot shape and size, skewing and field magnet design, has been discussed in [3], which shows some deviation of real back EMF to that of ideal trapezoidal back EMF. For getting the commutation signals, the ZCPs of phase back EMF are delayed to 30° which is done by filters and register circuits, and this reduces performance. Detection of line to line BEMF proves to be more appropriate when it is required to get zero difference points (ZDPs) of phase BEMFs [23, 24] or when direction independent U-function [25] has to be obtained for position sensorless commutation. The peak magnitude of the line to line back EMF is higher than that of the phase back EMF. So it is easy to detect at low speed. The ZDPs can be obtained by comparing the line to line detected BEMF to zero. Also, it is important in case when the virtual hall signal based commutation for sensorless control is required. Damodharan and Vasudevan [26] proposed a sensorless drive that detects back EMF from line voltage difference. Kim et al. [27] proposed another method based on terminal voltage difference. For getting BEMF, these line to line voltages appearing between two phase terminals are filtered. The simple and cost-effective digital low pass filtering inside controller are an appropriate alternative to bulky analog filters outside controllers so as to reduced space and component requirement. Field Programmable Gate Array (FPGA) provides an excellent solution for the development of new control tools to enhance the performance of highly complex control algorithms. FPGA has numerous advantages like programmable hardwired feature, fast time to market, shorter design cycle, embedding processor, low power consumption, and higher density for the implementation of the digital system. The FPGA based Least *P*th-norm IIR low pass filter has been realized for the line to line BEMF detection using line to line terminal voltage measurement. The HON_LPWM method has been used for the speed control of the BLDC motor at variable as well as fixed DC link voltage of 300 V.

2 Proposed Scheme of FPGA Based Line to Line BEMF Detection with Speed Control

The complete proposed system has been shown in Fig. 1. Single-phase AC supply is fed to the rectifier and then it is filtered through the DC like capacitor. This DC link voltage is fed to the six-switch three-leg VSI connected to motor. Hall sensor inputs and line current and line voltages of the BLDC motor are fed to the FPGA–WAVECT

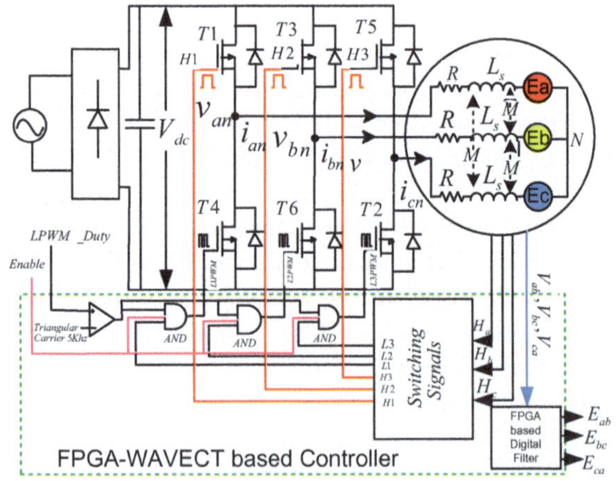

Fig. 1 Complete scheme of the proposed FPGA based back EMF detection with speed control

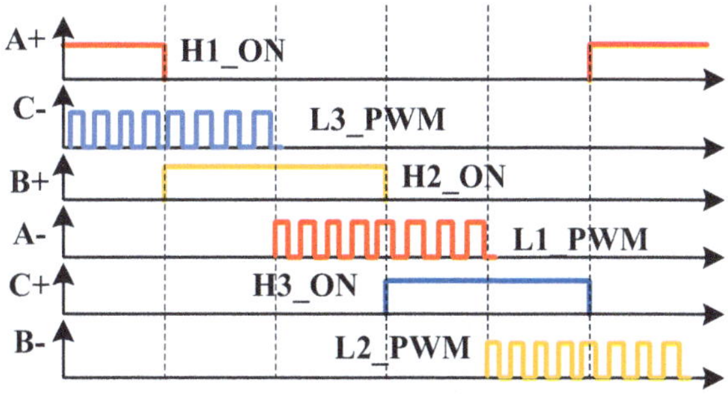

Fig. 2 HON_LPWM switching strategy for current control and back EMF extraction

controller. The PWM switching signals as output from FPGA–WAVECT controller are fed to the switches.

In the proposed system, the lower switches are operated with PWM of desired duty cycle as per required current for reference speed. Upper switches are fed with fundamental switching signals. The HON_LPWM strategy used in this system has shown in Fig. 2. Suppose hall sensors are giving the sector outputs such that A+ and C− have to be switched ON. Then the upper switch T1 is provided with H1_ON pulses as shown in Fig. 2 and T2 switch is provided with L3_PWM pattern. So the switch T1 and T2 are fed from HON _LPWM strategy. When motor rotates and next sector has given new hall signals levels, then B+ and C− phases have to be energized. For this, lower switch T2 is fed with L3_PWM while positive side switch T3 is fed with H2_ON. So, again the strategy for T2 and T3 is HON_LPWM. The sequences of the pulses are H1_ON, L3_PWM, H2_ON, L1_PWM, H3_ON, and L2_PWM which are fed to T1, T2, T3, T4, T5 and, T6, respectively.

3 FPGA–WAVECT Based Implementation

Field programmable gate array (FPGA) based control of systems is widely used in the applications such as industrial and space mechatronics systems demanding high performance, accuracy, and reliability. FPGA based controllers offer advantages such as high processing speed and high functional integration. A field programmable gate rray (FPGA) is a programmable logic device that supports implementation of relatively large logic circuit. FPGAs are quite different from simple programmable logic devices (SPLDs) and complex programmable logic devices (CPLDs). FPGAs provide logic blocks for the implementation of required functions.

3.1 FPGA–WAVECT Modules

FPGA based WAVECT is real-time digital simulator and controller which uses the Vivado 2014.4 design suite for Xilinx system generator integrated with MATLAB 2014b. ZynqTM-7000 SoC XC7Z020-CLG484-1 FPGA board is installed inside the WAVECT and default configuration WCU 200 of WAVECT release 2017.10 was used during the testing of proposed scheme. The WAVECT controller is appropriate when we need to capture large number of the output signals. It has the capability to make the algorithm schematically using the Xilinx based system generator toolbox in Simulink environment and transfer the code to the controller for hardware implementation. In this system up to 32 channels, virtual input and outputs together with speed encoder and hall sensor input interface are available for control and monitoring. The line to line back EMF detection scheme has been analyzed together with the speed control using FPGA–WAVECT system.

3.2 Complete Drive System

The schematic diagram representing the hardware realization of the scheme and real-time connection view has been shown in Fig. 3. The VSI is fed from DC link voltage which is rectified from the AC output from autotransformer the autotransformer. WAVECT module consists of four current and four voltage sensors. The converter output terminals connected through the shielded cables to the input of the WAVECT terminals. The output from the WAVECT unit is connected to three phase terminals. The current and the voltages are measured and the PWM is generated according to the

Fig. 3 FPGA based hardware implementation of BLDC drive system for HON_LPWM based speed control and line to line BEMF detection

Table 1 Motor parameters

S. No.	Parameters	Value	Unit
1.	Continuous stall torque	4	Nm
2.	Continuous current	3.16	A
3.	Electrical time constant	4.37	ms
4.	Rated speed at 300 V DC	1800	rpm
5.	Torque constant	1.58	Nm/A
6.	Voltage constant	135.4	VL-Lpk/krpm
7.	Resistance	7.88	Ω
8.	Inductance	34.4	mH
9.	Inertia	6.05	kg cm^2
10.	Mechanical time constant	1.92	ms
11.	Number of poles	8	No.
12.	FPGA simulation time	25×10^{-9}	s

control scheme. These six pulses are fed to IGBT switches. Details and parameters of the implemented motor have been given in Table 1. Figure 4 shows the MATLAB/Simulink and Xilinx based simulation model for hardware implementation.

3.3 *Commutation Scheme*

The commutation logic has been shown in Fig. 5 as per Table 2 for clockwise direction of rotation. Based on the logical value of hall sensor signals, the appropriate phases are turned ON or OFF. Suppose, the hall signal H_a is 1 and other hall signals H_b and H_c are 0, then switch T3 is energized for B+ and T2 is energized for C−. Similarly, all other switches are also operated based on the hall signals.

3.4 *HON_LPWM Scheme*

Xilinx based HON_LPWM scheme has been shown in Fig. 6 in which an enable signal is given to start or stop the motoring operation. A triangular carrier signal is generated using the counter. This triangular output is compared to appropriate duty cycle for reference current required for the reference speed.

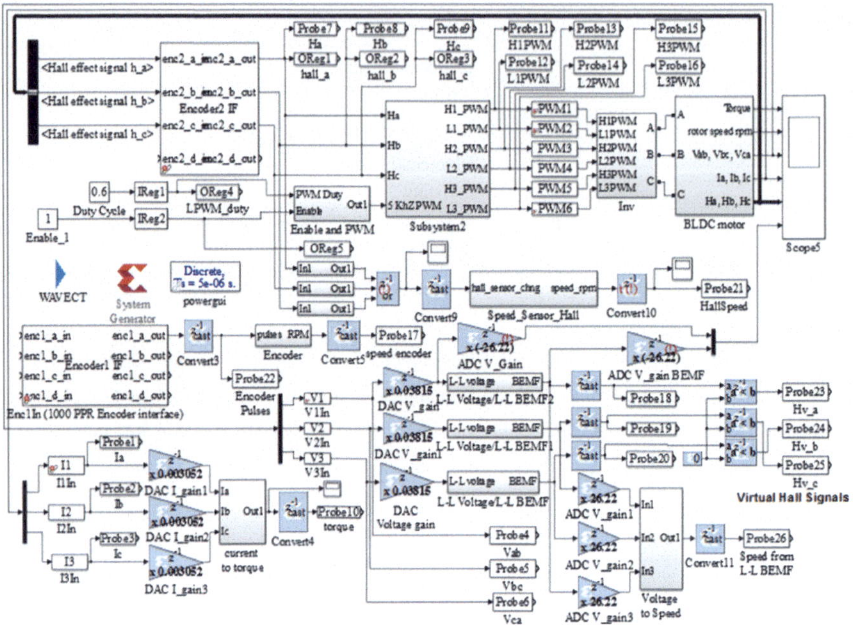

Fig. 4 MATLAB/Simulink and Xilinx based simulation model for hardware implementation of the HON_LPWM strategy with BLDC motor for speed control and BEMF detection

Table 2 Commutation logic for hall based operation

H_a	H_b	H_c	CW					
			A+	B+	C+	A−	B−	C−
1	0	1	0	0	1	1	0	0
0	0	1	0	0	1	0	1	0
0	1	1	1	0	0	0	1	0
0	1	0	1	0	0	0	0	1
1	1	0	0	1	0	1	0	0
1	0	0	0	1	0	0	0	1

3.5 Filtering of L–L Voltage for L–L BEMF Detection

For getting line to line BEMF with trapezoidal waveshape, the filter is implemented. Further, the need for filter to minimize the difference between actual and desired response of magnitude is solved using least *P*th design method. During the filtering, it is assured that there is least phase delay between the filtered back EMF and actual phase to phase measured voltage at two terminals. For reduced complexity of the filter with reduced order and considerable frequency to DB magnitude, the Least

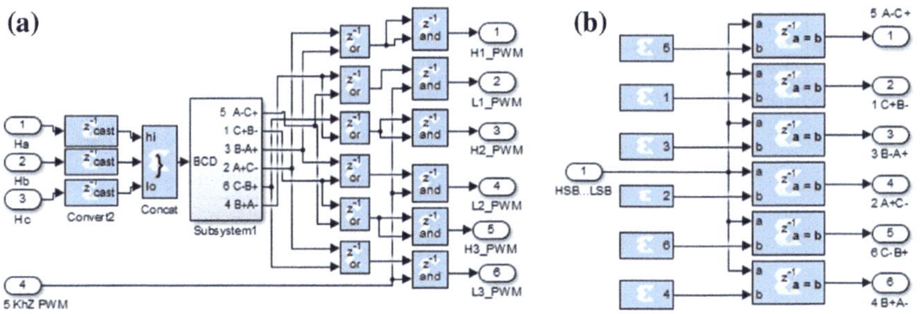

Fig. 5 **a** Phase commutation scheme based on the hall sensor signals, **b** sequence signal generation for switches

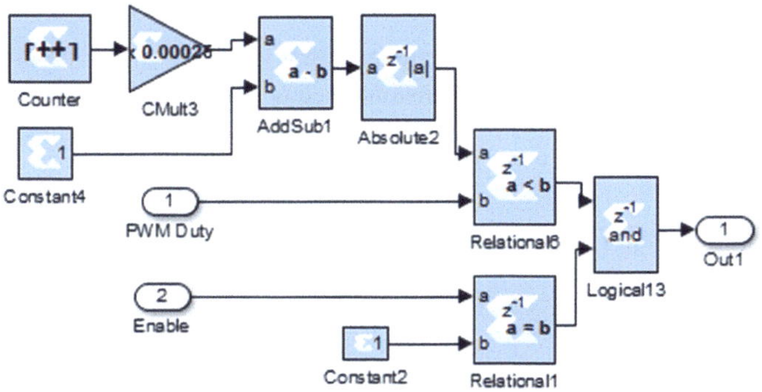

Fig. 6 Enable input and 5 kHz PWM generation based on input duty cycle for speed control and BEMF detection using HON_LPWM strategy

*P*th-norm IIR filter with numerator order 1 and denominator order 2 is used. The *P*th optimization as a design tool is not new. It was used quite successfully for the minimax design of IIR filters. The method does not need to update the weighting function, and it is an unconstrained convex minimization approach. The approach has advantages as filter quality, mathematical verification of the properties such as causality, stability, etc. using the pole zero and magnitude plots. The Least *P*th norm algorithm has a larger gradient driving it to converge faster when away from the optimum. Let, $H_d(w)$ be the desired frequency response over a frequency range which is a compact subset of $[0, \pi]$ and $H(w)$ be the *m*th-order transfer function given by [28],

$$H(w) = h0\frac{1 + b^{\mathrm{T}} f}{1 + a^{\mathrm{T}} f}, \tag{1}$$

$$f = \left[z^{-1} z^{-2} \ldots z^{-m}\right]^{\mathrm{T}} \tag{2}$$

$$a = [a_1\, a_2 \ldots a_m]^{\mathrm{T}},\ b = [b_1\, b_2 \ldots b_m]^{\mathrm{T}} \tag{3}$$

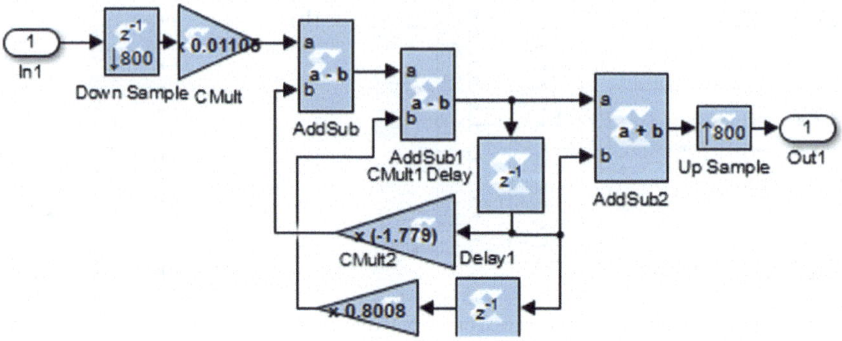

Fig. 7 Least *P*th-norm filter with numerator of first order and denominator of second order

and $z = e^{jw}$ with w the normalized frequency, that is $0 \leq w \leq \pi$. The filter design problem is to find real-valued M dimensional vector ($M = 2m + 1$).

$$X = [h_0 a_1 \ldots a_m b_1 \ldots b_m]^{\mathrm{T}} \tag{4}$$

That minimizes the *P*th norm objective function

$$E(x) = \int |H(w) - H_{\mathrm{d}}(w)|^P \mathrm{d}w \tag{5}$$

where $p \geq 2$ is an even integer and that the corresponding filter (6) is stable.

$$H(w) = \frac{B(w)}{A(w)} = \frac{\sum_{n=0}^{N} b(n)e^{-\mathrm{j}wn}}{1 + \sum_{n=1}^{M} a(n)e^{-\mathrm{j}wn}}. \tag{6}$$

where $B(w)$ is the Fourier transform of the forward coefficients, $A(w)$ is the Fourier transform of the backward coefficients, $b(n)$ is the set of forward coefficients, and $a(n)$ is the set of reverse coefficients.

The sampling frequency for the filter is kept 20 kHz. The pass frequency is kept at 300 and stop frequency is kept 800 Hz. Before filtering, the downsampling of 800 (filter sampling time is 800 times FPGA sampling time) has been used for matching to filter sampling time. After the filtering process through IIR Least *P*th-norm filter, upsampling of 800 has been done for matching to FPGA sampling time (Fig. 7).

3.6 Speed Calculation from Hall Sensors and Encoders

The schemes for speed measurement from detected line back EMF, hall signal changes and 1000 PPR optical encoder pulses have been shown in Figs. 8, 9 and 10 respectively. Total number of hall sensor signal changes is given by, $H_{\mathrm{sch}} = P \times$

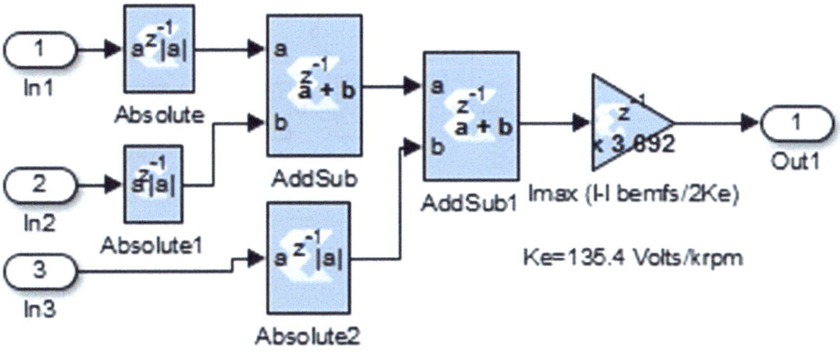

Fig. 8 Speed measurement from line to line back EMF

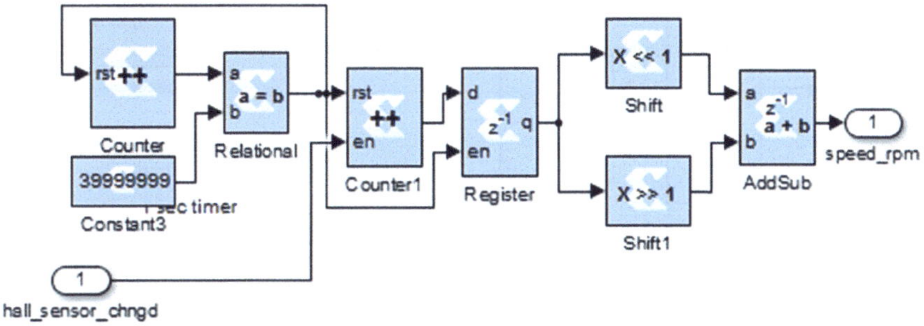

Fig. 9 Speed measurement from hall signal changes

$H_n \times 2$. Total 24 state changes in 1 min give one rpm. A counter is incremented to record time up to one minute. To calculate 1 s, counter should count up to 39,999,999 at FPGA sampling time of 25 ns. So each hall signal change will give $60 \times 25\text{e}{-9} \times$ 40,000,000/24 rpm. Also, the speed from 1000 PPR external encoder attached with motor is digitally detected using FPGA blocks as shown in Fig. 10. Each encoder pulse will count $60 \times 40{,}000{,}000/1000$, i.e., counter will increment up to 2,399,999 from 0 and then reset and during one cycle of count and reset, it will give 1/1000 rpm. Estimation of the speed from extracted line-to line back EMF uses the back EMF constant.

4 Simulation Results

Simulation result for operation at 300 V DC, 1.92 Nm load, duty cycle of LPWM 78.125%, operating speed of 1580 rpm and PWM frequency is 5 kHz (Figs. 11, 12, 13, 14, 15, and 16).

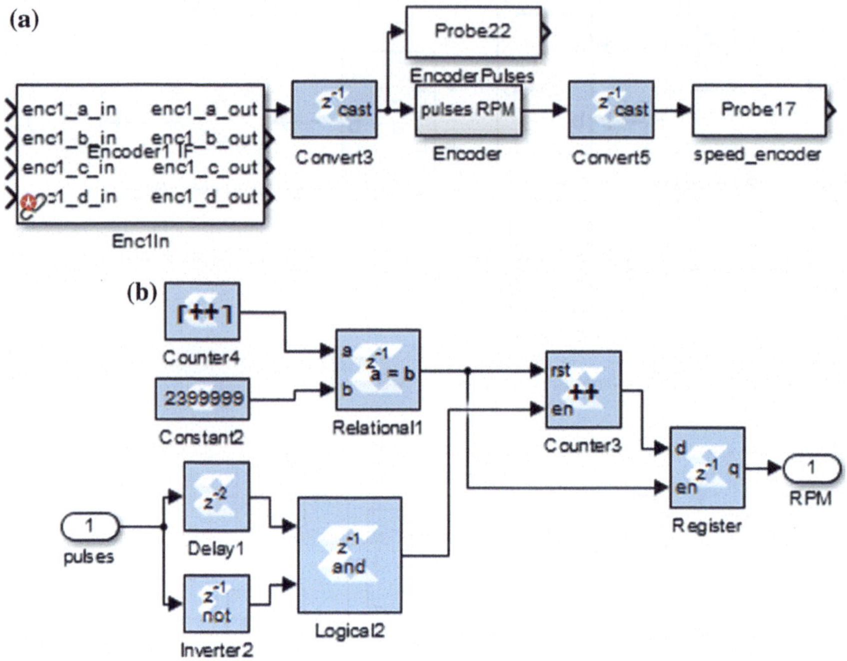

Fig. 10 **a** Encoder interface from hardware, **b** speed measurement from 1000 PPR encoder

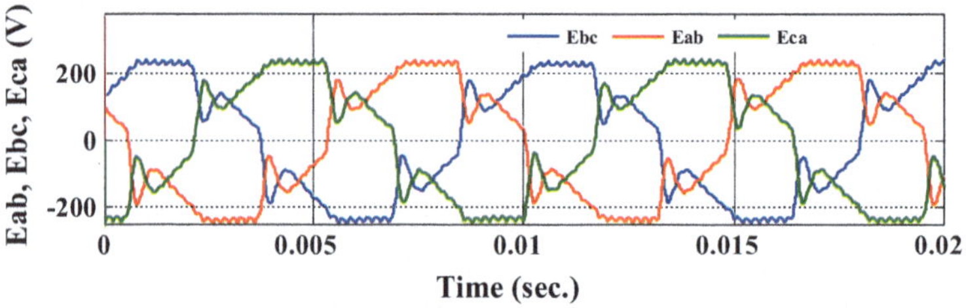

Fig. 11 Three line to line back EMFs at 1580 rpm

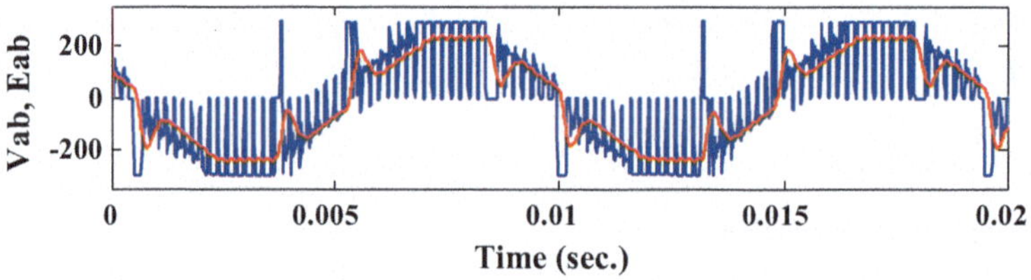

Fig. 12 Terminal voltage versus back EMF

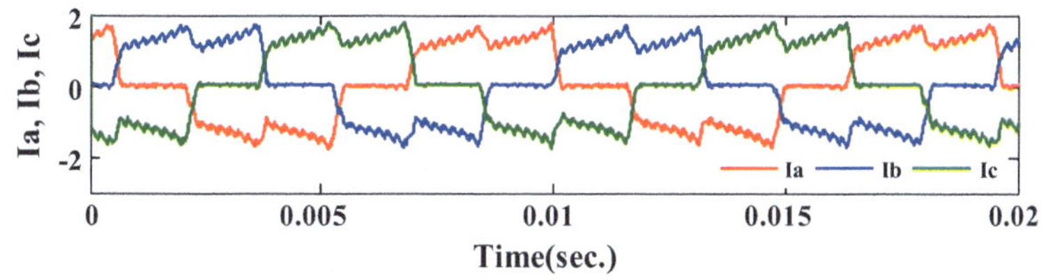

Fig. 13 Three-phase currents

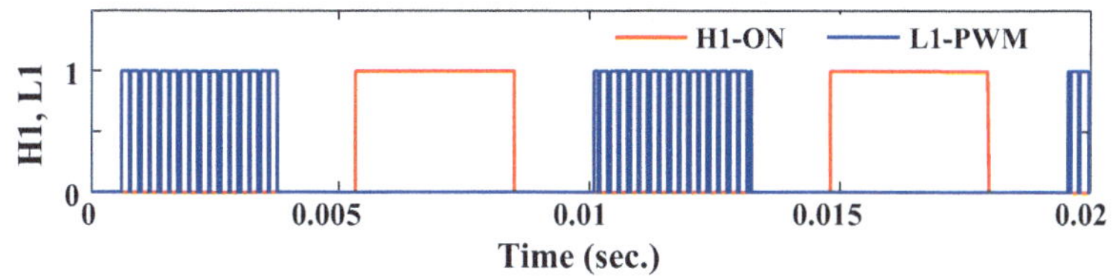

Fig. 14 H1_ON, L1PWM signals

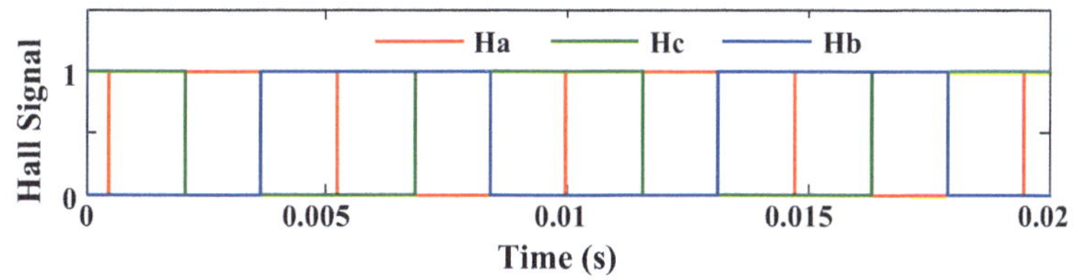

Fig. 15 Real hall sensor signals

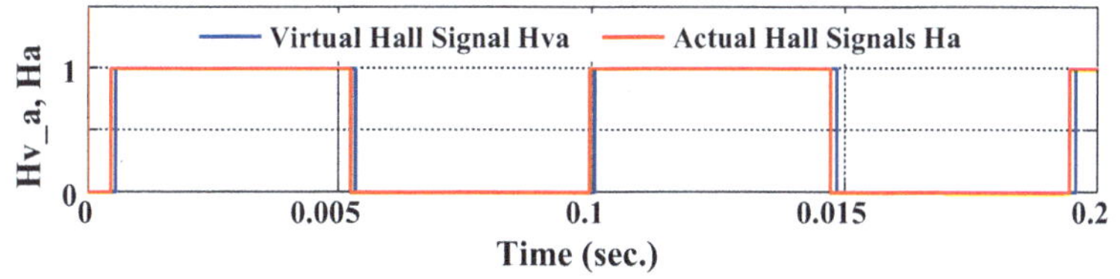

Fig. 16 Virtual hall signal versus actual hall sensor signals

5 Hardware Results

The hardware results of proposed line to line BEMF detection with the speed control has been discussed in the following sections. First, load and voltage are increased and duty cycle is kept 100%. Next, the duty cycle is varied with load and DC link voltage kept fixed. Results are shown in Figs. 17, 18, 19, 20 and 21.

Case 1: Fix DC voltage of 100 V DC (736 rpm), D_LPWM = 100%, no load (Fig. 17)

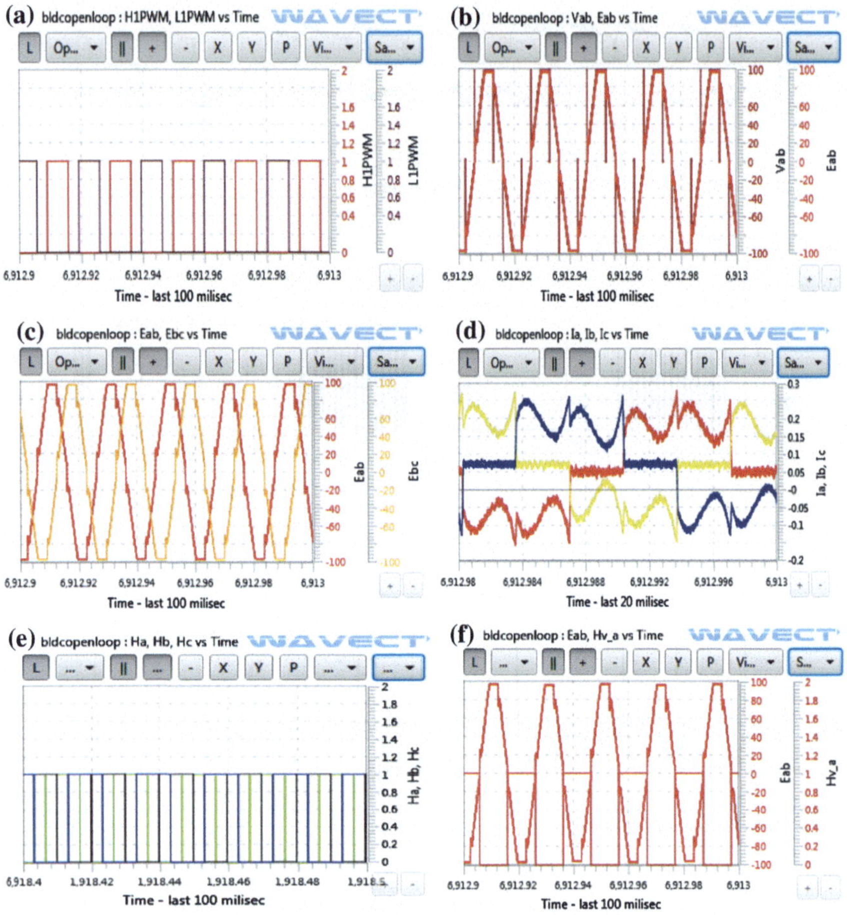

Fig. 17 **a** HON_LON strategy line to line voltage without speed control, **b** actual line to line voltage and line to line BEMF filtered from L–L voltages, **c** two line to line BEMFs, **d** three phase no load current, **e** hall sensor signals, **f** virtual hall sensor signal from one of line to line BEMF

Case 2: Vdc = 300 V, 50% rated load (1.92 Nm) load D_LPWM = 100%, operating speed = 2080 rpm (Fig. 18)
Case 3: 300 V DC, 1.92 Nm load and D_LPWM = 78.125%, operating speed = 1580 rpm, PWM frequency = 5 kHz (Fig. 19)
Case 4: 300 V DC, 1.92 Nm load and D_LPWM = 59.375%, operating speed = 1175 rpm, PWM frequency = 5 kHz (Fig. 20)
Case 5: 300 V DC, 1.92 Nm load and D_LPWM = 21.875%, operating speed = 394 rpm, PWM frequency = 5 kHz (Fig. 21)

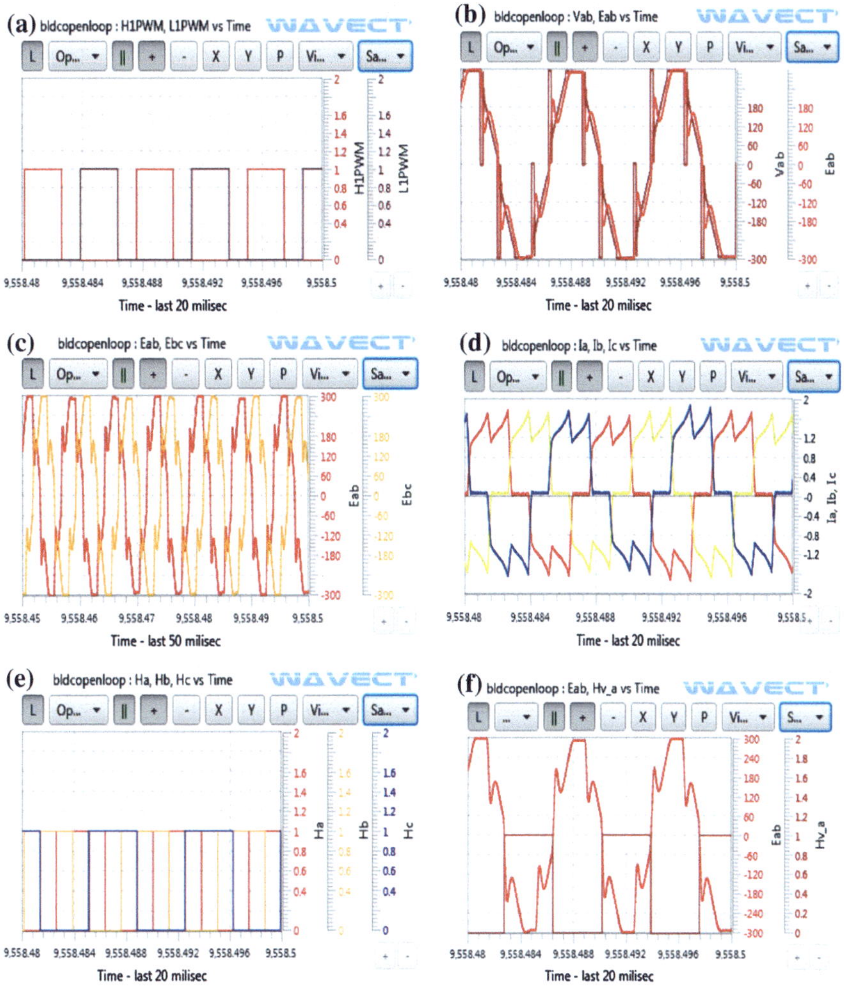

Fig. 18 At 300 VDC, 1.92 Nm load and D_LPWM = 100% **a** HON_LON strategy line to line voltage without speed control, **b** actual line to line voltage and line to line BEMF filtered from L–L voltages, **c** two line to line BEMFs, **d** three phase load current, **e** hall sensor signals, **f** virtual hall sensor signal from line to line BEMF

6 Analysis with Varying DC Link Voltage and Load

The outcome of comparative analysis regarding the effects of variation of DC link voltage and load torque on the phase current magnitude, BEMF magnitude, percentage current ripple, and speed at fixed PWM duty cycle of 100%, has been shown in Fig. 22. Current increases with increase in load but decreases with increase of DC link voltage. Line to line BEMF magnitude is proportional to the voltage. Speed is proportional to voltage. Phase current ripple reduces with the load.

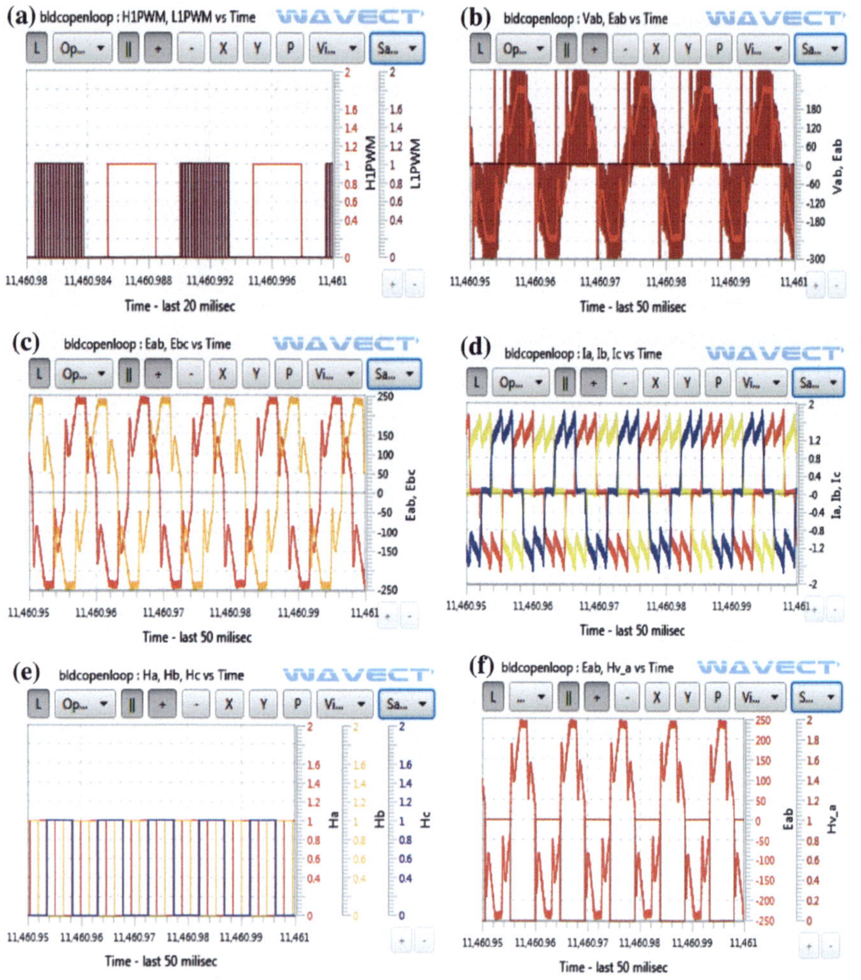

Fig. 19 At 300 VDC, 1.92 Nm (50%) load and D_LPWM = 78.125% **a** HON_LON strategy line to line voltage without speed control, **b** actual line to line voltage and line to line BEMF filtered from L–L voltages, **c** two line to line BEMFs, **d** three phase load current, **e** hall sensor signals, **f** virtual hall sensor signal from line to line BEMF

7 Analysis with Varying Duty Cycle at Fixed DC Link Voltage and Fixed Load

Analysis with variable duty cycle at fix DC link voltage and fix load is shown in Fig. 23. The intermittent peaks occur in the line to line filtered BEMF with certain range of duty cycles.

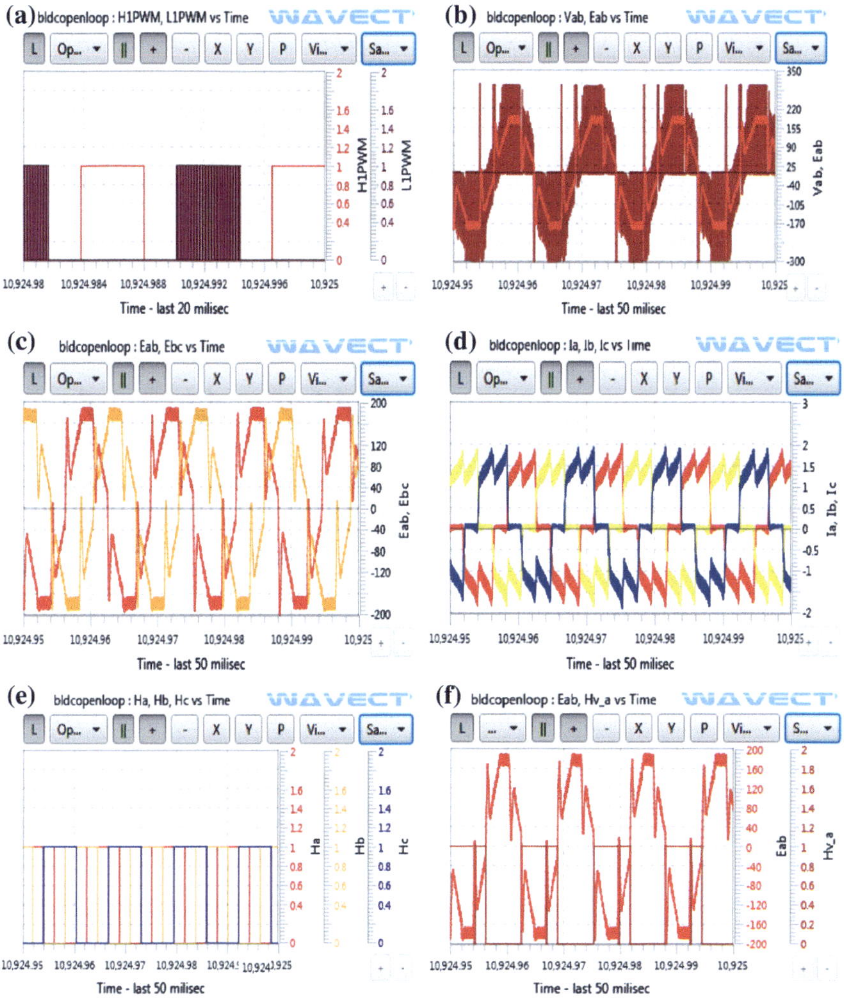

Fig. 20 At 300 VDC, 1.92 Nm (50%) load and D_LPWM = 59.375% **a** HON_LON strategy line to line voltage without speed control, **b** actual line to line voltage and line to line BEMF filtered from L–L voltages, **c** two line to line BEMFs, **d** three phase load current, **e** hall sensor signals, **f** virtual hall sensor signal from line to line BEMF

8 Conclusion

In proposed work, the FPGA based scheme of line-to-line back EMF detection from line-to-line terminal voltage was presented using the Least Pth-Norm IIR digital low pass filter with hall sensor based position control. It has been verified that near ideal trapezoidal waveshape of line-to- line back EMFs can be achieved if the proper selection of filters cutoff frequency and gain are exercised. It can be observed that

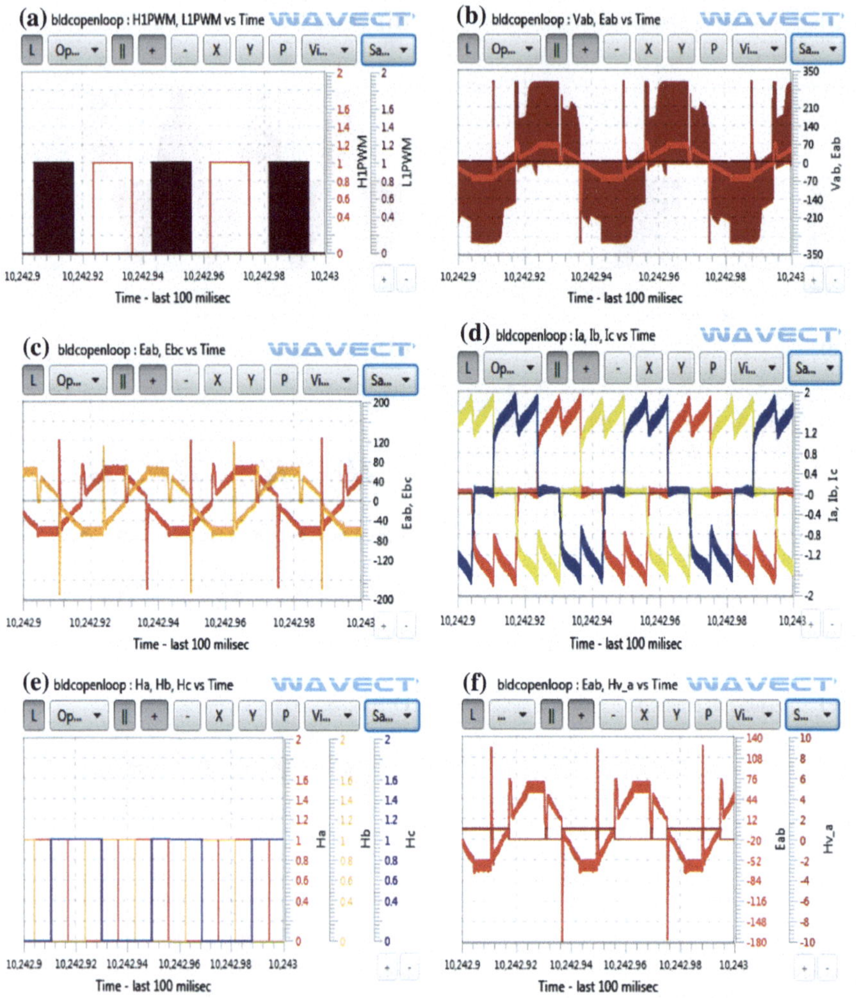

Fig. 21 At 300 VDC, 1.92 Nm (50%) load and D_LPWM = 21.875% **a** HON_LON strategy line to line voltage without speed control, **b** actual line to line voltage and line to line BEMF filtered from L–L voltages, **c** two line to line BEMFs, **d** three phase load current, **e** hall sensor signals, **f** virtual hall sensor signal from line to line BEMF

the speed varies linearly with LPWM duty cycle. Furthermore, the magnitude of flat portion of filtered BEMF from line to line BEMF also varies linearly with respect to duty cycle. For a fixed filtering frequency and system sampling, intermittent peaks due to PWM noise and spikes due to diode freewheeling as well multiple zero crossings of line back EMFs depend on the duty cycle. Up peaks in both positive and negative side do not cross the zero line at any duty cycle from 0 to 100%. However, down peaks become more prominent at certain duty cycles. Positive side down peak increases and crosses zero when duty ratio reduces below 40%, while the down peak in the

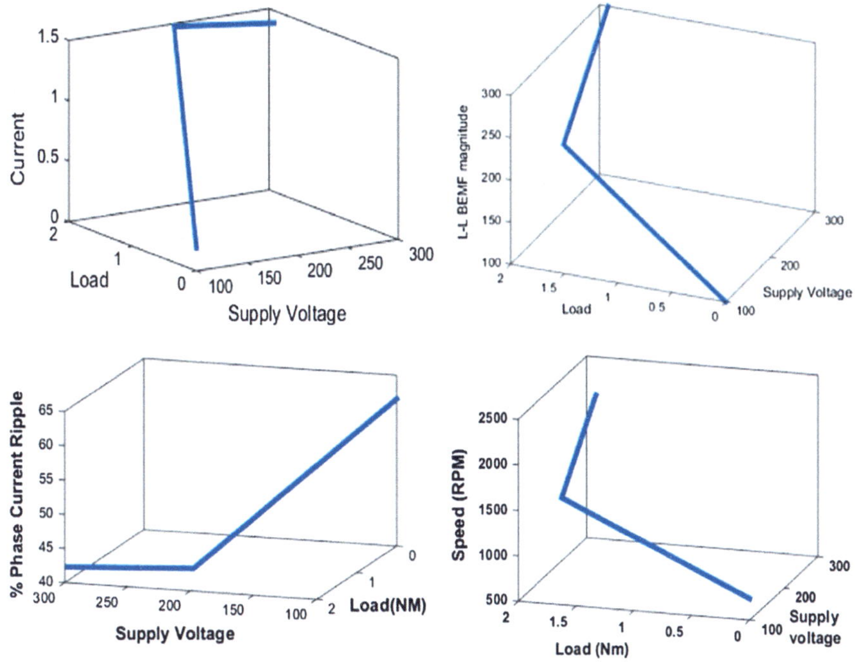

Fig. 22 Effect of variable DC link voltage on current, L–L BEMF phase current ripple at fix duty cycle of LPWM at 100%

negative side increase and crosses zero after about 62% duty cycle. This zero crossing behavior of up and down spikes can lead to improper commutation signals and may adversely affect the sensorless drive performance when back EMF is used for sensorless operation. So with this filter, a duty cycle limitation has got involved, i.e., 40–60%. This limitation can be removed if proper filtering with waveshape preservation is done together with least delay in the phase back EMF zero difference points or ZCPs of line to line back EMF. Various other strategies still remain to be investigated and tested experimentally with various filter configurations with FPGA based WAVECT controller for better waveshaping, less delay, with 0–100% duty cycle range. Lower sampling frequency and lower cutoff frequency may provide better filtering but can introduce phase delays between actual and extracted back EMFs. Coefficient optimization and phase delay compensation may provide right zero crossing points of line to line back EMF for sensorless application.

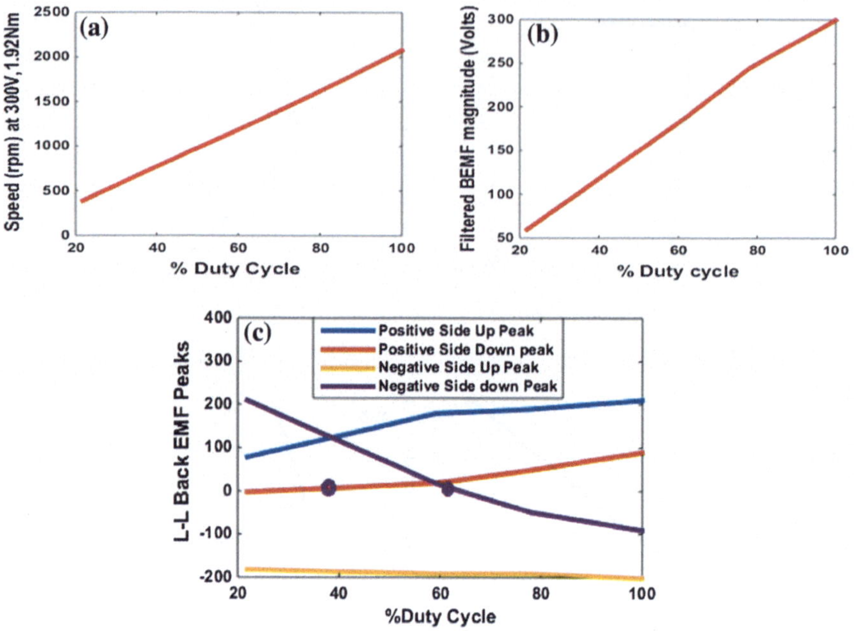

Fig. 23 Variation in **a** Speed, **b** line to line BEMF and **c** intermittent peaks in BEMF with respect to duty cycle

References

1. Kesava Rao MV (1934) Brush contact drops in DC machines. Bangalore Press, Accepted 25-6-1934
2. Tashakori A, Ektesabi M, Hosseinzadeh (2010) Characteristic of suitable drive train for electric vehicle. In: 2010, 3rd international conference on power electronics and intelligent transportation System (PEITS), 20–21 Nov 2010, China
3. Jeon YS, Mok HS, Choe GH, Kim DK, Ryu JS (2000) A new simulation model of BLDC motor with real back EMF waveform. In: 7th workshop on computers in power electronics (COMPEL 2000), pp 217–220
4. Kim TH, Ehsani M (2004) Sensorless control of BLDC motors from near zero to high speeds. IEEE Trans Power Electron 19(6):1635–1645. ISSN 0885-8993
5. Tashakori A, Ektesabi M, Hosseinzadeh N (2011) Modeling of BLDC motor with ideal back EMF for automotive applications. In: Proceedings of world congress on engineering, vol II, WCE 2011, 6–8 July 2011, London U.K.
6. Zwyssig C, Duerr M, Hassler D, Kolar JW (2007) An ultra-high-speed, 500000 rpm, 1 kW electrical drive system. In: IEEE power conversion conference-Nagoya 2007, PCC'07
7. Zwyssig C, Duerr M, Hassler D, Kolar JW (2008) An ultra high speed, low power electrical system. IEEE Trans Ind Electron 55(2)
8. Padmaja Y (2003) Brushless DC (BLDC) motor fundamentals. AN885, 2003 Microchip Technology
9. Tara S, Syfullah Khan Md (2013) Simulation of sensorless operation of BLDC motor based on the zero cross detection from the line voltage. Int J Adv Res Electr Electron Instrum Eng 2(12). ISSN 2320-3765

10. Lee J, Sathikumar S, Shrivastava Y (2008) A novel speed and position estimation of the brushless DC motor at low speed. In: Australian universities power engineering conferences, (AUPEC'08)
11. Kim T-S, Park B-G, Lee D-M, Ryu J-S, Hyun D-S (2008) A new approach to sensorless control method for brushless DC motors. Int J Control Autom Syst 6(4):477–487
12. Wen J-S (2008) Intelligent control of high speed sensorless brushless DC motor for intelligent automobiles. In: IEEE international conference on man machine and cybernatics SMC 2008, pp 3394–3398
13. Frus JR, Kuo BC (1976) Closed-loop control of step motors using waveform detection. In: Proceedings of International Conference on stepping motors and systems, Leeds, U.K., pp 77–84
14. Acarnely PP, Watson JF (2006) Review of position-sensorless operation of permanent magnet machines. IEEE Trans Power Electron 53(2):352–362
15. Matsui N (1996) Sensorless PM brushless DC Motor drives. IEEE Trans Ind Electron 43(2):300–308
16. Xin K, Zhan Q, Luo J (2006) A new simple sensorless control method for switched reluctance motor drives. KIEE J Electr Eng Technol 1(1):52–57
17. Huang Y, Xin Y, Zhang W (2008) An improved BEMF detection method for sensorless BLDC motor. In: IEEE international conference in Industrial technology, ICIT 2008
18. Iizuka K, Uzuhashi H, Kano M (1985) Microcomputer control for sensorless brushless motor. IEEE Trans Indus Appl 27:595–601
19. Erdman D (1987) Control system, method of operating an electronically commutated motor and laundering apparatus. US patent No. 4654566
20. Shao JW, Nolan D, Hopkins T (2003) Improved direct back EMF detection for sensorless brushless DC (BLDC) motor drives. In: Proceedings of IEEE APEC, pp 300–305
21. Zhang L, Xiao W, Qu W (2006) Sensorless control of BLDC motors using an improved low-cost back EMF detection method. In: 37th IEEE power electronics specialists conference PECS
22. Shao JW, Nolan D, Hopkins T (2003) A novel microcontroller based sensorless brushless DC(BLDC) motor drives for automotive fuel pumps. IEEE Trans Indus Appl 39(6):1734–1740
23. Soni UK, Tripathi RK (2017) Novel back EMF zero difference point detection based sensorless technique for BLDC motor. In: 2017 IEEE international conference on industrial technology (ICIT), Toronto, ON, pp 330–335
24. Soni UK, Tripathi RK (2017) Novel estimated back EMF ZDP based sensorless controlled BLDCM using unknown input observer. In: International seminar on intelligent technology and its applications (ISITIA), Surabaya, pp 205–210
25. Soni UK, Tripathi RK (2017) Direction independent U-function based sensorless control of BLDC motor. In: 2017 IEEE international conference on industrial and information systems (ICIIS), Peradeniya, pp 1–6
26. Damodharan P, Vasudevan K (2010) Sensorless brushless dc motor drive based on the zero-crossing detection of back electromotive force (EMF) from the line voltage difference. IEEE Trans Energy Convers 25(3):661–668
27. Kim T, Kim C, Lyou J (2011) A new sensorless drive scheme for BLDC motor based on the terminal voltage difference. In: Proceedings of IEEE industrial electronics society conference, pp 1710–1715
28. Gokhale MY, Khanduja DK (2008) Analysis and synthesis of speech using least Pth norm filter design. Int J Speech Technol 11(1):51–61

Umesh Kumar Soni has obtained B.E. degree form JEC Jabalpur (MP), India in 2003. He has three years of Industrial experience and more than 4 years of academic experience. He has completed his M.E. in 2011 in control System specialization from JEC, Jabalpur. Presently he is perusing the Ph.D. in MNNIT Allahabad, India. He interests in Control, Robotics, Fuzzy, Neuro-fuzzy, Wind Power System and Image Processing, Power Electronic Converters and Control, Superefficient Motors and Controls, Wireless Electricity.

Maloth Naresh received the Bachelor of Technology Degree in Electrical and Electronics Engineering from Indur Institute of Engineering and Technology, Siddipet, India, in 2012 and the Master of Technology degree in Power System from the MNNIT Allahabad, India, in 2014, currently working toward the Ph.D. degree in the area of Power Electronics application in Power System and renewable energy grid integration in NIT Allahabad. Research interests include Power flow control and power quality issues in distributed generation system (Wind, Solar energy system) in DC-microgrids, analysis of wind power plants.

Prof. Ramesh Kumar Tripathi has obtained Bachelor degree in Engineering from REC Durgapur, West Bengal in 1989, M.Tech. from IT-BHU, Varanasi in 1991 and Ph.D. from IIT Kanpur in 2002. He is Fellow IE and senior member IEEE. Presently he is working as Dean Academic and Professor in Department of Electrical Engineering, MNNIT Allahabad, India. He has more than 25 years of academic experience. He has many research publications in journals and conferences of international repute. His research interests are Power Electronics Switch-Mode Rectifiers and Power Supplies, Magnet Load Power Supplies, Power Factor Correction, Electrical Drives, Power Quality and Active Power Filters and Virtual Instrumentation.

Power Quality Enhancement Using FACTS Device in Transmission System with DPFC

Vandana Chaudhary, Krishna Dutt, Maloth Naresh and Ramesh Kumar Tripathi

Abstract Nowadays, power demand is increasing day by day and also ageing of the networks makes the system difficult for power flow control through the transmission lines. The voltage variation in transmission line should be limited which is due to the load changes; these problems are solved by using FACTS devices. A new device has been introduced called distributed power flow converter (DPFC), where distributed FACTS concept is utilized and common DC-link has been removed as present in the UPFC, and transmission line is used for the exchange of real powers between shunt and series converters. In this paper proposed carrier-less hysteresis band control (HBC) method has been used for controlling the pulses in series and shunt converters individually. The DPFC has same controllability as of the UPFC; a DPFC requires multiple numbers of one-phase series converters in place of a three-phase series converter of large rating. DPFC with single-phase series converters reduces cost and increasing reliability of the power; the whole research is implemented in MATLAB/SIMULINK.

Keywords UPFC · DPFC · Transmission lines · Series and shunt converters · Power transmission

1 Introduction

Due to the increasing demand and aging of networks, power has to be regulated fast and more reliably, and to achieve this objective, FACTS devices are used which are

V. Chaudhary · K. Dutt (✉) · M. Naresh · R. K. Tripathi
Department of Electrical Engineering, MNNIT Allahabad, Prayagraj, India
e-mail: dkrishna281093@gmail.com

V. Chaudhary
e-mail: vanu8392@gmail.com

M. Naresh
e-mail: ree0114@mnnit.ac.in

R. K. Tripathi
e-mail: rktripathi@mnnit.ac.in

S. N. Singh et al. (eds.), *Advances in Power and Control Engineering*, Lecture Notes in Electrical Engineering 609, https://doi.org/10.1007/978-981-15-0313-9_5

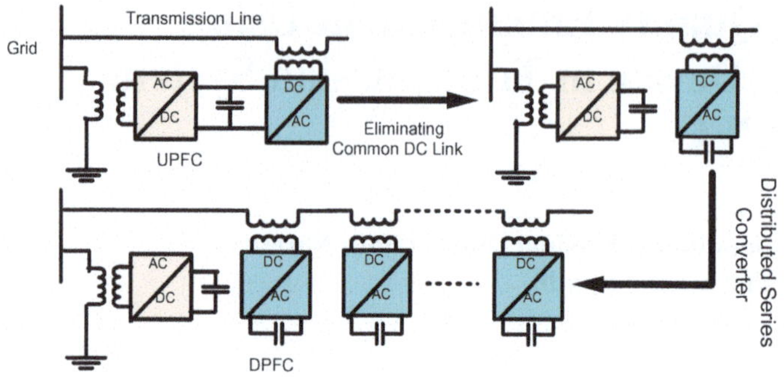

Fig. 1 Basic concept of UPFC to DPFC

high voltage and high power converters [1–3]. These FACTS devices are divided into four categories as per their usage and connections, namely: series, shunt, series-shunt and series-series [4–6].

A distributed FACTS (D-FACTS) device has been recently introduced which is less costly than UPFC and highly reliable as compared to other conventional FACTS devices and is known as DPFC [7–9]. DPFC is basically derived from UPFC and is much more advantageous than UPFC because it controls almost all power system parameters for transmitting the power such as transmission angle [10, 11], line impedance and magnitude of bus voltage for the availability of reliable power flow [12, 13]. DPFC is advantageous than UPFC because it removes common DC-link in between the shunt and series converters as in UPFC which is shown in Fig. 1. And, it uses the power line or transmission line for the exchange of real power at third harmonic frequency between the DPFC converters [14–17]. It employs multiple numbers of one-phase series converters instead of a large rating of three-phase series converter which reduces the cost and reliability of this controller, and it is the main and basic principle of this controller [18–20]. The DPFC can be referred as distributed FACTS device [21–23] because it uses the power line for the replace of real power between the converters when remove DC link in between DPFC converters by making them isolated with one another, whereas UPFC uses common DC link.

2 Operating Principle of Exchange Real Power Between DPFC Converters

The exchange real power between DPFC converters is derived from the method of 'power theory of non-sinusoidal components' [21]. As per the Fourier analysis, it can be said that the non-sinusoidal signals are expressed as summation of the sinusoidal signals at various frequencies having different magnitudes [22]. The multiplication of mean values of non-sinusoidal signals of currents and voltages results the real power. As the cross product of different frequency signals results zero, thus the real

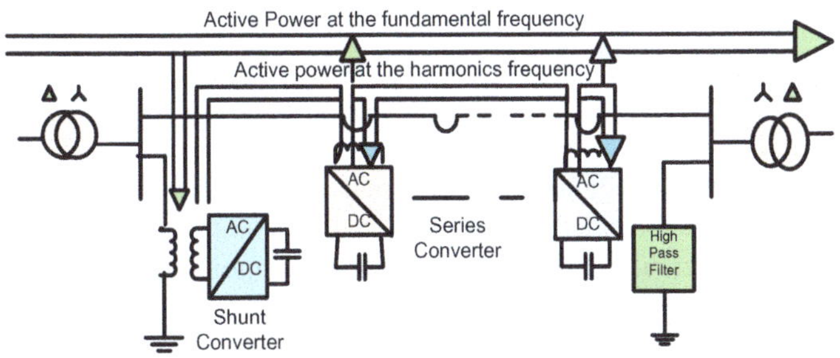

Fig. 2 Exchange of real power between DPFC converters

power can be given as:

$$P = \sum_{i=1}^{\infty} V_i I_i \cos \phi_i \tag{1}$$

Here I_i and V_i are currents and voltages at ith frequency and ϕ_i is the phase difference between currents and voltages at respective frequencies. Above expression represented as the power at various frequencies are independent as the product of voltages and currents of same frequency gives the real power at the same frequency [11]. The real power at certain frequency which can be absorbed power at some other frequencies. Using this method in DPFC [13], real power from the source at fundamental frequency can be absorbed and it can inject the same power back to the system at various harmonic components, thus harmonic power flows from the transmission lines connected with the series converters.

Figure 2 shows exchange of real power in between series and shunt converters of DPFC, and a HPF in the DPFC stops the fundamental component and supports harmonic frequency component. A closed loop path for harmonic current formed by the series converter, shunt converter, a HPF and ground.

3 Usage of Third Harmonic Component

As we know that the third harmonic components are identical in all the three phases; it means that they are equivalent to zero-sequence components and because of this property, only third harmonic component is selected for the exchange of real power in the DPFC [24]. As zero-sequence components are easily blocked by the star-delta-connected transformer, Δ does not have the neutral for the return path of third harmonic components (zero-sequence component), thus an extra filter is not required for the elimination of harmonic leakage [10, 11].

Since harmonic frequency is closer to cut-off frequency and voltage isolation is high, thus filter required would be costly. While using zero-sequence harmonic components, a cable is used in place of filter which connects as shown in Fig. 3. As Δ-winding will be acting as open circuited for third harmonic current, thus all the harmonic current passes through *Y*-winding and ground cable, thus a large HPF is eliminated.

Figure 4 shows, the path of harmonic current flow through grounding the star-delta transformer, as the transformer with floating neutral point without series converter which will be open circuited for the 3rd harmonic component thus there will be no 3rd harmonic component of current flow in the power line.

$$P_i = \frac{|V_{\mathrm{sh},i}||V_{\mathrm{se},i}|}{X_i}\sin(\theta_{\mathrm{sh},i} - \theta_{\mathrm{se},i}) \tag{2}$$

Above expression shows the real power exchange between *i*th frequency where X_i is the line impedance, $|V_{\mathrm{sh},i}|$, $|V_{\mathrm{se},i}|$are magnitudes of voltages and $(\theta_{\mathrm{sh},i} - \theta_{\mathrm{se},i})$ are the phasor difference among the two voltages at the *i*th frequency. The impedance of transmission line is almost inductive which is proportional to the harmonic frequency, the higher voltages in the converters [13, 14, 16]. Thus, zero-sequence components, i.e. third harmonic frequency is chosen for the operation of DPFC.

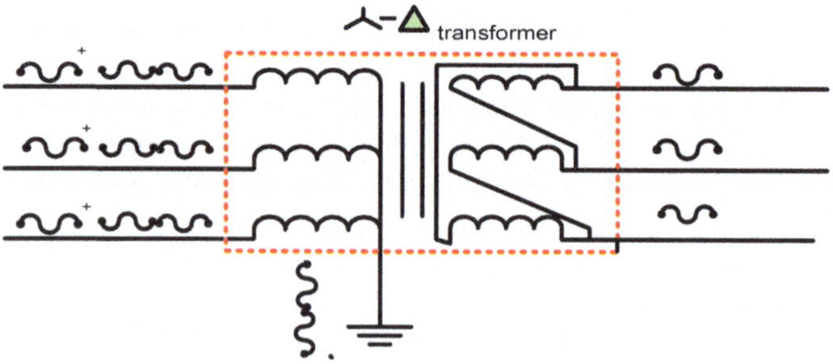

Fig. 3 Utilization of the grounded star-delta transformer for the path to zero-sequence third harmonic

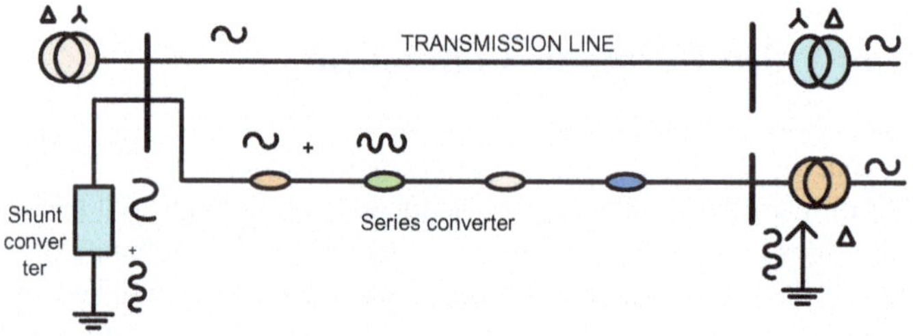

Fig. 4 Routing of harmonic current with grounding status of star-delta transformer

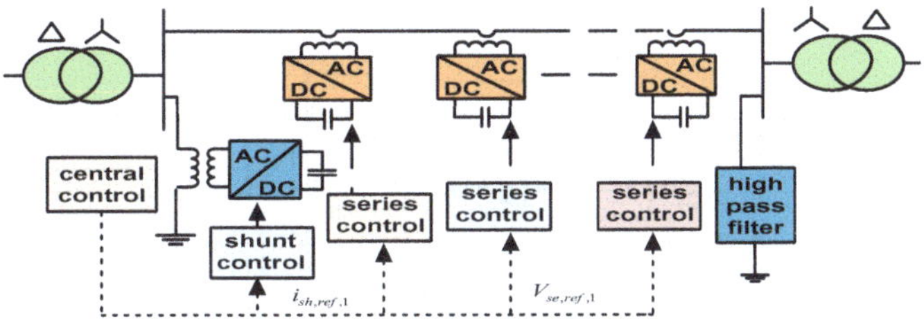

Fig. 5 Control scheme diagram

4 DPFC Control

DPFC is having basically three types of controllers for the controlling of its converter, these are: series control, central and shunt control is provide for the controlling of the DPFC. There are used for controlling their own converters which are localized.

4.1 Central Control

This controller is used for signals related to the fundamental frequency component and its controlling is based on the functions specified which is to be controlled by the use of DPFC's application such as balancing of asymmetrical components, low-frequency power oscillations, and power flow control. This control provides respective current signal for shunt converter (Fig. 5).

4.2 Shunt Control

It is used for the injection of third harmonic frequency current into the neutral of transformer to supply real power for shunt converter as shown in Fig. 6. Sequentially, it helps the capacitor for maintaining the constant DC voltage by absorbing real power at fundamental frequency from the source and injects the reactive current required at fundamental frequency in grid.

4.3 Series Control

The series converter are having their individual series control, and it is used to make the DC capacitor voltage at constant level of their individual converters with the help of third harmonic components; additionally, they generate series voltage signal in quadrature with line required as shown in Fig. 7.

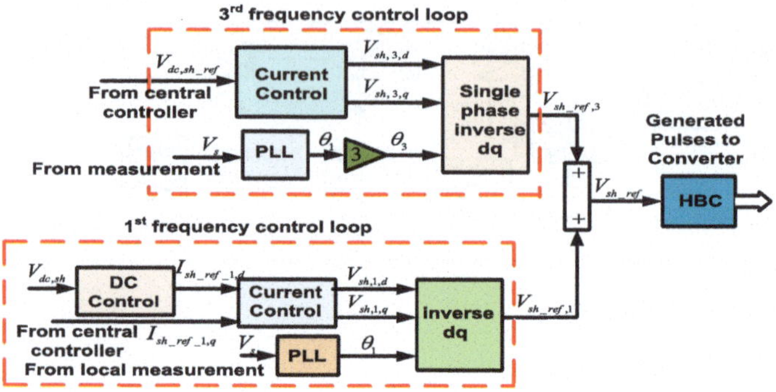

Fig. 6 Method for the control of shunt converter

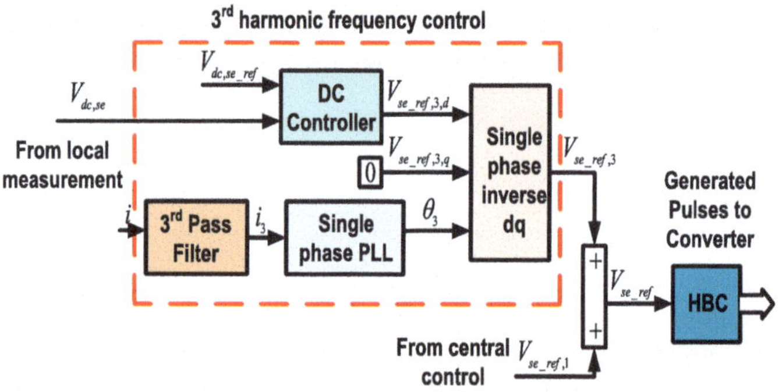

Fig. 7 Method for the control of series converter

4.4 Carrier-less Hysteresis Band Control in DPFC

This control is used for fundamental frequency because it does not require any carrier and gives a band for traction of signals within the specified band. It can be used for the control of both voltage and current signals by making a controlled VSI and CSI. This control technique provides good dynamics, good accuracy and conditional stability as shown in Fig. 8.

By limiting the signals into certain band, it controls the switches of VSI so that actual signal follows reference signals more accurately within the band. When the actual signal is less than the reference signal, it forces the signal to increase that amount with respect to the reference signal; similarly, when it is more than the reference signal, then it forces the signal to reduce that amount accordingly gate pulses are being generated for both the converters. Voltage across the capacitor used in shunt converter must be greater than 2.45 times the line voltage, and the capacitor voltage across series converter must be almost equal to line-to-neutral voltage so that continuous variations can be corrected by tracking the reference signal more accurately.

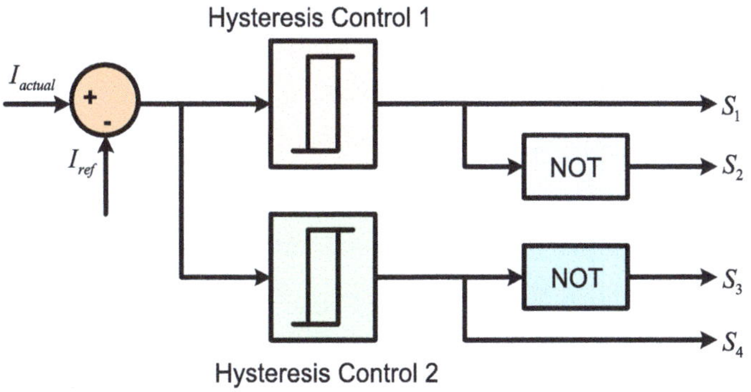

Fig. 8 Proposed carrier-less hysteresis band control for pulse generation

DPFC is equivalent to the UPFC which utilizes the concept of distributed FACTS (D-FACTS) and power exchange with the help of harmonics. Thus, DPFC consumes all the advantages of D-FACTS and UPFC; these are:

- ***High Control Capability***: DPFC is capable of handling all the power system parameters, i.e. impedance of line, bus voltage magnitude and transmission angle. By removing the common DC link, the DPFC converters can be installed and controlled effectively and individually. With this high control capability, it is capable of improving quality of power and stability of system like as restoration of voltage sag, asymmetry balancing and damping of low-frequency power oscillations.
- ***High Reliability***: More reliable due to the availability of redundant series converters. Additionally, all series and shunt converters are independent of each other, thus fault on any one of the converter does not affect the others but still there will be little influence due to this.
- ***Low Cost***: Due to redundant series a converter no ph-to-ph voltage isolation is required; converter rating is also small which are manufactured easily making it economical.

5 Simulation and Results

Here, simulation results of DPFC model are studied and analysed, where DPFC comprises the two converters, i.e. shunt and series converters. Modelling of DPFC has been done in the software named as MATLAB/SIMULINK as in Fig. 9. Series converter is used for the compensation of voltage profile disturbances, whereas shunt converter is utilized for the reactive power compensation for maintaining DC capacitor voltage at constant level (Table 1; Fig. 10).

Above simulation is done on the three-phase source voltage of 380 V as in Fig. 11 in per unit, as per the rectification of step change in the transmission line the third

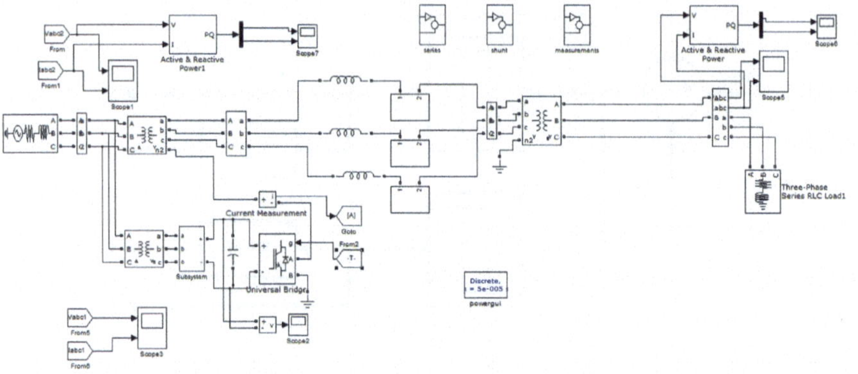

Fig. 9 DPFC simulation model

Table 1 Simulation parameters

Parameters	Values
V_s (3 phase V_{LL})	380 V
V_r (3 phase V_{LL})	380 V
Line impedance (L)	6 mH
C_{sh}	2 mF
C_{se}	2 mF
Frequency	50 Hz
Transformer ($T1$)	250 MVA, 1:1
Transformer ($T2$)	250 MVA, 25:1
Transformer ($T3$)	250 MVA, 1:5

harmonic current is being injected by DPFC at bus 1, respective voltage and current profiles can be seen in Fig. 12. Since third harmonic current has been introduced, change is observed only in current profiles, accordingly V_{THD} and I_{THD} can be observed in Figs. 13 and 14.

After synchronizing the third harmonic frequency component with fundamental frequency component by using above-explained scheme, improved voltage and current and voltage waveforms are shown in Fig. 15, where the I_{THD} has been improved from 33.90 to 0.02% as in Fig. 16 and V_{THD} has been improved from 0.23 to 0.11% as in Fig. 17.

Above all results are encompassed with a step change in the reference signal (as it can be the condition of a faulty line) at $t = 0.29$ s of magnitude 1.5 to -1.5 (in pu) which is in Fig. 18, accordingly the voltage injected by the one-phase series converter for its enhancement is shown in Fig. 19 thus the changed line current injected by shunt converter at third harmonic frequency is shown in Fig. 20.

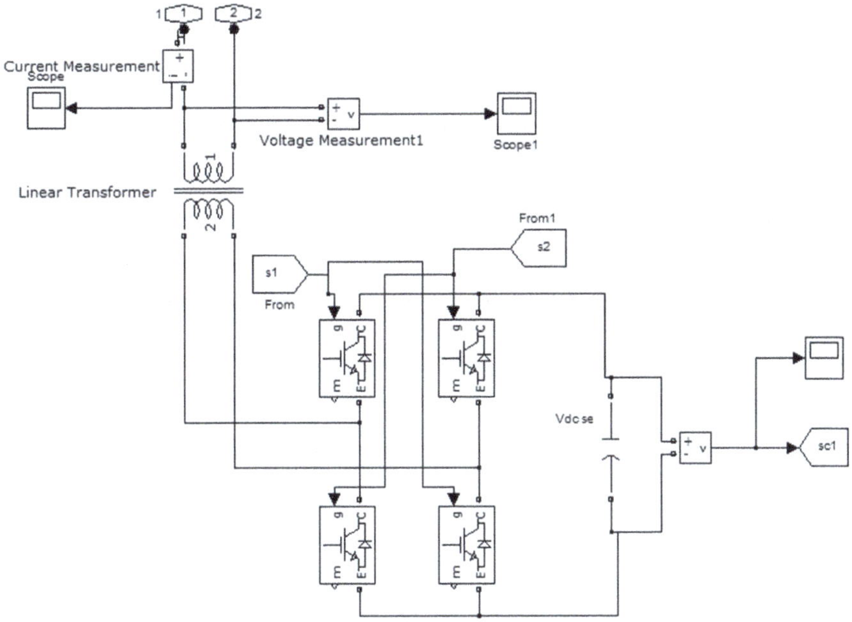

Fig. 10 One-phase series converter

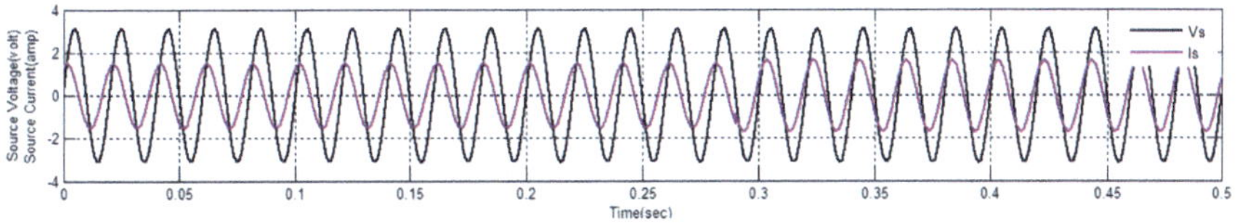

Fig. 11 Source voltage and current

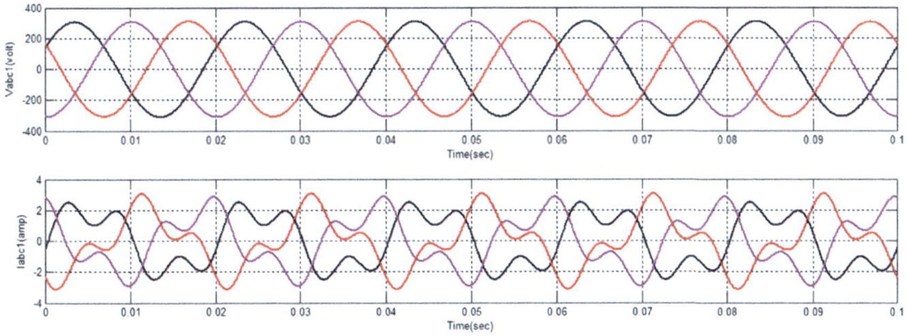

Fig. 12 Voltage and current waveform of bus 1 when injecting the third harmonics frequency current into the neutral of the transformer

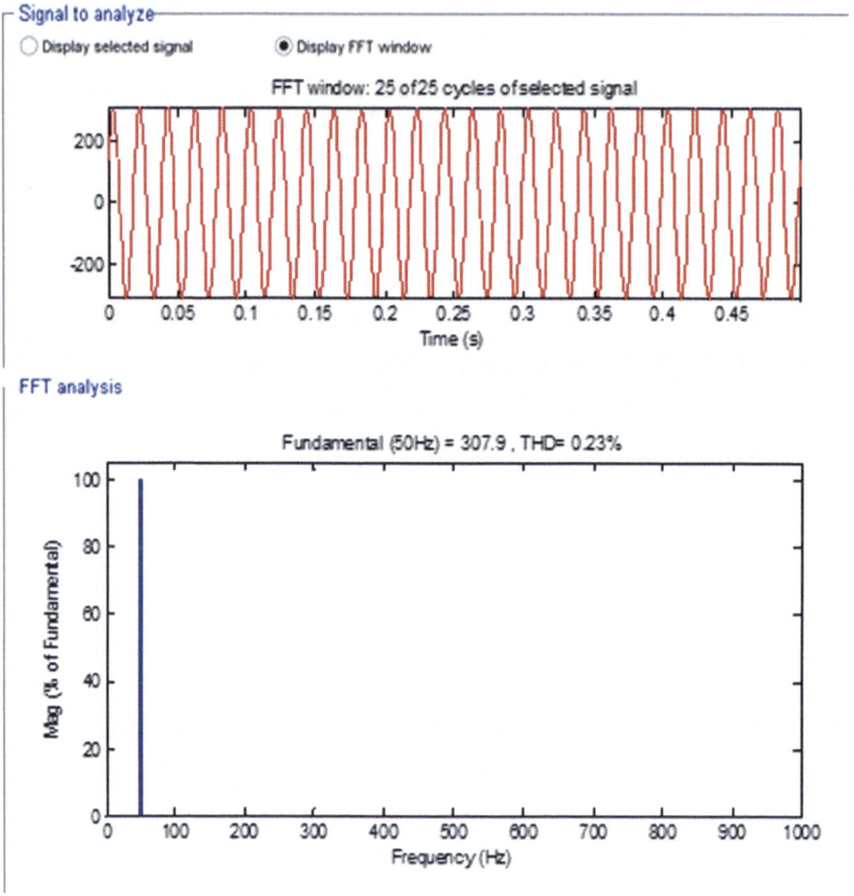

Fig. 13 Voltage THD = 0.23% of Bus 1

As per the step-change provided to the reference signal of the converter after controlling according to the HBC methodology used, the pulses for shunt and series converters are observed in Figs. 21 and 22.

DC capacitor voltages of series and shunt converters for supplying the energy can be seen in Figs. 23 and 24, respectively, where DC voltage profiles are being settled at 0.1 s accordingly power factor, active and reactive power profiles can be observed in Fig. 25 having very less oscillations in their profiles and the effect of addition and removal of the step change in the transmission line.

For controlling DPFC, conventionally SPWM control technique was used, which shows improvement in current and voltage at their extreme limits, can be seen in terms of THDs which limit its applicability. But Hysteresis Band Control (HBC) technique improves current and voltages, power flow also significantly, giving wide

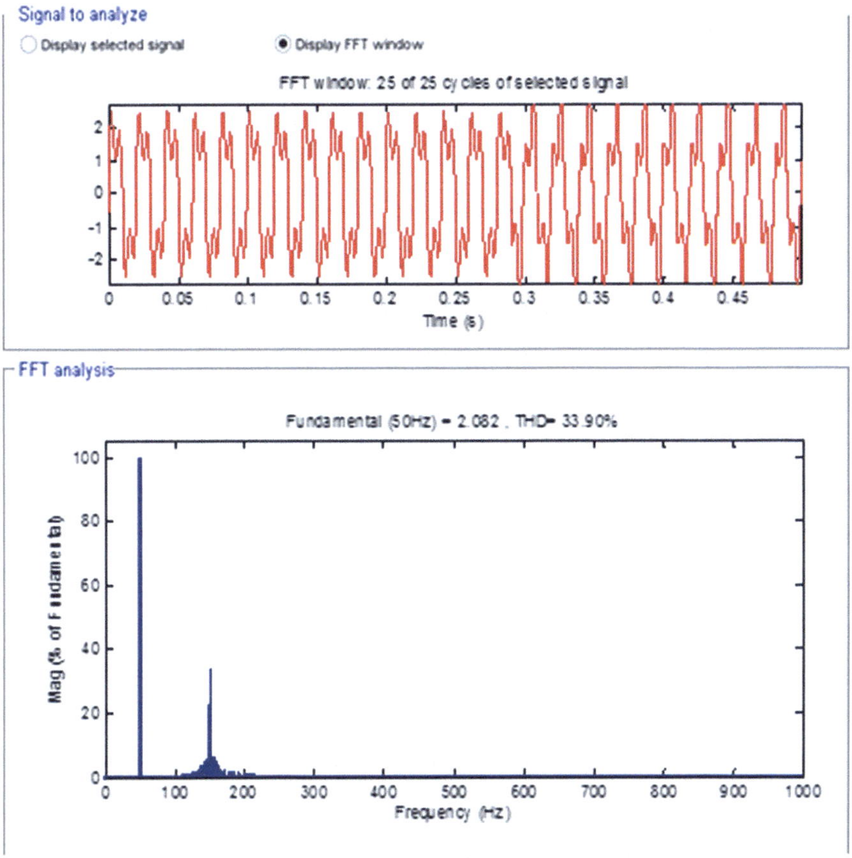

Fig. 14 Current THD = 33.90% of Bus 1

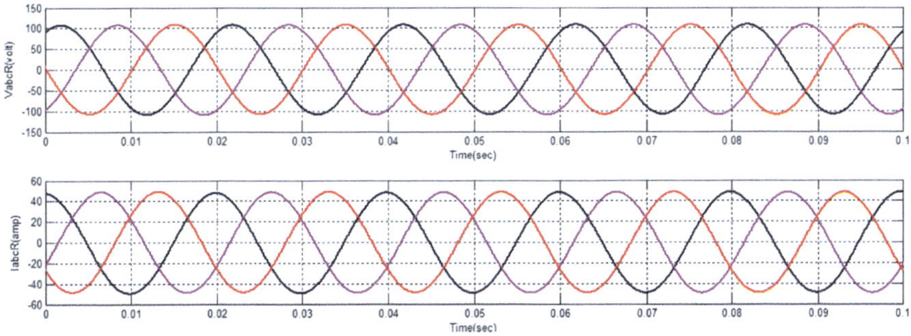

Fig. 15 Voltage and current waveform at receiving end after synchronizing the third harmonics frequency component with fundamental frequency component

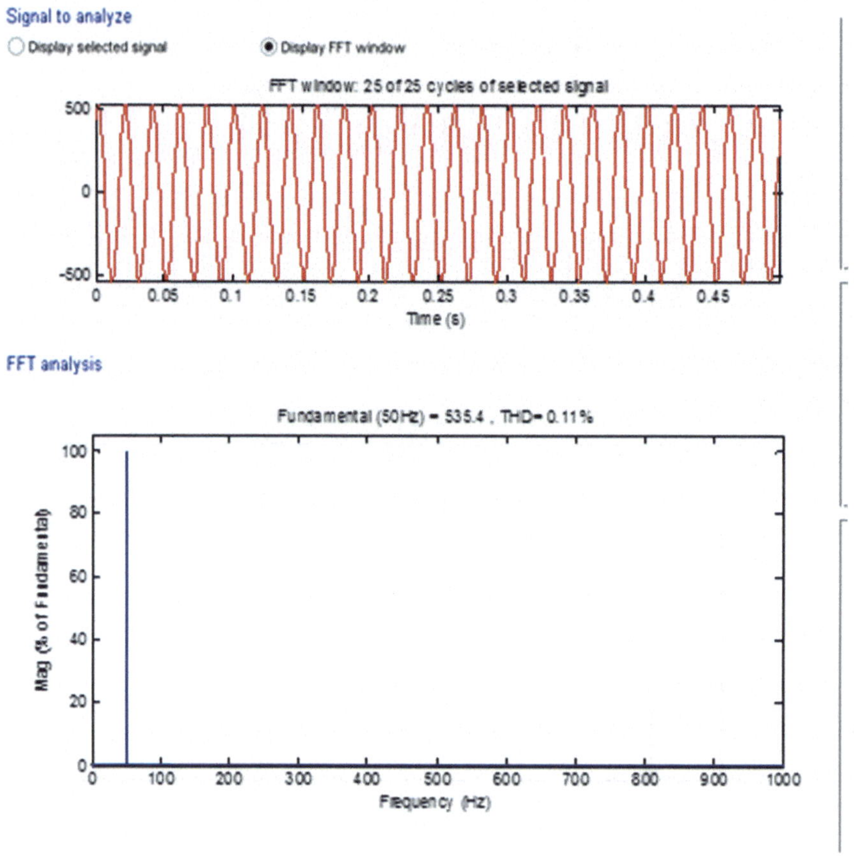

Fig. 16 Voltage THD = 0.11% of receiving end

its application for the rectification of PQ problems. Thus compared results of both SPWM and HBC technique can be seen from Table 2.

6 Conclusion

In this paper, a new control concept of UPFC has been taken out which is named as DPFC. The control of transmission angle, magnitude of bus voltage, the DC-link used in UPFC for the exchange of real and reactive powers is eliminated in DPFC it utilizes the transmission line for exchange of these powers at 3rd harmonic frequency level. DPFC employs distributed FACTS concept. Three-phase series converter, which reduces the necessity of high-voltage isolation, high cost and is easy in production which is of low rating. The $1-\phi$ converters inject the controllable reactive and real power derived at fundamental frequency.

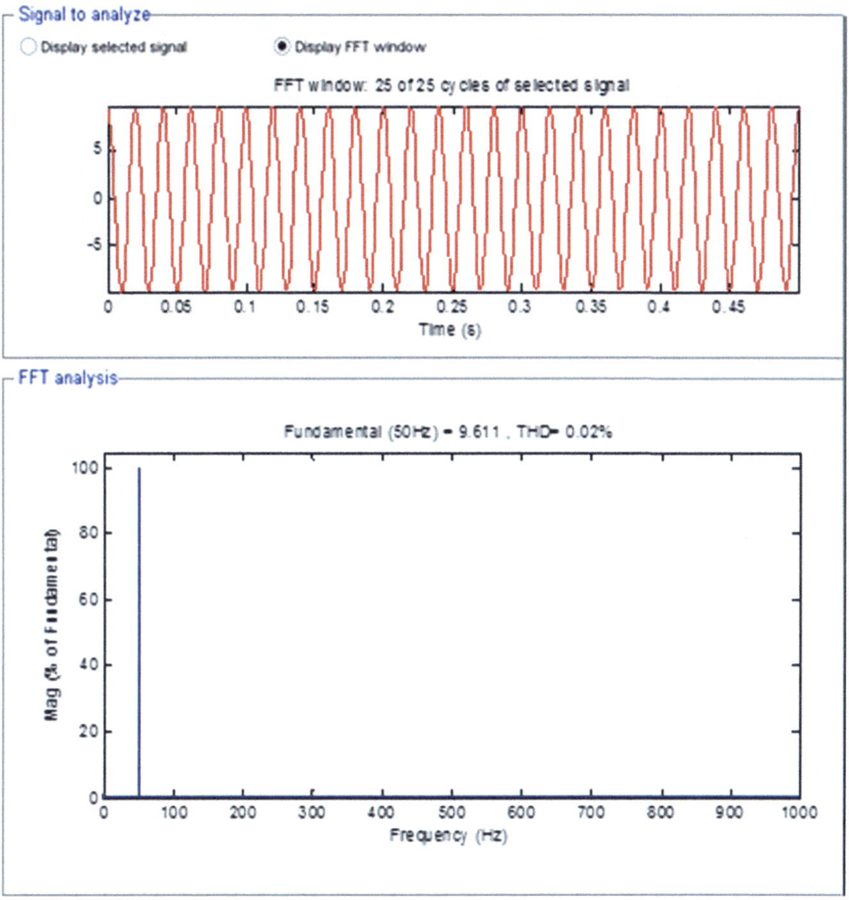

Fig. 17 Current THD = 0.02% of receiving end

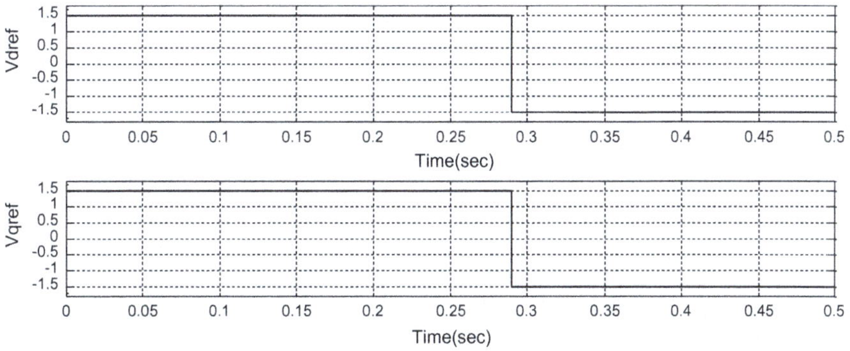

Fig. 18 Step change in reference signals

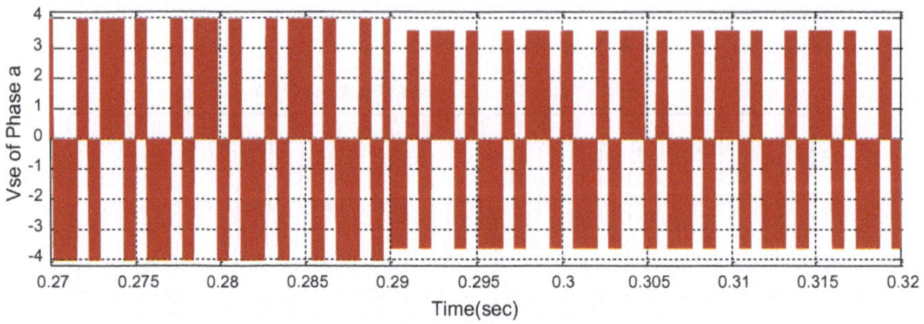

Fig. 19 Voltage injected by one-phase series converter

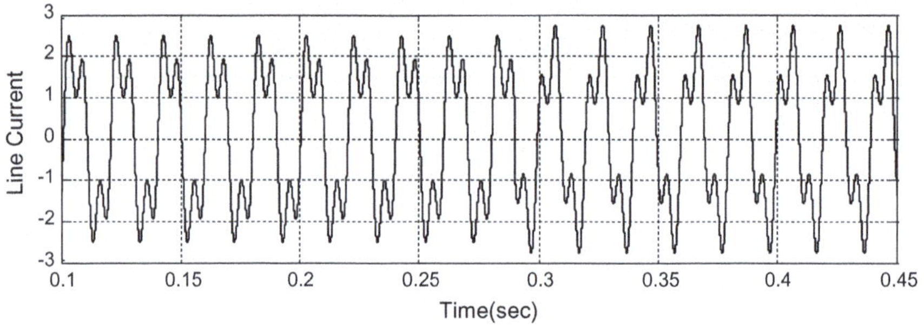

Fig. 20 Line current

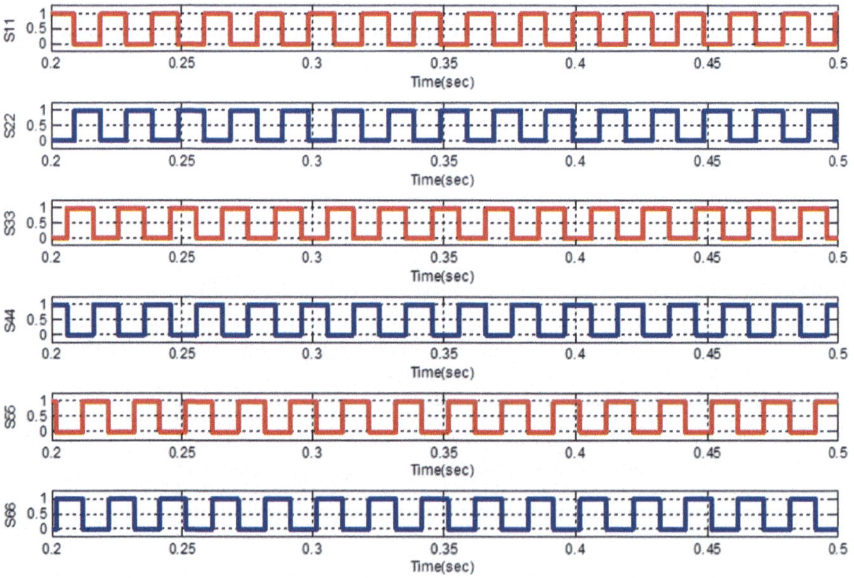

Fig. 21 Pulses for three-phase shunt converter

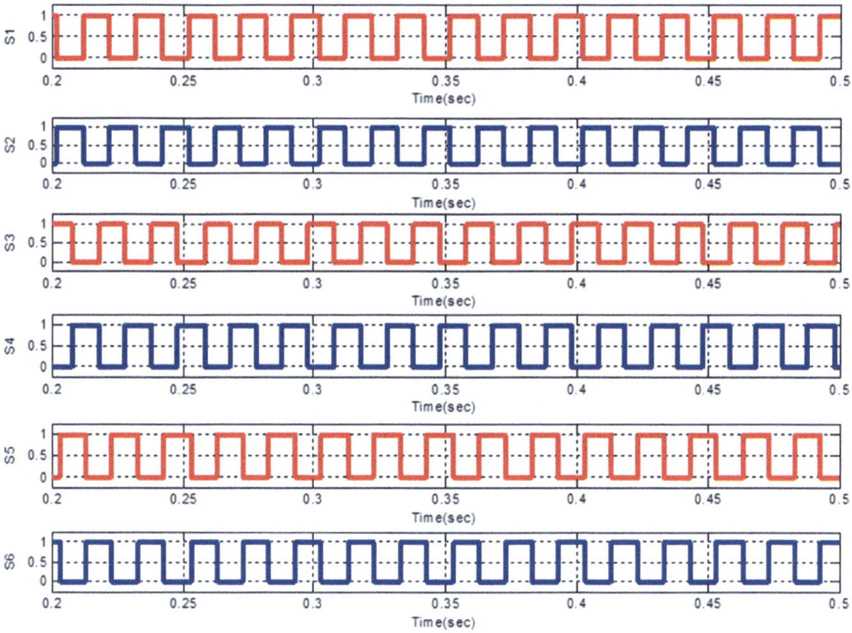

Fig. 22 Pulses for three one-phase series converters in all three phases

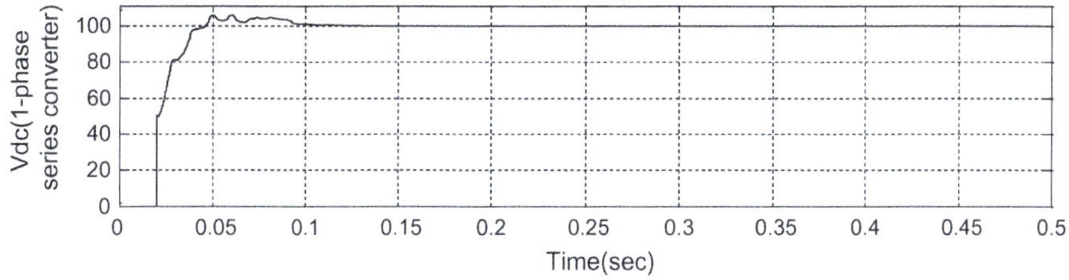

Fig. 23 DC capacitor voltage across one-phase series converter

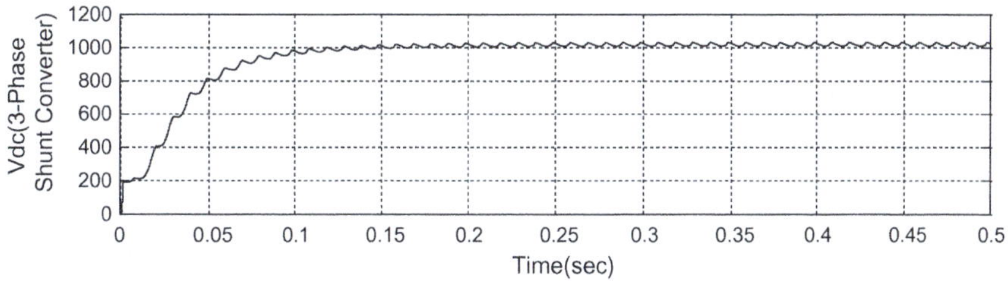

Fig. 24 DC capacitor voltages across three-phase shunt converter

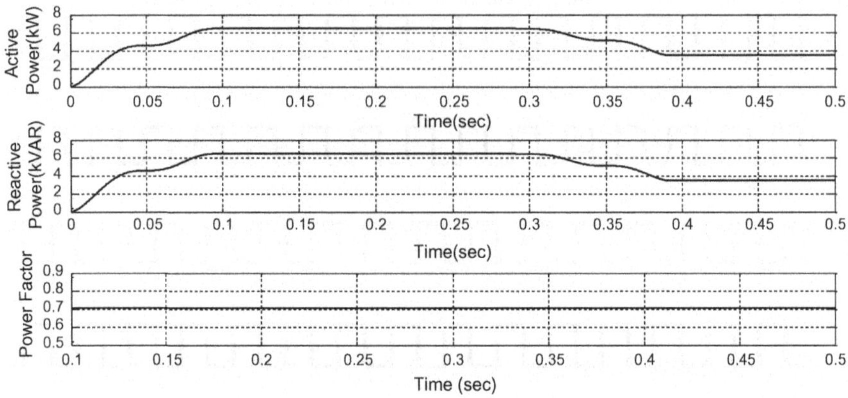

Fig. 25 Power flow and power factor observed at receiving end

Table 2 Comparison of THDs from two different control strategies of DPFC

	SPWM control		Hysteresis band control	
	V_{thd} (%)	I_{thd} (%)	V_{thd} (%)	I_{thd} (%)
Sending end $V_{s(abc)}$	0.40	4.85	0.16	0.76
Bus 1	5.16	33.91	0.23	33.90
Bus 2	5.38	4.50	0.11	0.02
Receiving end $V_{r(abc)}$	5.32	4.45	0.11	0.02

References

1. Hingorani G, Gyugyi L (2000) Understanding FACTS: concepts and technology of flexible AC transmission systems. IEEE Press, New York
2. Gyugyi L (1992) Unified power-flow control concept for flexible AC transmission systems, generation, transmission and distribution. In: IEE proceedings-generation, transmission and distribution, IEE Proceedings C
3. Song YH, Johns AT (1999) Flexible ac transmission systems (FACTS). Institution of Electrical Engineers, London
4. Gyugyi L, Sen KK, Schauder CD (1999) The interline power flow controller concept: a new approach to power flow management in transmission systems. In: IEEE transactions on power delivery
5. Strzelecki R, Benysek G, Fedyczak Z (2002) Interline power flow controller-probabilistic approach. In: Digital object identifier
6. Divan D (2004) A distributed static series compensator system for realizing active power flow control on existing power lines. In: Power systems conference and exposition
7. Chandrasekaran K, Vengkatachalam PA, Karsiti MN, Rama Rao KSM (2009) Mitigation of power quality disturbances. J Theor Appl Inf Technol 8(2):105–116
8. Yuan Z, de Haan SWH, Ferreira JB, Cvoric D (2010) A FACTS device: distributed power-flow controller (DPFC). IEEE Trans Power Electron 25(10)
9. Yuan Z, de Haan SWH, Ferreira B (2010) DPFC design procedure—a case study using the KEPCO UPFC as an example. IEEE

10. Lin X, Han Y (2011) Effective controller design for the cascaded Hbridge multilevel DSTATCOM for reactive compensation in distribution utilities. Elektrotehniski Vestnik 78(4):229–235
11. Suresh S, Devarajan N, Geetha M, Rajasekaran V (2011) Investigation on D-STATCOM operation for power quality improvement in a three phase three wire distribution system with a new control strategy. Control Theory Inf 1(2)
12. Reddy GP, Reddy KR (2012) Power quality improvement using neural network controller based cascaded HBridge multilevel inverter type D-STATCOM. In: International conference on computer communication and informatics
13. Rai A (2012) Enhancement of voltage stability and reactive power control of distribution system using facts devices. Int J Sci Res Eng Technol 1(9):1–5
14. Deshmukh SM, Dewani B (2012) Overview of dynamic voltage restorer (DVR) for power quality improvement. Int J Eng Res Appl 2(6):1372–1377
15. Boudaghi A, Tousi B (2012) DSTATCOM based five-level cascade HBridge multilevel inverter for power quality improvement. Int J Tech Phys Prob Eng 4(3):110–117
16. Kumbha V, Sumathi N (2012) Power quality improvement of distribution lines using DSTATCOM under various loading conditions. Int J Mod Eng Res 2(5):3451–3457
17. Jamshidi A, Barakati SM, Ghahderijani MM (2012) Impact of distributed power flow controller to improve power quality based on synchronous reference frame method. IACSIT Int J Eng Technol 4(5)
18. Paul S, Sarkar S, Sarkar S, Saha PK, Panda GK (2013) By dynamic voltage restorer for power quality improvement. Int J Eng Comput Sci 2(1):234–239
19. Gaigowal SR, Renge MM (2013) Some studies of distributed series FACTS controller to control active power flow through transmission line. ICPEC
20. Chhabra P (2013) Study of different methods for enhancing power quality problems. Int J Curr Eng Technol 3(2):403–410
21. Mohan Reddy D, Gowrimanohar T (2013) Cascaded multilevel inverter based DSTATCOM for restructured power systems to compensate the reactive power and harmonics using shift carrier techniques. IOSR J Electr Electron Eng 4(3):39–48
22. Sateesh Kumar Reddy M, Narasimhulu N (2013) A new distributed power flow controller to improve the power quality by mitigating voltage sag and swell. IJERT 2(9)
23. Bindal RK (2014) A review of benefits of FACTS devices in power system. IJEAT
24. Pande SV, Ghute AA (2015) Implementation of distributed power flow controller (DPFC) for power quality improvement. Int J Adv Eng Technol
25. Song W, Zhou X, Liang Z, Bhattacharya S, Huang AQ (2009) Modelling and control design of distributed power flow controller based-on per-phase control. IEEE
26. Singh B, Yadav I, Kumar D (2011) Mitigation of power quality problems using FACTS controllers in an integrated power system environment: a comprehensive survey. Int J Comput Sci Artif Intell 1(1):1–12
27. Divya S, Shyamala U (2015) Power quality improvement in transmission systems using DPFC. In: IEEE sponsored 2nd international conference on electronics and communication system (ICECS 2015)
28. Ghosh A, Ledwich G (2002) Power quality enhancement using custom power devices. Kluwer Academic Publishers
29. Renduchintala UK, Pang X, Maddukuri SVPK, Aravinthan V (2016) Smooth shunt control of a fuzzy based distributed power flow controller to improve power quality. In: 2016 IEEE international conference on information and automation for sustainability (ICIAfS), Galle, pp 1–5
30. Sharma M, Bhargava A, Yadav P (2016) Oscillation damping with DPFC using optimization techniques. In: 2016 international conference on micro-electronics and telecommunication engineering (ICMETE), Ghaziabad, pp 343–348
31. Zhang Y, Lu G, Khan WA, Zhang Y, Zhu Q Direct power flow controller—a new concept in power transmission. IEEE Trans Power Electron
32. Xiaolei Q, Kaiyang Z, Shijing Z, Xiong C (2018) Architecture design of distributed power flow controller system based on centralized control. In: 2018 IEEE 8th annual international

conference on CYBER technology in automation, control, and intelligent systems (CYBER), Tianjin, China, pp 1209–1213
33. Renduchintala UK, Pang C, Kumar Maddukuri SP, Aravinthan V (2018) Fuzzy theory based distributed power flow conditioner to enhance power system dynamic stability. In: 2018 international conference on power energy, environment and intelligent control (PEEIC), Greater Noida, India, pp 1–5
34. Mewara B, Bhargava A, Jain K (2018) DPFC performance for improvement of power quality in power system undergoing unbalance faulty condition. In: 2018 international conference on inventive research in computing applications (ICIRCA), Coimbatore, pp 704–708
35. Xiao M et al (2018) Two-level control method for DPFC series units based on PLC communication. In: 2018 2nd IEEE conference on energy internet and energy system integration (EI2), Beijing, pp 1–5
36. Wu M, Tang A, Wang Z, Zheng X, Xu Q (2018) Study on an exact equivalent model of DPFC. In: 2018 IEEE 3rd advanced information technology, electronic and automation control conference (IAEAC), Chongqing, pp 2626–2629

Optimal Placement of Resistive Superconducting Fault Current Limiters in Microgrid

Sneha Rai and Mala De

Abstract The basic feature of the future smart grid is redistribution of the conventional grid into a number of smaller grids recognized as microgrid. The extensive incorporation of distributed energy resources (DER) with conventional power grids results in the rise of fault current by several times during the occurrence of fault or any abnormality which has to be solved for the successful execution of microgrid. One of the most important utilization of superconducting fault current limiters for upcoming microgrid is related to the curtailment of abnormal rise in fault current and the viable position in microgrid. This paper analyzes the implementation of resistive superconducting fault current limiter (RSFCL) for designed microgrid system, which is an extension of the current power grid. RSFCL model of resistive type has been modeled in Simulink and its performance has been studied at various positions in the designed system by integrating a wind power plant of 10 MVA capacity with traditional power plant, to minimize the fault current in the microgrid which is caused by the occurrence of different types of faults at various locations in the grid. Consequently, the viable location of RSFCL which has no adverse effect on wind farm and is also cost-effective has been analyzed for all types of faults at discrete positions in the microgrid.

Keywords Microgrid · Smart grid · Resistive superconducting fault current limiters · Fault current

1 Introduction

Smart grid is the improvement of future power grid which combines communication technology and renewable energy resources for the present grid to transmit the electrical power which is reliable, resilient, cleaner, and responsive than the present power grid [1, 2]. The basic element of the smart grid is redistribution of traditional grid networks in smaller grids, called as microgrid which have DERs combined with them [3, 4]. The increased assimilation of DERs with the traditional power plant to

S. Rai (✉) · M. De
National Institute of Technology Patna, Patna 800005, India
e-mail: sneharai0212@gmail.com

S. N. Singh et al. (eds.), *Advances in Power and Control Engineering*, Lecture Notes in Electrical Engineering 609, https://doi.org/10.1007/978-981-15-0313-9_6

cater the rising electricity demand results in the enormous fault current which is a severe problem and needs to be taken care for the proper functioning of the grid [5].

The traditional devices used for protection like relays and circuit breakers allow some initial fault current to pass through them before their initialization [6, 7]. But the evolution of resistive RSFCL results in the replacement of traditional protective devices as it limits the fault current within its first cycle and hence results in transient stability enhancement of the designed system [8, 9].

The nonlinear characteristic of superconductor is responsible for the fault current limitation of the RSFCL [10, 11]. RSFCL offers negligible resistance to the system during the normal operating condition but the resistance increases rapidly during fault condition as the current surpasses its critical value, and as a consequence, the device changes its state from low resistance to high resistance and reduces the fault current to the permissible level [12]. In this paper, the whole microgrid system has been developed by interconnecting wind farm of 10 MVA capacity with the traditional power plant. The effects of RSFCL on the wind farm have been analyzed as well as its feasible position in the microgrid that results in the maximum limitation of fault current from both power sources has been proposed.

2 Modeling and Simulation

The entire microgrid has been developed using Simulink/SimPowerSytem in MATLAB. A RSFCL has also been designed in it.

2.1 Microgrid Model

The microgrid model consists of a conventional power plant of 100 MVA capacity using three-phase synchronous generator which generates electricity at 20 kV which is then stepped up to 154 kV by a step-up transformer TR1 for transmission purpose as shown in Fig. 1. A 200 km long transmission line is used to transmit power to the receiving station and the transformer TR2 has been connected to step down the voltage level from 154 to 22.9 kV. A 100 MW load is connected through bus B1 before TR1. An industrial load of 6 MW is also connected through bus B8. A reactive load of 330 MVAR is supplied via bus B4 of the developed system. Three distribution transformers TR4, TR5, and TR6 are also being connected to step down the voltage from 22.9 kV to 400 V in the customer side. Three household loads of 10 KW each are also connected by the three transformers via 5 km long transmission line. The wind farm of capacity 10 MVA is connected through bus B7 via transformer TR3 and it consists of five parallel-connected induction-type wind turbines having 2 MVA capacity each.

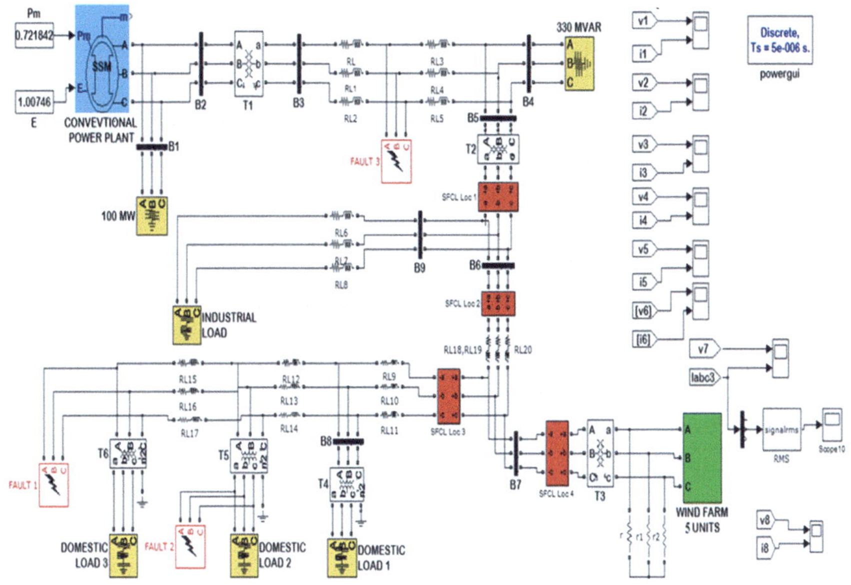

Fig. 1 Microgrid model showing various fault locations and RSFCL positions

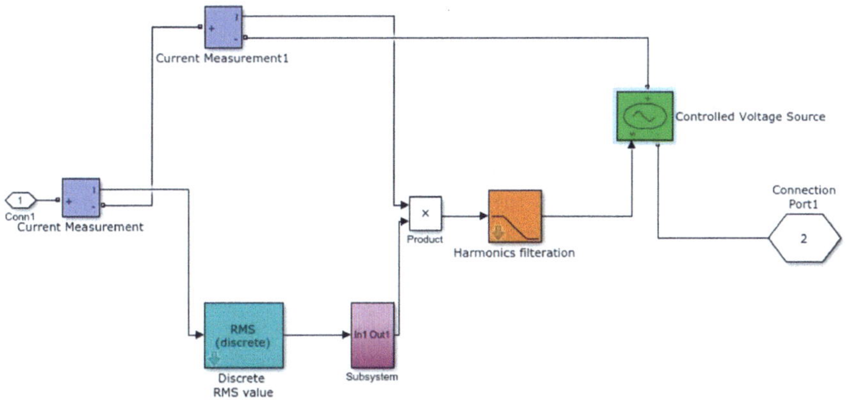

Fig. 2 RSFCL model

2.2 *RSFCL Model*

The three-phase RSFCL model has been designed in Simulink (as shown in Fig. 2) by taking 22.9 kV as base operating voltage. The four basic parameters of RSFCL are as follows:

- Activation current which has been taken as 550 A according to the requirement of the designed system.

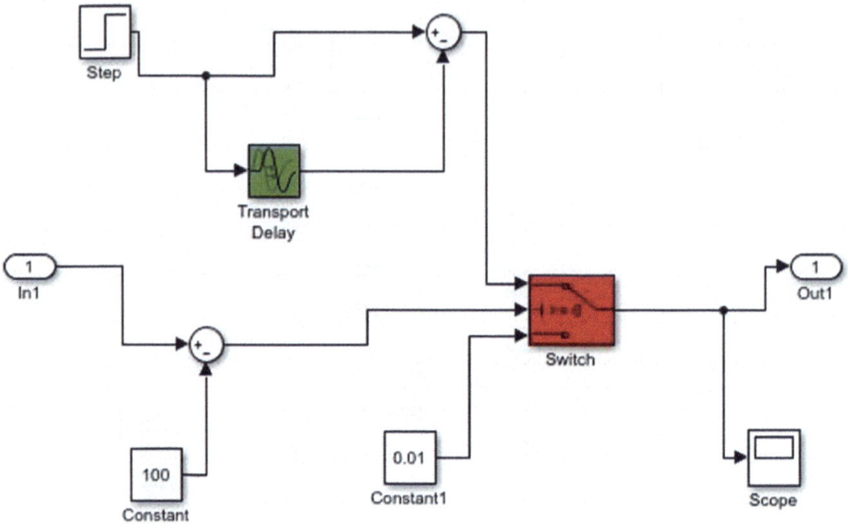

Fig. 3 RSFCL characteristic table

- Response or transformation time has been taken as 2 ms.
- The minimum impedance of the system has been taken as 0.01 Ω and the maximal impedance as 20 Ω.
- The readjustment time has been taken as 10 ms.

Firstly, the RMS block calculates RMS value of the flowing current through input port and then this value is compared with RSFCL characteristic table. Figure 3 demonstrates the subsystem of the RSFCL characteristics table.

Now, if the value of flowing current surpasses the activation current of designed RSFCL, the transition of its state occurs from low resistance (superconducting state) to high resistance (normal state) in the predetermined time which is also called as response time or triggering time. After the fixation of fault when the flowing current drops below its activation current level, the RSFCL takes its recovery time to go to the normal operating state. The harmonics occurring in the designed system can be reduced by the use of first-order low-pass filter. To reimburse the voltage sag caused by the increased fault current level, controlled voltage source has been connected to the overall circuit.

3 Results and Discussion

All the distinct faults, including three-phase-to-ground (LLLG), LG, LL, and LLG faults, have been created at different prone locations of the microgrid and the change in the level of fault current has been observed for the four different locations of RSFCL in the designed system. The first placement of RSFCL has been considered to be at the substation which is denoted by position 1. After that, the RSFCL has been located at the branch network known as position 2 and at the integration of the conventional power source and the wind farm known as position 3 of the RSFCL. At last, two RSFCLs located at position 1 and position 4 have been considered which are near the wind farm.

3.1 Fault Near Distribution Grid

As it has been observed from the below mentioned results, the distribution grid fault current increases for the placement of RSFCL at position 1 and position 2 as due to the occurrence of fault the impedance level of the complete system has increased which results in the reduction in the current coming from the traditional power source, but as the incorporated wind farm is the another power source near the fault location, therefore it supplies the increased fault current to the fault position which is more than the case without RSFCL. Figure 4 shows the wind farm fault current comparison for four RSFCL locations in case of three-phase-to-ground fault occurring near distribution grid.

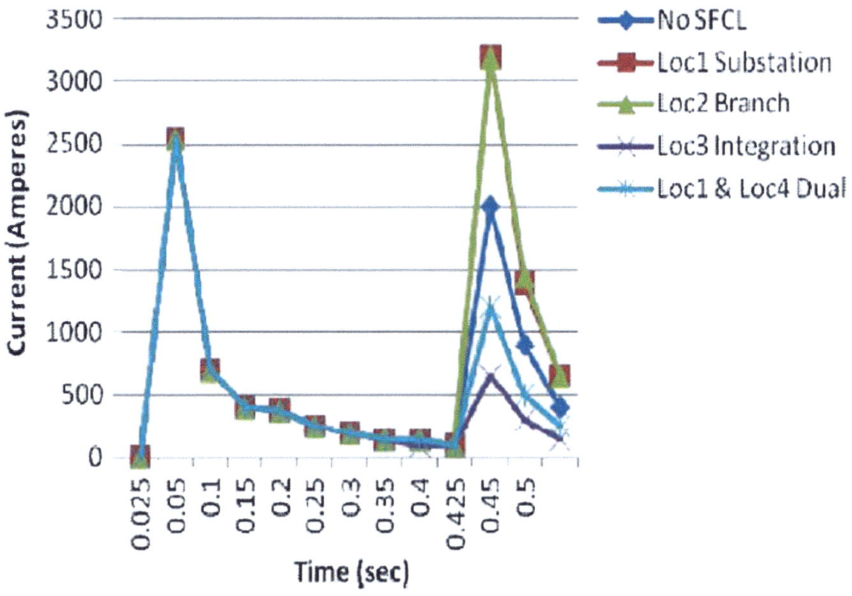

Fig. 4 Wind farm fault current comparison considering four RSFCL positions for the LLLG fault occurring near distribution grid

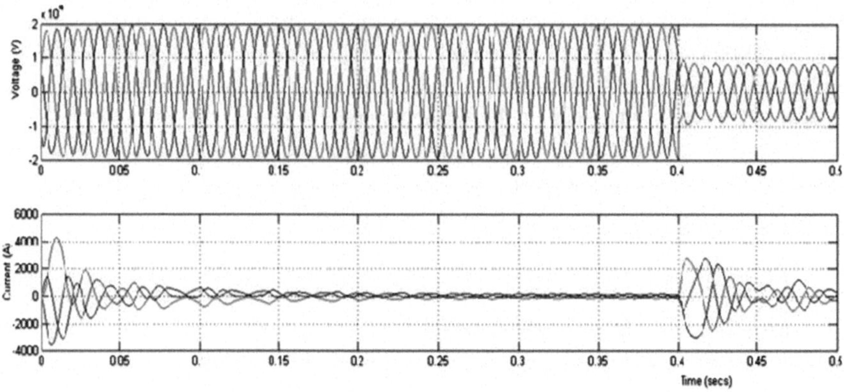

Fig. 5 Fault current of wind farm without RSFCL

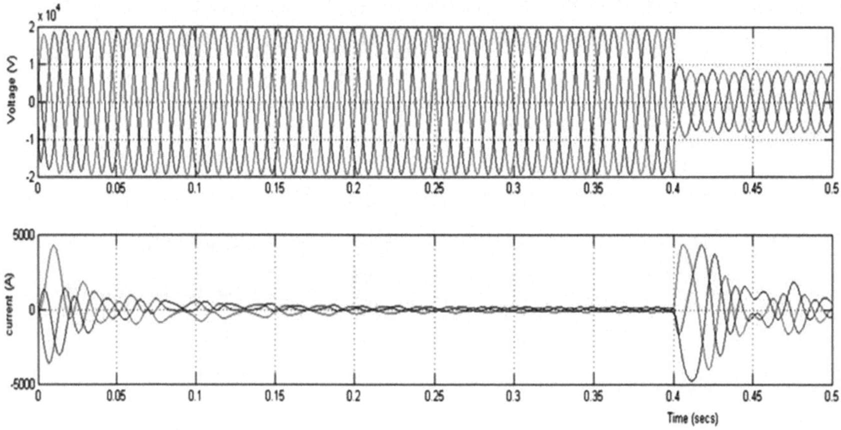

Fig. 6 Fault current of wind farm with RSFCL placed at position 1

Figures 5, 6, 7, 8, and 9 represent the wind farm voltages and currents for different RSFCL locations when the fault has occurred in distribution grid side.

3.2 *Fault Near Customer Grid*

It has been observed from the result obtained for the customer grid fault that the best position of RSFCL which limits the maximum fault current came to be position 3 as similar to the previous distribution grid case. Since this fault is very minor in nature, it has appeared in distribution side of the network which is generally of low voltage.

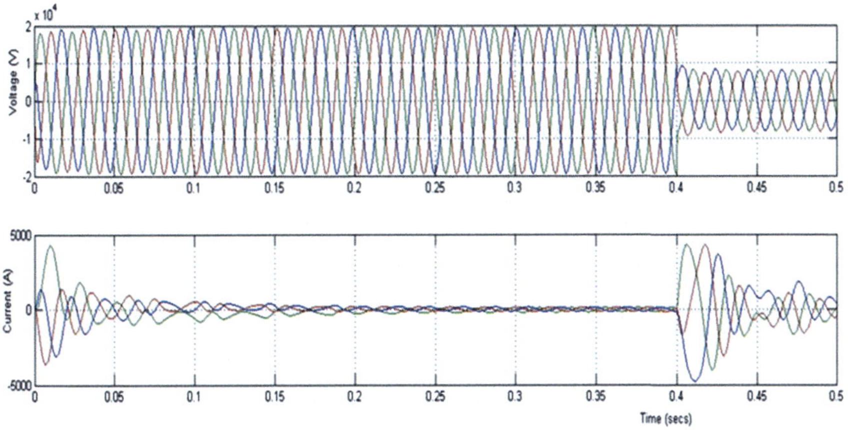

Fig. 7 Fault current of wind farm with RSFCL placed at position 2

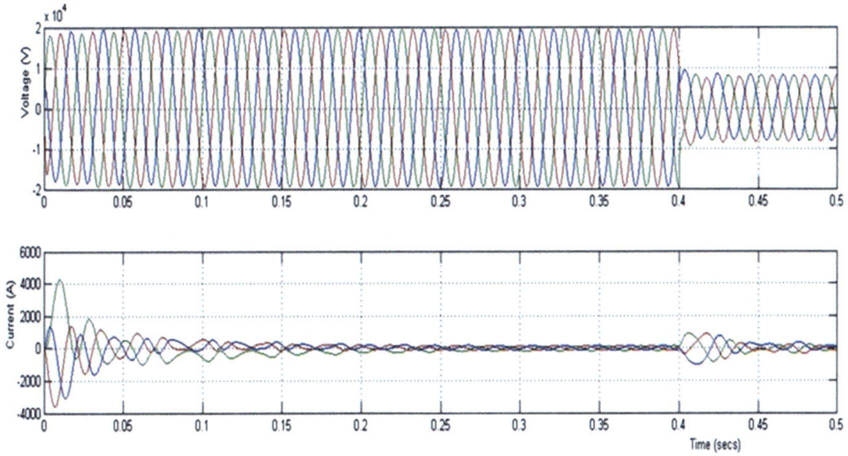

Fig. 8 Fault current of wind farm with RSFCL placed at position 3

Figure 10 shows the wind farm fault current comparison for four RSFCL locations in case of three-phase-to-ground faults occurring near customer grid.

3.3 Fault Near Transmission Line

This fault is very unlikely occurring fault that causes the abrupt rise in fault current from the wind power plant which is measured from bus B7 in Fig. 1. In this case, as in the contrary to the above-mentioned cases, position 1 and position 2 of the

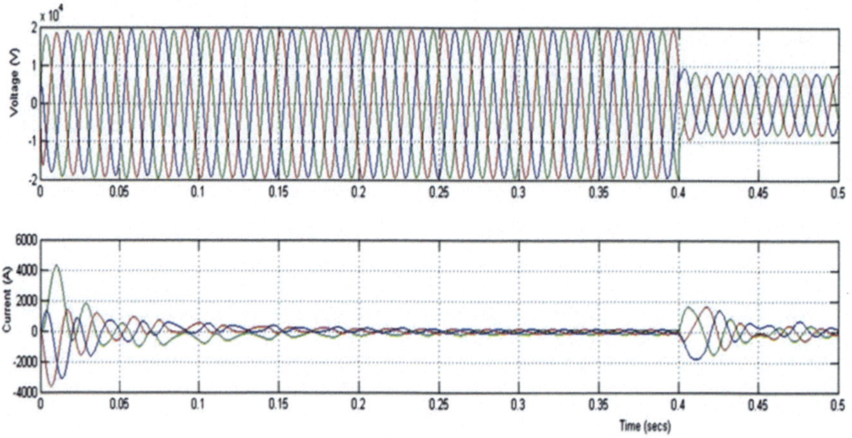

Fig. 9 Fault current of wind farm with RSFCL placed at position 1 and 4

Fig. 10 Wind farm fault current comparison considering four RSFCL positions for the LLLG fault occurring near customer grid

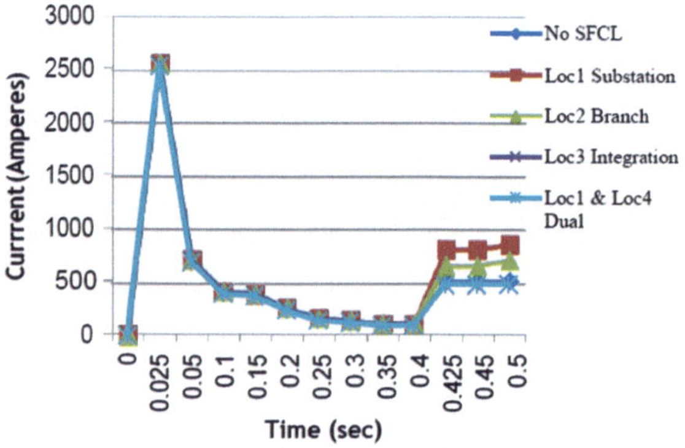

RSFCL result in the fault current minimization as these two locations of RSFCL comes directly in the path of the conventional source as well as wind farm fault current. Figure 11 shows the wind farm fault current comparison for four RSFCL locations in case of three-phase-to-ground faults occurring near transmission line.

3.4 Fault Near Distribution Grid and Customer Grid

When two faults occur simultaneously at the distribution grid and customer grid, there is an abrupt rise in the fault current originating from wind farm for RSFCL located at position 1 (substation) and position 2 (branch network) as the wind power plant supplies more current to the fault location since it is the other power source. But the RSFCL placed at position 3 which is the interconnecting junction of the

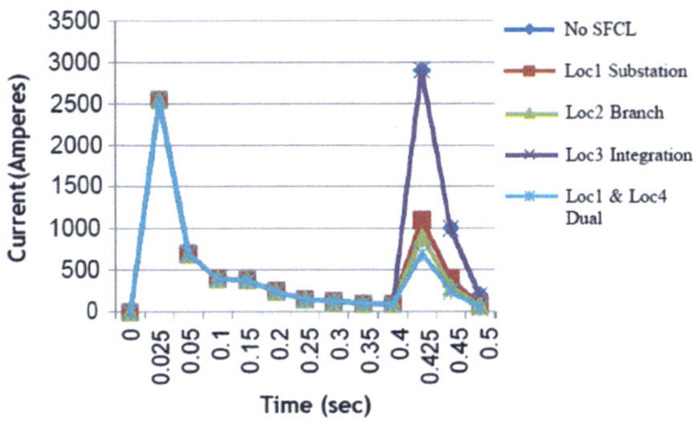

Fig. 11 Wind farm fault current comparison considering four RSFCL positions for the LLLG fault occurring near transmission line

traditional power plant and wind farm results in the maximum minimization of peak fault current as here RSFCL is directly located in the way of the wind farm fault current.

3.5 Fault Near Transmission Line and Distribution Grid

For the faults occurring simultaneously near distribution grid and the transmission line, the best position of the RSFCL which causes maximum minimization in the fault current is position 3, i.e., the interconnecting point of wind farm and conventional source.

3.6 Fault Near Customer Grid and Transmission Line

When two faults occur simultaneously near the customer grid and the transmission line, position 1 and position 2 of RSFCL also mark the minimization of fault current as in both the cases RSFCL is directly positioned in the path of the wind farm current going to transmission line fault, but the maximum fault current reduction has been observed for the RSFCL placed at position 3 only that is the interconnecting junction of wind farm and the conventional source.

Table 1 shows the percentage change in the fault current of wind farm due to the RSFCL positions for three-phase-to-ground faults occurring at different locations in the microgrid. Table 2 represents the percentage change in the wind farm fault current due to RSFCL locations for two three-phase-to-ground (LLLG) faults occurring simultaneously at different positions in the microgrid.

Table 1 Reduction in fault current for various RFCL positions considering three-phase-to-ground faults

RSFCL position	LLLG fault on distribution side	LLLG fault on customer side	LLLG fault on transmission line
	Change (%)	Change (%)	Change (%)
Position 1	63 (rise)	59 (rise)	61 (reduced)
Position 2	62 (rise)	31 (rise)	66 (reduced)
Position 3	71 (reduced)	22 (reduced)	0.3 (reduced)
Positions 1 and 4	43 (reduced)	3 (reduced)	73 (reduced)

Table 2 Minimization in fault current for various RFCL positions in case of two (LLLG) faults occurring simultaneously

RSFCL position	LLLG faults on distribution side and customer side	LLLG faults on transmission line and distribution side	LLLG faults on customer side and transmission line
	Fault current change (%)	Fault current change (%)	Fault current change (%)
Position 1	55 (rise)	26 (rise)	24 (reduced)
Position 2	48 (rise)	22 (rise)	19 (reduced)
Position 3	63 (reduced)	31 (reduced)	32 (reduced)
Positions 1 and 4	42 (reduced)	14 (reduced)	16 (reduced)

4 Conclusion

The implementation of resistive RSFCL for microgrid has been presented in this paper. The microgrid has been modeled by interconnecting wind farm of 10 MVA capacity as the distributed energy resource with the conventional power source. A RSFCL model is formulated and the transient analysis has been performed by placing RSFCL at distinct positions in the designed microgrid for various faults (including LLLG, LLG, LL, and LG) occurring at three different locations. The consequences of RSFCL have also been analyzed for two faults occurring simultaneously at two different locations. It has been observed that the direct installation of RSFCL at the two prone positions which are substation point and branch network causes the fault current to rise abruptly as compared to the no RSFCL case. The optimal position of RSFCL which causes the maximum fault current reduction and has no adverse impacts on distributed energy resource is the interconnecting junction of wind farm with the conventional source. The analysis is performed for all the types of faults and the same result has been obtained for LLG, LL, and LG faults.

References

1. Lee H-Y, Asif M, Park K-H, Lee B-W (2018) Feasible application study of several types of superconducting fault current limiters in HVDC grids. IEEE Trans Appl Superconduct 28(4)
2. Angeli G, Bocchi M, Serri L, Martini L (2018) Short-circuit current limitation through 2G YBCO resistive-type RSFCL devices: a model for technical and economic comparison with traditional air-core reactors. IEEE Trans Appl Supercond 28(4)
3. Sadeghkhani I, Golshan MEH, Guerrero JM (2017) A current limiting strategy to improve fault ride-through of inverter interfaced autonomous microgrids. IEEE Trans Smart Grid 8(5):2138–2148
4. Ramadan A (2017) Permanent magnet fault current limiter for the power grid. Energy Procedia 134:768–775
5. Chen L, Zheng F, Deng C, Li Z, Guo F (2015) Fault ride-through capability improvement of DFIG-based wind turbine by employing a voltage-compensation-type active RSFCL. Can J Electr Comput Eng 38(2):132–142
6. Meerovich V, Sokolovsky V (2015) High-temperature superconducting fault current limiters (FCLs) for power grid applications. In: Superconductors in the power grid, pp 283–324
7. Salvadori F, Gehrke CS, de Oliveira AC, de Campos M, Sausen PS (2013) Smart grid infrastructure using a hybrid network architecture. IEEE Trans Smart Grid 4(3)
8. Kim J-C, Cho S-M, Shin H-S (2013) Advanced power distribution system configuration for smart grid. IEEE Trans 4(1)
9. Khan UA, Shin WJ, Seong JK, Oh SH, Lee BW (2011) Feasibility analysis of application and positioning of DC HTS FCL in a DC microgrid through modeling and simulation using Simulink and SimPowerSystem. Phys C Supercond Appl 471(21–22):1322–1326
10. Ye L, Juengst K-P (2004) Modelling and simulation of high temperature resistive superconducting fault current limiters. IEEE Trans Appl Supercond 14(2)
11. Dessaint L, Al-Haddad K, Le-Huy H, Sybille G, Brunelle P (1999) A power system tool based on simulink. IEEE Trans Ind Electron 46(6):1252–1254
12. Sugimoto S, Kida J, Arita H, Fakui C, Yamagiwa T (1996) Principle and characteristics of a fault current limiter with series compensation. IEEE Trans Power Delivery 11(2):842–847

Design of Optimal PID Controller for Control of Twin Rotor MIMO System (TRMS)

Sumit Kumar Pandey, Jayati Dey and Subrata Banerjee

Abstract This paper presents the design of proportional-integral-differential (PID) controller to control twin rotor multi-input multi-output system (TRMS). The TRMS is a nonlinear test rig associated with cross-coupling. Initially, a decoupling method is used in the linear model of TRMS to eliminate the effect of cross-coupling. Further, two PID controllers are designed by using three different optimization techniques: particle swarm optimization (PSO), real-coded genetic algorithm (RCGA), and bacterial foraging optimization (BFO) for the two resulting decoupled systems which, in turn, control the pitch and yaw motions. A relative interpretation conducted from the obtained responses in which it is recognized that responses of BFO optimized PID controller are superior.

Keywords PID · TRMS · PSO · RCGA · BFO

1 Introduction

TRMS [1] is the control apparatus similar to the helicopter in some respect. The control of this system is the critical job by cause of nonlinearity and cross-couplings associated with it. Some studies have [2–13] revealed that several control techniques have been applied for control and improvement of the performance of TRMS. TRMS is controlled by model predictive control technique in [2]. The unscented Kalman filter is applied here for observing the system's state variables. The state-space method is used to linearize the nonlinear part adaptively. In [3], the authors described that the effect of coupling acts as a disturbance to TRMS. The system model is developed

S. K. Pandey (✉) · J. Dey · S. Banerjee
Department of Electrical Engineering, National Institute of Technology (NIT), Durgapur, Durgapur, West Bengal 713209, India
e-mail: skpdmk@gmail.com

J. Dey
e-mail: deybiswasjayati@gmail.com

S. Banerjee
e-mail: bansub2004@yahoo.com

S. N. Singh et al. (eds.), *Advances in Power and Control Engineering*, Lecture Notes in Electrical Engineering 609, https://doi.org/10.1007/978-981-15-0313-9_7

by adopting the identification technique. Each loop of the two single-input single-output (SISO) system is controlled through the robust deadbeat controller. In [4], the authors applied two-degree-of-freedom (DOF) SISO control technique to control the TRMS. A pre-compensator is applied to decouple the MIMO system. To implement this, the nonlinear model is linearized around the origin. Some works are reported in [5–7] for control of TRMS by using intelligent techniques based PID controllers.

The authors proposed a novel method by using evolutionary optimization and PID controller to control a MIMO nonlinear system [5]. Real-value-type genetic algorithm (RGA) is used for designing of the PID controller. The neural network based PID controller is applied for control of TRMS in [6]. Due to the application of single neuron in PID controller, on-line learning and real-time control of MIMO system are feasible. The computing time of the system reduces in the proposed method. The authors describe an inverse-model control method to control the motion of TRMS by applying intelligent techniques [7].

The TRMS is controlled by adaptive second-order sliding mode (SOSM) control method in [8]. The authors have described dynamic modeling and linear quadratic Gaussian control technique to control TRMS in [9, 10]. In [10], black-box identification technique is applied to model the TRMS. The control signal induced to the plant is reduced by using a command path pre-filter.

A derivative filter coefficient is added in the PID controller for control of TRMS in [11]. The authors added the derivative filter coefficient with PID controller to remove the effect of the noise and improve the system performance. In [12], an optimal state feedback controller is used for control of TRMS based on the linear quadratic regulator (LQR) method. The optimal controllers are used for control of pitch and yaw motion of TRMS. Tuning of the PID controller is done by using fuzzy logic and genetic algorithm [13] for control of TRMS. The authors propose a PSO-based PID control technique for controlling the TRMS in [14]. The aim of the designed controller is to minimize the error. The detail process of the basic BFO algorithm is described by authors in [15] in which the analysis of chemotaxis elimination and dispersal operation is depicted.

In this work, an optimal PID controller is designed by using different optimization techniques known as PSO, RCGA, and BFO to control the decoupled TRMS. In [16], the methods for decoupling of dynamic MIMO systems are described. In this, both static and dynamic decoupling of systems for open-loop control is discussed. A novel classical control approach is proposed to control the highly nonlinear coupled dynamics parallel-type motion platforms [17]. In this method, the MIMO nonlinear system is divided into several SISO systems due to decoupling, and the controller is designed for these SISO systems. A diagonal decoupling method is described for linear multivariable systems in [18]. The condition for the implementation of the decoupling controller is described. The authors presented a generalized theory of decoupling control by state feedback method for linear and nonlinear systems within the same framework [19].

The paper is organized as below: Mathematical modeling and design of decoupler for TRMS are depicted in the next section. PID control technique is explained in Sect. 3. Results are analyzed in Sect. 4 followed by the conclusion section.

2 Mathematical Modeling and Decoupling of TRMS

TRMS aero-dynamical diagram is portrayed in Fig. 1. It shows the two-degree-of-freedom rotation, first one is concern to horizontal, and the other one is with respect to vertical axis.

The obtained nonlinear equations of TRMS are linearized to origin by considering the numerical value from [1] to get the linearized plant system represented as [4]

$$\dot{x} = Ax + Bu, \quad y = Cx + Du \tag{1}$$

$$A = \begin{bmatrix} -0.8333 & 0 & 0 & 0 & 0 & 0 & 0 \\ 0 & 0 & 1 & 0 & 0 & 0 & 0 \\ 1.246 & -4.706 & -0.08824 & 0 & 0 & 0 & 0 \\ 0 & 0 & 0 & -1 & 0 & 0 & 0 \\ 0 & 0 & 0 & 0 & 0 & 1 & 0 \\ 1.482 & 0 & 0 & 3.6 & 0 & -5 & 18.75 \\ -0.01694 & 0 & 0 & 0 & 0 & 0 & -0.5 \end{bmatrix}, \quad B = \begin{bmatrix} 0 & 1 \\ 0 & 0 \\ 0 & 0 \\ 1 & 0 \\ 0 & 0 \\ 0 & 0 \\ 0 & 0 \end{bmatrix},$$

$$C = \begin{bmatrix} 0 & 1 & 0 & 0 & 0 & 0 & 0 \\ 0 & 0 & 0 & 0 & 1 & 0 & 0 \end{bmatrix}, \quad D = \begin{bmatrix} 0 & 0 \\ 0 & 0 \end{bmatrix}$$

The plant transfer function matrix is obtained as

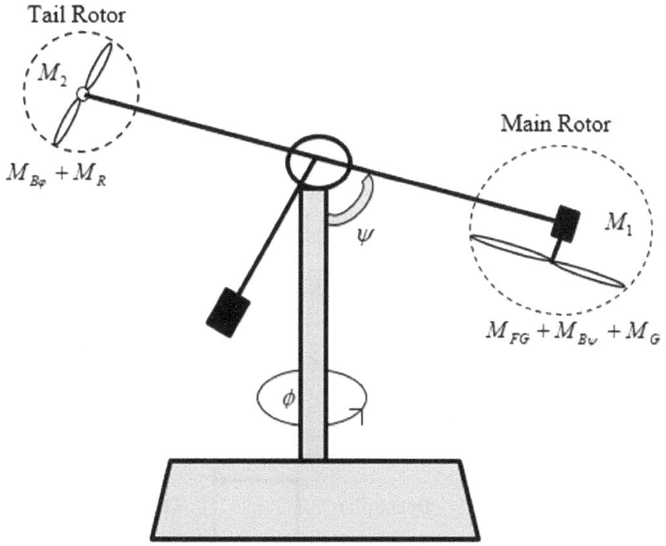

Fig. 1 TRMS aero-dynamical model

$$G(s) = \begin{bmatrix} \frac{1.246}{s^3+0.9215s^2+4.77s+3.918} & 0 \\ \frac{1.482s+0.4234}{s^4+6.33s^3+7.07s^2+2.08s} & \frac{3.6}{s^3+6s^2+5s} \end{bmatrix} \tag{2}$$

It is clearly observed from Eq. (2) that strong interaction is present in between $u_1 - y_2$, but in between u_2 and y_1, no interaction is seen. The main rotor and tail rotor transfer function is shown by G_{11}, G_{22}, respectively.

The coupling effects are abolished by the decoupler design. The generalized decoupling technique is applied as below.

$$D(s) = G^{-1}(s)G_{\mathrm{R}}(s) \tag{3}$$

where

$D(s)$ decoupler matrix,
$G^{-1}(s)$ inverse of $G(s)$, and
$G_{\mathrm{R}}(s)$ diagonal matrix of $G(s)$.

$$D(s) = \begin{bmatrix} 1 & 0 \\ \frac{-0.4492s^6-3.143s^5-7.083s^4-17.06s^3-22.28s^2-9.60s+5.51e^{-14}}{s^6+6.906s^5+15.22s^4+37.09s^3+47.03s^2+19.44s+4.749e^{-14}} & 1 \end{bmatrix} \tag{4}$$

3 PID Controller Design

The control scheme of TRMS is portrayed in Fig. 2.

The controller performance is mainly dependent on its gain values. Therefore, to get the optimum performance, the gains are tuned well. Different optimization techniques are applied for tuning of controller gains. PID controller transfer function of main and tail rotor is represented through Eqs. (5) and (6), respectively.

$$G_{\mathrm{cm}}(s) = K_{\mathrm{pm}} + K_{\mathrm{im}}/s + K_{\mathrm{dm}}s \tag{5}$$

$$G_{\mathrm{ct}}(s) = K_{\mathrm{pt}} + K_{\mathrm{it}}/s + K_{\mathrm{dt}}s \tag{6}$$

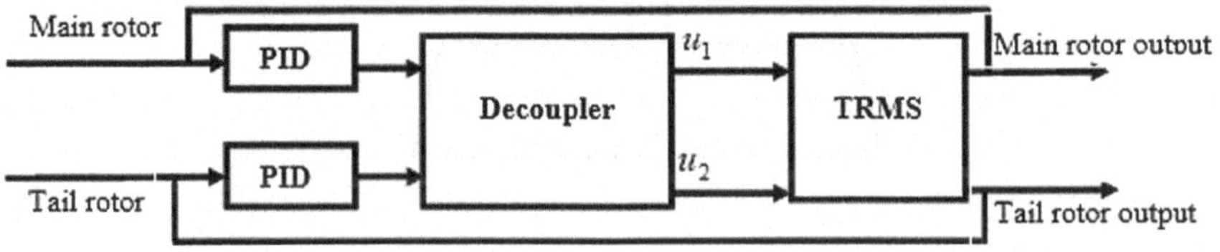

Fig. 2 Decoupled PID control TRMS

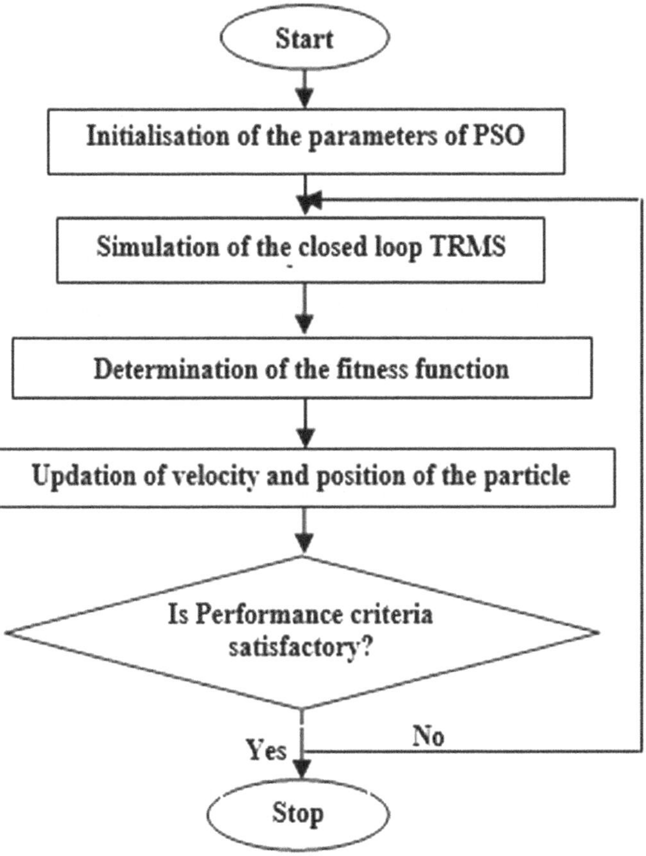

Fig. 3 Steps of PSO algorithm

Three different optimization techniques PSO, RCGA, and BFO are applied for tuning. The reason behind selecting these optimization techniques is their effectiveness in solving difficult real-world optimization problems in a manner such that less computational time and more convergence accuracy are attained (Figs. 3, 4, and 5).

3.1 Particle Swarm Optimization (PSO) Technique to Tune PID Controller Parameters

PSO is an optimization technique that optimizes an objective function [14]. The velocity and position formula is calculated by Eqs. (7) and (8).

$$V_{\text{id}}^{n+1} = V_{\text{id}}^{n} + c_1 \,\text{rand}\,() \cdot \left(P_{\text{id}}^{n} - X_{\text{id}}^{n}\right) + c_2 \,\text{rand}\,() \cdot \left(P_{\text{gd}}^{n} - X_{\text{id}}^{n}\right) \tag{7}$$

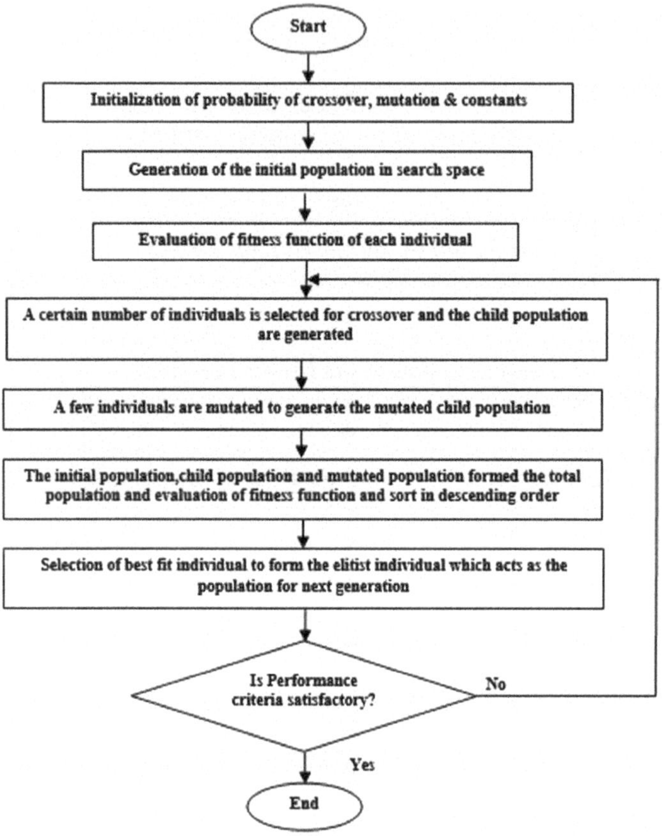

Fig. 4 Steps of RCGA optimization

$$X_{\mathrm{id}}^{n+1} = X_{\mathrm{id}}^{n} + V_{\mathrm{id}}^{n+1} \tag{8}$$

The cost function described by Eq. (9) for calculating the PID controller gains.

$$W = \left(1 - e^{-\beta}\right) \cdot \left(M_{\mathrm{p}} + E_{\mathrm{ss}}\right) + e^{-\beta}(t_{\mathrm{s}} - t_{\mathrm{r}}) \tag{9}$$

where

M_{p} overshoot,
t_{r} rise time,
t_{s} settling time, and
E_{ss} steady-state error.

β (weighting factor) is suitably selected as 0.5. The steps of PSO technique are portrayed in Fig. 3. The initialization of population is done at first, and then, the TRMS is simulated to determine the fitness function. The parameter of this optimization

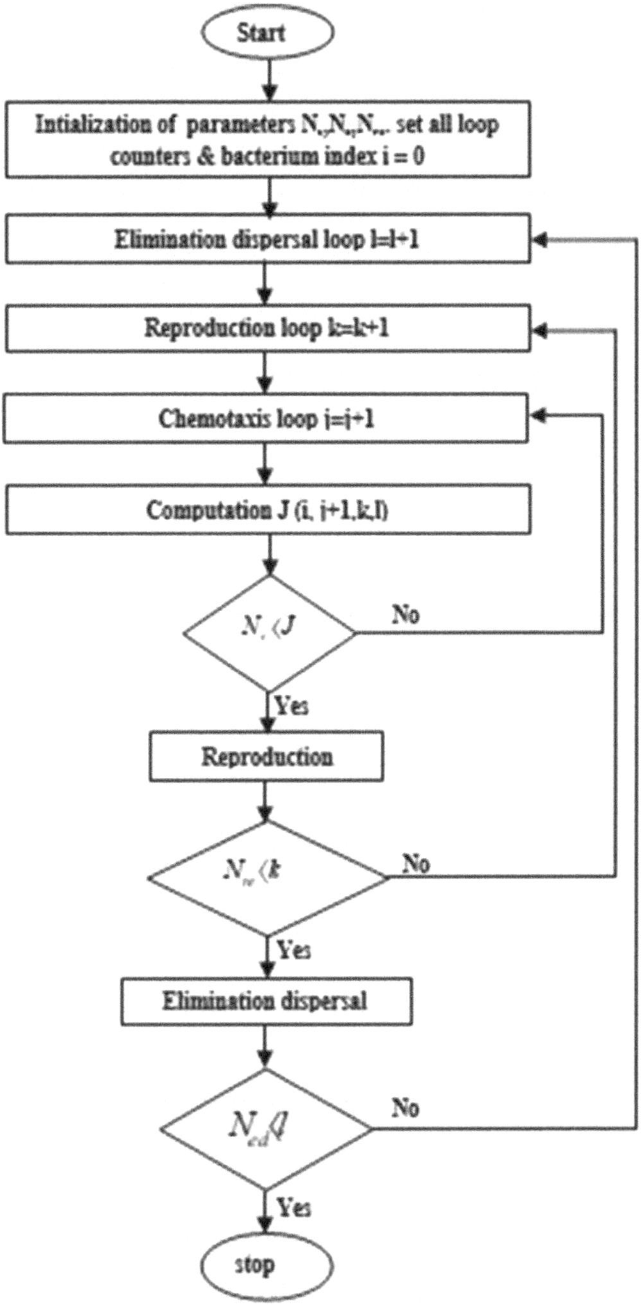

Fig. 5 Steps of BFO algorithm

Table 1 Parameters of PSO optimization

S. no.	Parameter	Value
1	Cognitive constant (C_1)	1.50
2	Group constant (C_2)	1.50
3	Number of initial population	1500
4	Number of iteration	40

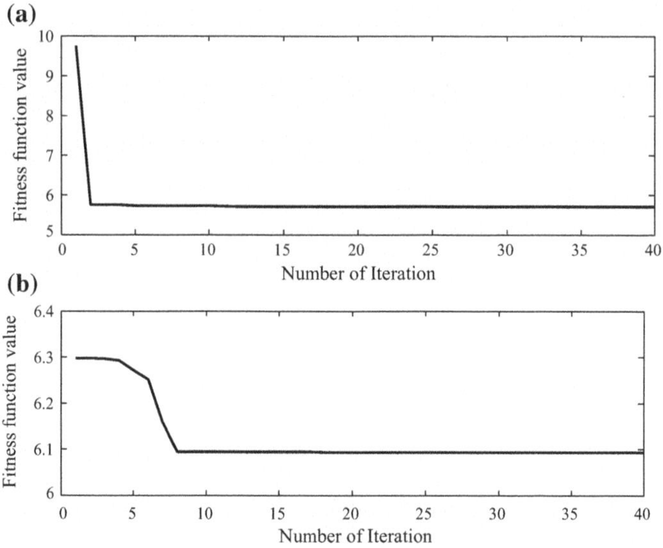

Fig. 6 **a** Convergence characteristics of the main rotor, **b** convergence characteristics of the tail rotor

technique is taken as described in Table 1. The convergence characteristics of the main and tail rotor are shown in Fig. 6.

3.2 *Real-Coded Genetic Algorithm (RCGA) Technique to Tune PID Controller Parameters*

RCGA started from an initial point and generates randomly throughout the search space [5]. After the initialization, it enters into the main loop to achieve the optimum solution and works as a global optimization. The steps of the RCGA optimization are described in Fig. 4. In this, three genetic operators are selection, crossover, and mutation. In RCGA, ith chromosome for the nth population gene (g) can be described as

$$\text{chromosome} = g^{n,i} = \left[g1^{n,i}\ g2^{n,i} \cdots gN^{n,i} \right] \tag{10}$$

The selection operator selects favorable chromosomes based on their objective function and generates an interim population commonly known as mating pool. The common strategy used for the selection is roulette wheel, ranking, and stochastic binary tournament. The convergence of the algorithm is possible due to the selection operator. The crossover operator is known as main search device of the algorithm. Different type of the crossover is possible such as one-point, two-point, and random crossover. After crossover, some genes are inverted in the candidate offspring due to mutation operation.

A real crossover is based on average of two-parent genes. The child population represented as

$$g_q^j(k+1) = g_q^j(k) + \alpha v_{qr}^j \omega_p \in_q^j \tag{11}$$

$$g_r^j(k+1) = g_r^j(k) + \alpha v_{rq}^j \omega_p \in_r^j \tag{12}$$

where $g_q^j(k)$, $g_r^j(k)$ are the qth and rth inputs of the population and v_{rq}^j is the crossing vector represented as

$$v_{rq} = g_r(k) - g_q(k) \tag{13}$$

$\in$ is a random number to determine the amount of gene crossing, and when ω (gene selection vector) = 1, the crossover takes place. α is the crossover range. The child formed by the mutation is described as

$$g_q^j(k+1)' = g_q^j(k+1) + \beta\phi \tag{14}$$

β is step size and ϕ is the random value that decides the amount of individual mutation. The parameter of RCGA optimization technique is taken as described by Table 2.

Table 2 Parameters of RCGA optimization

S. no.	Parameter	Value
1	Population	1500
2	Dimension	3
3	Maximum iteration	40

3.3 Bacterial Foraging Optimization Algorithm to Tune PID Controller Parameters

BFO technique is based on the Escherichia coli bacteria performance behavior [15]. This method is used for real-time optimization problems. The steps of the BFO algorithm are shown in Fig. 5. The complete BFO algorithm is divided into four basic steps. The first one is chemotaxis in which the motion of a bacteria coli bacterium by swimming and tumbling is simulated. Next step is swarming, where some of the bacterium will attract each other and move with higher density. In the reproduction step, the weak bacteria will die, while other healthy bacteria are divided into two parts and will take their place. In the elimination and dispersal step, either group of bacteria gets killed or dispersed to a new position. The specifications of the BFO method are written in Table 3.

4 Simulation Results

The optimally selected values of the PID controller parameters, as discussed in Sect. 3, are tabulated (Table 4).

With these gain values, the designed PID controller is implemented with simulation model of TRMS. Figures 7 and 8 show the output response and control signal of both rotors. The results clearly show that the PID controller with the BFO technique

Table 3 Specifications of BFO algorithm

S. no.	Specifications
1	Dimension for searching ($n = 3$)
2	Bacteria population (S) = 1500
3	Chemotactic step (N_c) = 5
4	Swim step (N_s) = 2
5	Reproductive step (N_{re}) = 2
6	Elimination and dispersal step (N_{ed}) = 1
7	Reproductions of bacteria ($S_r = S/2$)
8	Elimination probability for each bacteria (P_{ed}) = 0.3

Table 4 PID controller parameters determined by optimization techniques

Controller parameters	Main rotor			Tail rotor		
	K_{pm}	K_{im}	K_{dm}	K_{pt}	K_{it}	K_{dt}
PSO	0.08	0.8	0.08	2.91	0.5	1.2
RCGA	0.09	1.05	0.51	2.93	1.55	1.7
BFO	0.08	0.51	0.45	1.75	0.51	1.6

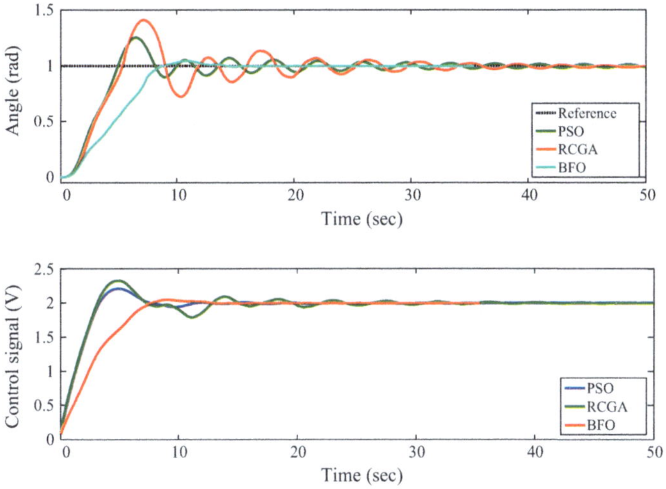

Fig. 7 **a** Tracking performance of main rotor, **b** control signal of main rotor

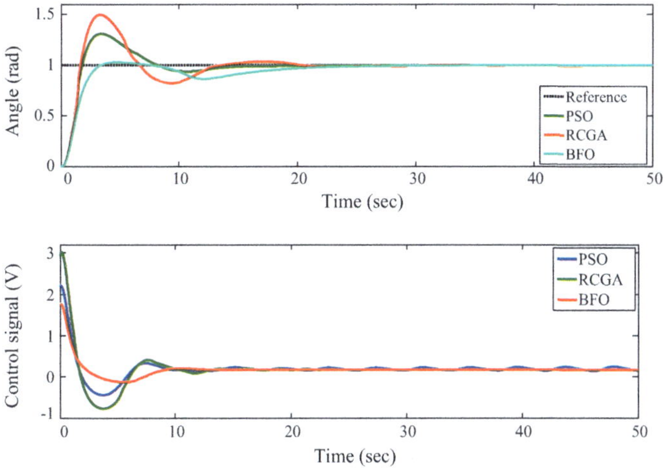

Fig. 8 **a** Tracking performance of tail rotor, **b** control signal of tail rotor

shows superior performance than others. Table 5 depicts the comparison of three optimization methods. It is shown in Table 5 that t_s, t_r, and $\%M_p$ all three specifications are much less in the PID controller tuned by BFO technique than the other optimization method. The control signal is also considered as an important aspect in the present work in concern with the real-time implementation. The rated limit of the control signal for both the rotors in the experimental setup is ± 2.5 V. It is visible from Figs. 7b and 8b that the control voltages are within the prescribed limit for all three optimization methods, whereas the BFO method shows the minimum control

Table 5 Comparison of time-domain performance by different techniques

Different methods	Main rotor			Tail rotor		
	t_s (s)	t_r (s)	$\%M_p$	t_s (s)	t_r (s)	$\%M_p$
PSO	20	5	30	10	1.6	30
RCGA	20	6	40	9	1.4	45
BFO	10	8	5	13	2	4

signal. Therefore, the proposed work can be implemented to the real-time setup of the TRMS without any complexity.

4.1 Disturbance Rejection

This property is analyzed by adding the pulse signal to the output of the system. The pulse signals of 20% magnitude of pitch and yaw are inserted to the system output of the BFO optimized PID controller. It is seen from Fig. 9a, b that the

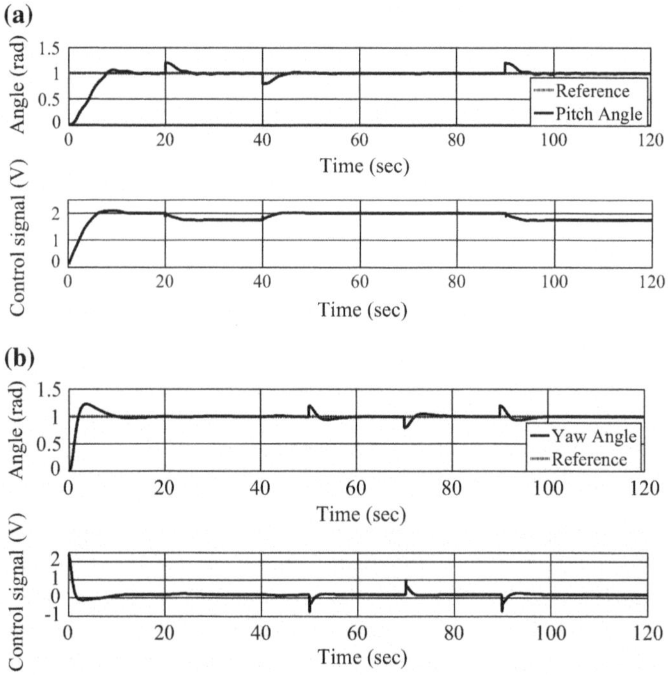

Fig. 9 **a** Output response and control signal of main rotor with disturbance, **b** output response and control signal of tail rotor with disturbance

results corresponding to these inputs settle within 5 s which justify that the designed controller is capable of disturbance rejection.

5 Conclusion

In this work, the decoupled twin rotor MIMO system (TRMS) is controlled by optimized PID controller. The generalized decoupling method is adopted to eradicate the coupling effects of the MIMO system. Three different optimization techniques: PSO, RCGA, and BFO are applied for the fine-tuning of the controller, and the simulation results are also portrayed for all the methods. The simulated responses show satisfactory performance, whereas BFO optimized PID controller shows superior response than others. The disturbance elimination property of the BFO optimized PID controller is also checked by adding the output pulse disturbance to the system. The responses of the system with the disturbances are impressive as the output tracks the reference input smoothly even in the presence of disturbances.

References

1. TRMS 33-220 (1998) TRMS 33-220 user manual, feedback company. E. Sussex, UK
2. Rahideh A, Shaheed MH (2012) Constrained output feedback model predictive control for nonlinear systems. Control Eng Pract 20:431–443
3. Wen P, Lu TW (2008) Decoupling control of a twin rotor MIMO system using robust deadbeat control technique. IET Control Theory Appl 2(11):999–1007
4. Pradhan JK, Ghosh A (2013) Design and implementation of decoupled compensation for a twin rotor multiple-input and multiple-output system. IET Control Theory Appl 7(2):282–289
5. Juang JG, Huang MT, Liu WK (2008) PID control using presearched genetic algorithms for a MIMO system. IEEE Trans Syst Man Cybern Part C Appl Rev 38(5):716–727
6. Juang JG, Liu TK (2011) Real-time neural network control of a twin rotor system. Adv Mater Res 271–273:616–621
7. Toha SF, Tokhi MO (2011) PID and inverse model based control of a twin rotor system. Robotica 29:929–938
8. Mondal S, Mahanta C (2012) Adaptive second-order sliding mode controller for a twin rotor multi-input–multi-output system. IET Control Theory Appl 6(14):2157–2167
9. Ahmad SM, Chipperfield AJ, Tokhi MO (2003) Dynamic modeling and linear quadratic Gaussian control of a twin-rotor multi-input multi-output system. Proc I Mech E Part-I J Syst Control Eng 217(13):203–227
10. Ahmad SM, Chipperfield AJ, Tokhi MO (2000) Dynamic modeling and optimal control of a twin rotor MIMO system. In: National aerospace and electronics conference
11. Pandey SK, Laxmi V (2014) Control of twin rotor MIMO system using PID controller with derivative filter coefficient. In: IEEE student's conference on electrical, electronics and computer science, India
12. Pandey SK, Laxmi V (2015) Optimal control of twin rotor MIMO system using LQR technique. Comput Intell Data Min 1:11–21
13. Juang JG, Liu WK, Lin RW (2011) A hybrid intelligent controller for a twin rotor MIMO system and its hardware implementation. ISA Trans 50:609–619

14. Biswas P, Maiti R, Kolay A, Sharma KD, Sarkar G (2014) PSO based PID controller design for twin rotor MIMO system. In: International conference on control, instrumentation, energy and communication (CIEC)
15. Li J, Dang J, Bu F, Wang J (2014) Analysis and improvement of the bacterial foraging optimization algorithm. J Comput Sci Eng 8(1):1–10
16. Banka S, Dworak P (2007) On decoupling of MIMO systems with guaranteed stability. PAK 53(6):46–51
17. Aminzadeh M, Mahmoodi A, Menhaj MB (2010) A novel decoupling controller design for parallel motion platforms. In: IEEE international conference on control and automation Xiamen, China
18. Park K (2011) Parameterization of decoupling controllers in the generalized plant model. IEEE Trans Autom Control 57(4):1067–1070
19. Sinha PK (1977) State feedback decoupling of nonlinear systems. IEEE Trans Autom Control 22:487–489

Fractional-Order (FO) Control of DC–DC Buck–Boost Converter

M. K. Sameer Kumar, Jayati Dey and Reetam Mondal

Abstract DC–DC converter is required to sustain a steady output DC voltage even under variation in input voltage, load current or any other uncertainties which might arise. Here in this proposed work, it is accomplished by fractional-order (FO) controller designed in such a manner that it not only provides robust stability but maintains voltage regulation successfully even under perturbation in different operating conditions.

Keywords DC–DC buck–boost converter · Fractional-order (FO) controller · Robust stability · Voltage regulation

1 Introduction

One of the fundamental topics to study in the sphere of power electronics is DC–DC converter. It mainly focuses on Boost and buck–boost converters. The initial approach is to perform detailed analysis of the structure and the working principle of these converters [1–3]. After this, it is imperative to develop the small-signal model of these converters to develop the output to control transfer function [3] utilizing which controllers will be designed for the plants. There are diverse approaches for the design methodology of a controller for a boost converter. The fundamental controller by which it is mostly implemented is the PI controller [4, 5]. Another approach can be the implementation of sliding mode control theory to design the controller for a boost converter [6]. Likewise, these two approaches can be amalgamated into a single controller, comprising of a voltage and a current loop to generate the switching pulse to operate the converter [7]. Design on the basis of frequency response of the

M. K. Sameer Kumar · J. Dey · R. Mondal (✉)
Department of Electrical Engineering (EE), National Institute of Technology (NIT), Durgapur, Durgapur, West Bengal 713209, India
e-mail: reetammondal2008@gmail.com

M. K. Sameer Kumar
e-mail: mksameer3293@gmail.com

J. Dey
e-mail: deyjayatibiswas@gmail.com

S. N. Singh et al. (eds.), *Advances in Power and Control Engineering*, Lecture Notes in Electrical Engineering 609, https://doi.org/10.1007/978-981-15-0313-9_8

converter, selecting the position of the zero and the pole, is also another approach [8]. While persuasive control approaches are promptly and conveniently obtainable, the PID controller is still favored due to its reasonable comparative lucidity with coherence and the quality of being most relevant [9]. A frequently conceded fact is that only a fragment of the prevailing PI/PID controller-based loops are modulated to accomplish the best or most favorable behavior [10]. Effective tuning techniques of the five parameters of FO $PI^{\lambda}D^{\mu}$ controllers have been reported in the past years. The progress of fractional calculus and its use in control system including their design and tuning using variegated rendition of software tools associated with the formulation and execution of FO $PI^{\lambda}D^{\mu}$ controllers have been reviewed in [11]. The non-integer controllers are thoroughly implemented to earn and acquire robust performance of plant under control [12]. Two sets of tuning rules, i.e., Ziegler–Nichols rule for tuning of plants with S-shaped unit step response and analytical tuning methods using internal model control (IMC) and tuning by minimization using frequency response, have been dealt with in [13]. A universal integer-order PI–PD controller as a stand-alone unit to control various industrial processes has been produced in [14]. An extended work on reformed and remodeled PI–PD Smith predictor leading to augmentation in the control of processes which are quite slow and sluggish or inherently having zero at the origin or plants with RHP poles is presented in [15]. Synthesis of feedback systems with various techniques has been discussed in [16]. An FOPID controller formulated and executed to land with the performance intended, for a separately excited DC motor put into its operation by these converters [17]. The application of FO $PI^{\lambda}D^{\mu}$ controllers for speed modulation in a DC motor drive is described in [18]. Diverse epitome of FO $PI^{\lambda}D^{\mu}$ controllers in the velocity control of a servo system has been investigated in [19]. The results were juxtaposed to differentiate with the classical PID controllers and scrutinized the virtue of benefit of the proposed design algorithm. A few years back, classical PID controllers were conceived for linear time-invariant (LTI) fractional-order plants in [20]. It is quite familiar that a large part of the control loops in industries are making practical use of the PI or PID controllers. Therefore, it is of sufficiently great appeal to explore the problem of upgrading the common PID controllers to FO $PI^{\lambda}D^{\mu}$ controller leading to change in the dynamical properties of the existing control loops.

In this proposed work, a mathematical output to control transfer function of buck–boost converter has been taken up for the performance study of these non-minimum phase plants which exhibits right half plane (RHP) zero. The disadvantage of a non-minimum phase system is that it produces undershoot in the transient response due to the existence and appearance of right half plane (RHP) zero in the s-plane, thus rendering the output response slow. FO $PI^{\lambda}D^{\mu}$ controller has been therefore designed to control the aforementioned converter in continuous conduction mode. In this approach, the plant transfer function has an RHP zero in the complex s-plane which makes it a non-minimum phase system. This is not the case of discontinuous conduction mode. Therefore, here the research interest is predominantly fascinated on the control aspects of the non-minimum phase plant employing the non-integer controller and to study its transient performance in different operating conditions. The fractional order (FO) allows an additional degree of freedom so as to meet the desired performance criteria accurately. The converter has been subjected to various conditions such as input voltage variation, reference variation and load variation to ensure stability and robustness under all these conditions. Simulation results in MATLAB have been furnished to ratify the design method proposed. It is contemplated that the fractional-order (FO) controller could attain desired response quite satisfactorily.

2 Buck–Boost Converter

Buck–boost converters can be used for both step-down (buck) and step-up (boost) operations of the input DC voltage. It consists of a switch (S), an inductor (L), a diode and a capacitor (C), parallel to which the load (R) is applied. It must be noted that the output here is inverted. The circuit diagram is given in Fig. 1.

The transfer function for the converter operated in continuous conduction mode adopted from [2, 3] is derived as

$$\frac{\hat{v}_O(s)}{\hat{d}(s)} = \frac{\frac{1}{LC} - \frac{sD}{RC(1-D)^2}}{s^2 + \frac{s}{RC} + \frac{(1-D)^2}{LC}} V_s \tag{1}$$

The numerical measurable values in Eq. (1) are enlisted in Table 1.

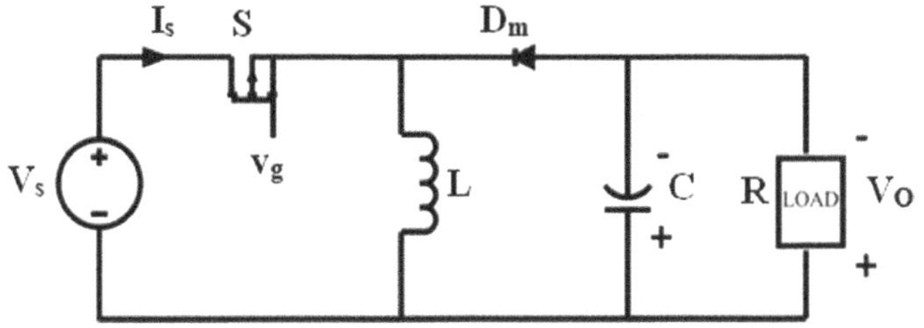

Fig. 1 Buck–boost converter

Table 1 Technical quantities of the DC–DC buck–boost converter [2, 3]

Sl. no.	Elements	Symbols	Parameter values
1.	Load	R	48 Ω
2.	Inductor	L	220×10^{-6} H
3.	Capacitor	C	80×10^{-6} H
4.	Duty cycle	D	2/3
5.	Supply voltage	V_s	12 V
6.	Switching frequency	f_s	5 kHz

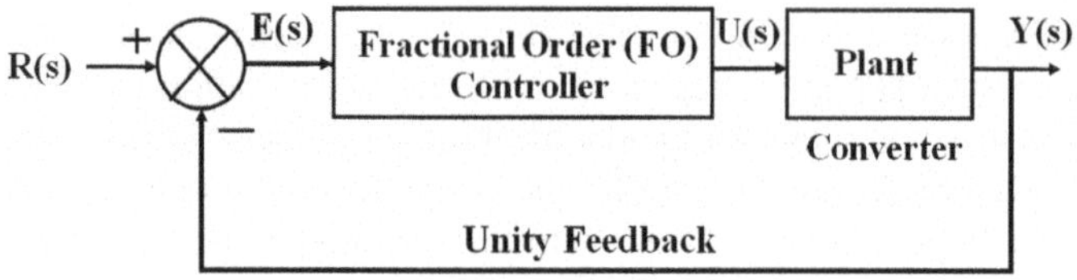

Fig. 2 Framework overview of fractional-order (FO) control

Thus, the transfer function in Eq. (1) takes the form as

$$\frac{\hat{v}_0(s)}{\hat{d}(s)} = \frac{1562.5(s - 36,363.637)}{s^2 + 260.416s + 6,313,131.313} \tag{2}$$

It is seen here in Eq. (2) that it is inherent with a zero in the RHP of s-plane at 36,363.637 and a stable complex conjugate pole pair at $-130.208 \pm 2509.2185i$. The output voltage (V_0) and the output load current (I_0) to be maintained are -24 V and -0.5 A, respectively.

3 Fractional-Order (FO) Controller Synthesis for Buck–Boost Converter

Contemporarily, the blooming arrival of fractional calculus has made feasible the transmutation from classical controllers to those generalized controllers described by differential equations with its non-integer orders. A PID controller is a common feedback control framework extensively operated and deployed in industries for more than a decade [11]. In the extension of its derivative and integrator orders from integer to non-integer numbers it allocates and avails supplemental adaptable adjusting scheme of the $PI^{\lambda}D^{\mu}$ controllers and consequently an elementary approach to achieve control requisites in contrast to its integer equivalent [12].

The control action affects the plant behavior in Fig. 2 by increasing the pace of the dynamic reaction by slashing error in steady state quickly. The control law thus sustained here in the form as

$$C_{PI^{\lambda}D^{\mu}}(s) = K_P + \frac{K_I}{s^{\lambda}} + K_D s^{\mu} \tag{3}$$

with K_P, K_I and K_D the gains of the controller in Eq. (3) and the non-integer orders λ and μ with $\lambda, \mu \in (0, 2)$ [11].

This section describes a novel design method of this non-integer-order controller for the non-minimum phase plant in Eq. (2). Since, the fractional order (FO) controller is the generalized form of the classical integer order (IO) controller having the following pole-zero form:

$$C_{PID}(s) = \frac{K_D s^2 + K_P s + K_I}{s} = \frac{K(s + z_1)(s + z_2)}{s} \tag{4}$$

the controller gains can be estimated by root locus technique. Now, if all the three gains are nonzero, selection of the two finite zeros in Eq. (4) above with $-z_1$ and $-z_2$ is suitably placed at -2142 and -283, which guides the locus of the roots to the stable half of the complex s-plane as shown in Fig. 4a. Integer-order controller parameters in Eq. (4) are thus obtained taking $K = K_D = -9.9 \times 10^{-5}$, leading to damped output reaching the desired quickly. The insertion and inclusion of zero in forward path transfer function thus draw the locus of the roots to the left, managing the system to be stable further by securing faster step response. Utilizing the same values of the controller gains, the orders of the proposed fractional-order (FO) controller $PI^{\lambda}D^{\mu}$ in the form of Eq. (3) have been varied and adjusted between 0 and 2 to consummate the desirable positive gain margin (GM), phase margin and peak sensitivity margin tuned graphically as exhibited in Fig. 3.

On fulfilling satisfactory frequency domain criteria, the fractional-order (FO) compensator according to the form in Eq. (3) is formulated for the plant in Eq. (2) as

$$C_{PI^{\lambda}D^{\mu}}(s) = 0.24 + \frac{60}{s^{1.2}} + 9.9 \times 10^{-5} s^{0.99} \tag{5}$$

Otherwise, repeating the entire design procedure in similar way with new values of $-z_1$ and $-z_2$ until desired sensitivity peak magnitude and stability margins is accomplished. A Bode diagram of buck–boost converter with nominal value of the load resistor of 48 Ω and the FO $PI^{\lambda}D^{\mu}$ compensator in forward path is shown in Fig. 4b, which has gain margin (GM) of 17.8 dB at 280,000 rad/s and phase

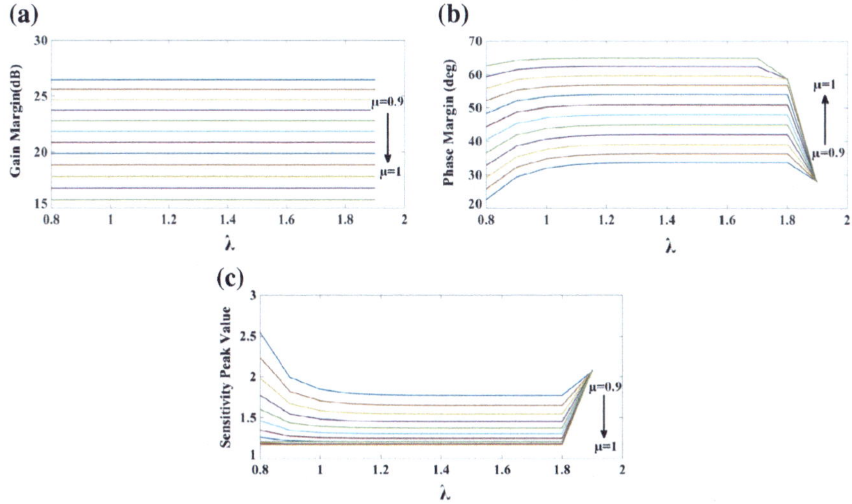

Fig. 3 **a** Gain margin (dB), **b** phase margin (PM) and **c** sensitivity peak varied with λ and μ

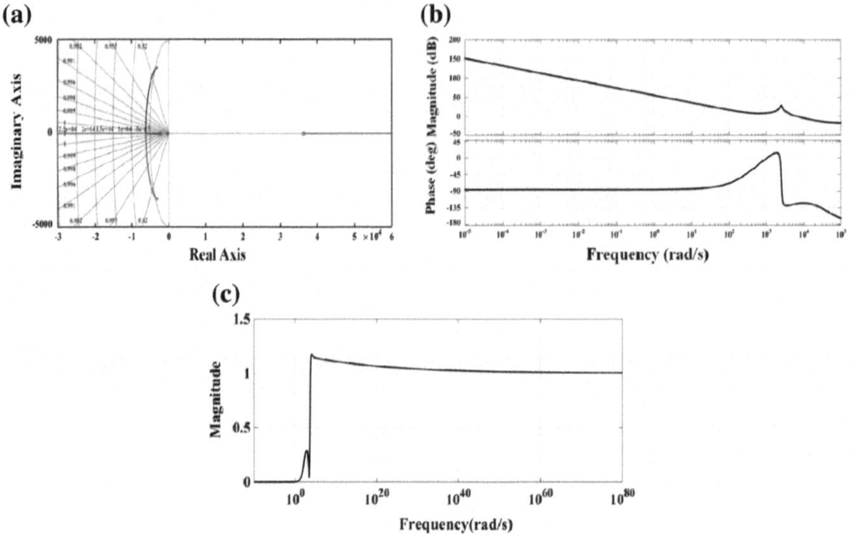

Fig. 4 **a** Root locus plot of the compensated system, **b** characteristic of the loop transfer function with frequency, **c** peak magnitude attained in sensitivity function

margin (PM) of 59.3 at 6550 rad/s which is reasonably adequate and acceptable. The sensitivity plot applying the FO $PI^{\lambda}D^{\mu}$ controller is shown in Fig. 4c. It is examined here that the maximum magnitude of the sensitivity function has < 2 in the mid-frequency region. This makes certain that the robust stability during plant parameter variations can also be fulfilled by concurrent appeasement of GM > 2 (or, 6 dB) and PM >30° [7, 8].

4 Stability Interpretation of FO-LTI System

The state of being stable in the FO-LTI system has been studied based on the Riemann surface. The closed-loop characteristic equation obtained by implementing the FO $PI^{\lambda}D^{\mu}$ controller in Eq. (5) for the buck–boost converter is

$$s^{3.2} - 0.15469s^{3.19} - 114.58s^{2.2} + 5625s^{2.19} + 1.9949 \times 10^7 s^{1.2} + 93{,}750s + 3.4091 \times 10^9 = 0 \tag{6}$$

Equation (6) in s-plane can be transformed into w-plane by $w = s^{1/m}$. Here, $m = 100$ indicates the quantity of sheets within the Riemann surface (Fig. 5a).

$$w^{320} - 0.15469w^{319} - 114.58w^{220} + 5625w^{219} + 1.9949 \times 10^7 w^{120} + 93{,}750w^{100} + 3.4091 \times 10^9 = 0 \tag{7}$$

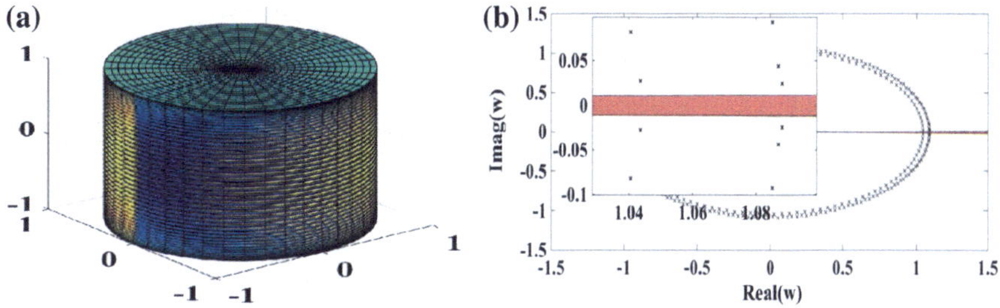

Fig. 5 **a** Riemann surface of function $w = s^{1/m}$ with $m = 100$ Riemann sheets and **b** closed-loop pole positions of buck–boost converter in complex w-plane

The stable region in the s-plane $\phi_s > \frac{\pi}{2}$ transforms to the sector in $\phi_w > \frac{\pi}{2m}$ in w-plane. Stability is confirmed if the principle sheet poles in w-plane lie within this segment. It is clearly noticed from diagram in Fig. 5b that the absence of poles lying within the unstable segment or sector confirms that the arguments have $|\phi_w| < \frac{\pi}{2m} < 0.01574$ which reasserts stability.

5 MATLAB Results and Performance Analysis of Buck–Boost Converter Using Fractional-Order (FO) Controller

The DC–DC converter employed with the fractional-order (FO) $PI^{\lambda}D^{\mu}$ controller designed to regulate the DC output voltage following the scheme in Fig. 6.

The voltage variation of the converter in Eq. (2) has been given in Fig. 7a with percentage overshoot within 0%. Settling time is 19 ms. Similarly, the responses for inductor current, output current and duty ratio are depicted in Fig. 7b–d with same settling time for all responses.

Fig. 6 Fractional-order (FO) control scheme of buck–boost converter implemented in MATLAB

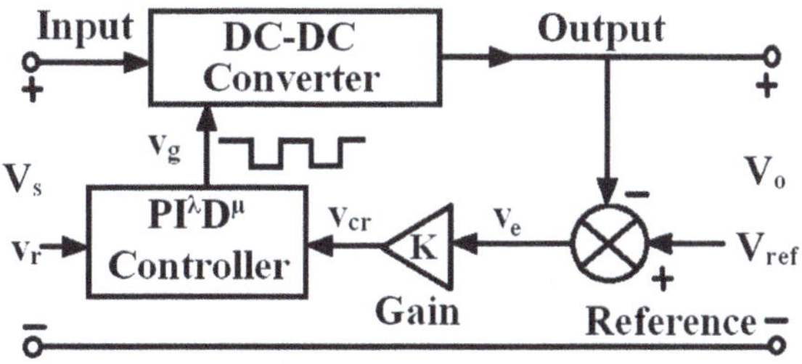

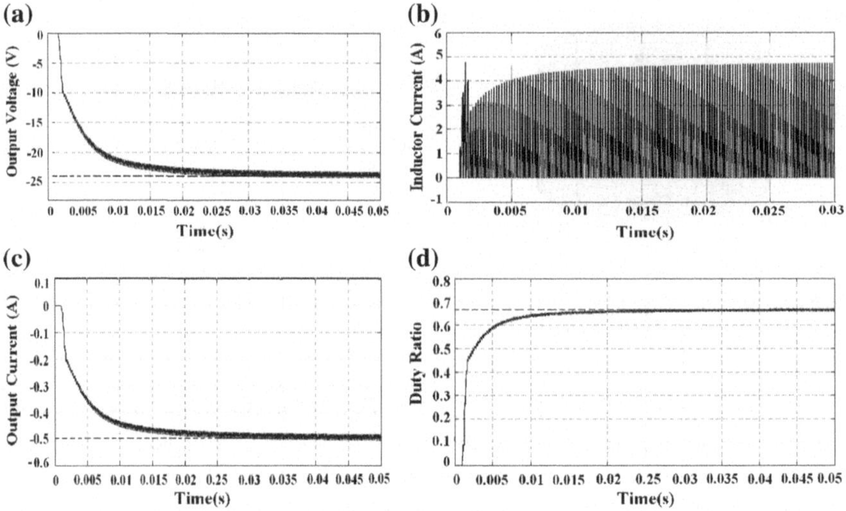

Fig. 7 **a** Load voltage variation, **b** inductor current variation, **c** output current response, **d** duty ratio response of buck–boost converter using FO controller

5.1 *Effect on Reference Variation*

Reference voltage has been varied from −24 to −30 V (25%) with step variation −6 V. The variation in output voltage, inductor current, output current and duty ratio due to variation in reference voltage is portrayed in Fig. 8 which conveys that with this variation, the percentage overshoot is within 0% and settling time is 17 ms for all responses.

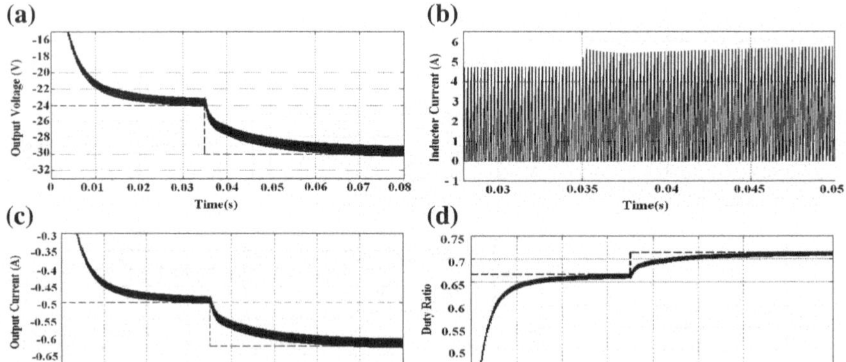

Fig. 8 **a** Load voltage variation, **b** inductor current variation, **c** output current response, **d** duty ratio response of buck–boost converter with reference variation

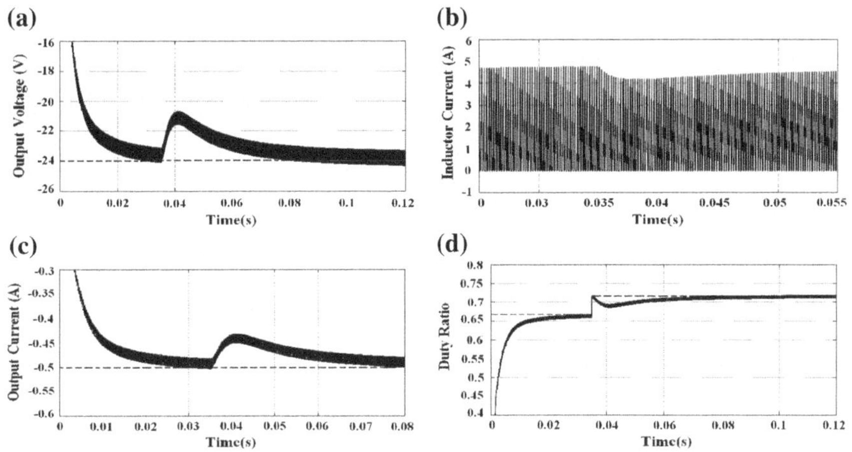

Fig. 9 **a** Load voltage variation, **b** inductor current variation, **c** output current response, **d** variation of duty ratio with input voltage variation

5.2 *Effect on Input Voltage Variation*

Input voltage has been varied from 12 to 9 V (25%) with step variation −3 V. The variation in output voltage, inductor current, output current and duty ratio due to variation in input voltage is displayed in Fig. 9 indicating that with this variation, the percentage overshoot induced in the plant is 18.33% and settling time is 28 ms for all responses.

5.3 *Effect on Load Variation*

Load has been varied from 48 to 36 Ω (25%) and again reaches 48 Ω. The variation in output voltage, inductor current, output current and duty ratio responses of buck–boost converter due to variation in load is depicted in Fig. 10 which claims that when the load is varied at the instance of 0.03 s and 36–48 Ω at the instance of 0.06 s, the percentage overshoot during both the variations is within 6.25% and settling time is 13 ms for first interval and 9 ms for second interval for all responses.

6 Conclusions

The converter under investigation is controlled with fractional-order (FO) $PI^{\lambda}D^{\mu}$ controller. A novel control scheme has been delineated to obtain the desired output voltage. The controller parameters are selected by root locus technique with the non-integer orders graphically tuned to reach desired frequency domain characteristics

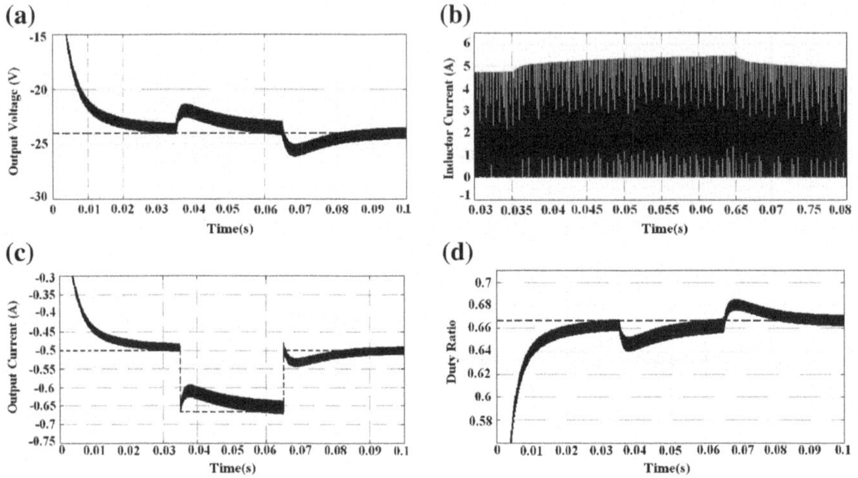

Fig. 10 **a** Load voltage variation, **b** inductor current variation, **c** output current response, **d** duty ratio response of buck–boost converter with load variation

and sensitivity peak. Therefore, cascading the fractional-order controller with it, satisfactory stability margins were obtained. Stability evaluation of the non-minimum phase system compensated with the non-integer-order controller is affirmed utilizing the notion of Riemann surface. Moreover, the fractional-order (FO) controller could provide satisfactory robust performance. The converter has been subjected to various test conditions, such as reference variation, load variation and input voltage variation. It has been observed that under all such variable conditions, the system remained stable and quick response was achieved. Thus, this investigation has thoroughly verified the plant performance with RHP zero in MATLAB employing the fractional-order (FO) controller.

References

1. Shaffer R (2007) Fundamentals of power electronics with MATLAB. Charles River Media, Boston, Massachusetts, USA
2. Mohan N, Undeland TM, Robbins WP (1995) Power electronics converters, applications and design, 2nd edn. Wiley, New York
3. Erickson RW, Maksimovic D (2001) Fundamentals of power electronics, 2nd edn. Kluwer Academic Publishers, New York
4. Fereshtehpoor V, Almasi ON, Monfarad M (2014) A bump-less multi-level PI controller for a boost converter. In: 5th power electronics, drive systems and technologies conference (PEDSTC 2014), IEEE, pp 518–523
5. Zhang J, Dorrell DG, Li L, Argha A (2015) A novel sliding mode controller for DC-DC boost converters under input/load variations. In: 41st annual conference of industrial electronics society (IECON), IEEE, pp 1698–1703

6. Abbas G, Farooq U, Gu J, Asad MU (2015) Controller design for low-input voltage switching converters having non-minimum phase characteristics. In: Canadian conference on electrical and computer engineering (CCECE), IEEE, pp 1294–1298
7. Dey J, Saha TK (2013) Design and performance analysis of two degree of freedom (2DOF) control of DC-DC boost converter. In: International conference on industrial technology (ICIT), IEEE, pp 493–498
8. Dey J, Saha TK, Mahato SN (2014) Robust voltage regulation of DC-DC PWM based buck–boost converter. In: International conference on industrial technology (ICIT), IEEE, pp 1374–1379
9. Kushwaha BK, Narain A (2012) Controller design for Cuk converter using model order reduction. In: International conference on power, control and embedded system, IEEE, pp 1–5
10. O'Dwyer A (2009) Handbook of PI and PID controller tuning rules, 3rd edn. Imperial College Press
11. Shah P, Agashe S (2016) Review of fractional PID controller. Mechatronics 38:29–41
12. Tepljakov A, Gonzalez EA, Petlenkov E, Belikov J, Monje CA, Petras I (2016) Incorporation of fractional-order dynamics into an existing PI/PID DC motor control loop. ISA Trans 60:262–273
13. Valerio D, Costa JS (2006) Tuning of fractional PID controllers with Ziegler Nichols type rules. Sig Process 86:2771–2784
14. Siemens (1999) https://www.rgbautomatyka.pl/.../RKN88T+SIEMENS+MANUAL+DATASHEET.pdf
15. Kaya I (2003) Auto tuning of a new PI-PD Smith predictor based on time domain specifications. ISA Trans 42:559–575
16. Horowitz IM (1963) Synthesis of feedback systems. Academic Press, Colorado
17. Bhaumik A, Srivastava S, Kumar Y, Mafizul Sk (2016) Performance studies of a separately excited DC motor speed control fed by a buck converter using optimized $PI^{\lambda}D^{\mu}$ controller. In: International conference on circuit, power and computing technologies (ICCPCT), IEEE, pp 1–6
18. Biswas A, Das S, Abraham A, Dasgupta S (2009) Design of fractional order PID controller with an improved differential evolution. Eng Appl Artif Intell, Elsevier 22:343–350
19. Barbosa R, Machado JAT, Jesus IS (2008) On the fractional PID control of a laboratory servo system. In: Proceedings of the 17th world congress international federation of automatic (IFAC). IFAC, Seoul, Korea, pp 15273–15278
20. Ozbaya H, Bonnet C, Fioravanti AR (2012) PID controller design for fractional-order systems with time delays. Syst Control Lett 61:18–23

Comparison of Optimal DG Placement in Radial Distribution System Using Centrality Index

Dibya Bharti and Mala De

Abstract Distribution system usually functions at low voltage and high currents, which consequently develops excess power loss along with poor voltage profile. Various methods have been suggested to resolve the problems of distribution system like loss reduction, voltage profile enhancement, and curtailment of network operational cost. Optimal distributed generation (DG) allocation is one of the pre-eminent strategies to maintain the operational security of distribution system. In this article, optimal placement of two dissimilar types of DGs is compared in radial distribution network for voltage profile improvement. Electrical centrality indices are applied for identifying the optimal location of DGs and capacity of DGs is optimized to improve the voltage profile, which consequently reduces system loss. By using electrical centrality indices, two dissimilar types of DGs are placed in network separately and system performance is compared. The proposed method evaluates the placement of different types of DGs in 34-bus radial distribution network as well as 69-bus radial distribution network.

Keywords Distributed generation (DG) · Radial distribution network · Voltage profile improvement · Average electrical centrality index

Nomenclature

P_{gi_slack}	real power generation at slack bus
P_{li}	active power of load
P_k	real power instilled on bus '*k*'
P^i_{DG}	real power production by DG at bus '*i*'
Q_{gi_slack}	reactive power output of slack bus
Q^i_{DG}	reactive power produced by DG fixed at bus '*i*'
Q_{li}	reactive load
Q_k	reactive power infused at bus '*k*'

D. Bharti (✉) · M. De
Department of Electrical Engineering, National Institute of Technology Patna, Patna 800005, India
e-mail: dibya_minu1@rediffmail.com

S. N. Singh et al. (eds.), *Advances in Power and Control Engineering*, Lecture Notes in Electrical Engineering 609, https://doi.org/10.1007/978-981-15-0313-9_9

V_{gi_slack}	slack bus voltage
V_k	bus voltage for 'k' bus
Y_{ki}	admittance matrix element related to buses 'k' and 'i'
δ_k	phase angle of voltage for 'k' bus
θ_{ki}	phase angle for admittance matrix element related to nodes 'k' and 'i'
LF_l	line flow
$P_{loss\_w\_DG}$	real power loss in network with DG
$P_{loss\_wo\_DG}$	real power loss in network without DG

1 Introduction

The incorporation of distributed generations (DGs) is exceedingly frequent in decentralized power system. The DGs are installed within electrical network with various objectives like network loss reduction (active and reactive), voltage profile improvement, peak load curtailment, and system security enhancement [1]. The DG placement in distribution system provides both technical and economical advantages, as immense quantity of produced power is dissipated as line loss in distribution level. DG impacts the network loss similar to placement of capacitors banks with a major difference that several DGs influence both active and reactive power flow in the system. The selection of inappropriate location/size for DGs in distribution system results into increased system loss like capacitors [2, 3].

In recent years, various schemes have been recommended for deciding the optimal site and enormity of DGs (single or multiple) in radial distribution system [4–13]. Analytical methods [14–18], optimal power flow [19, 20], evolutionary algorithms [21–30], differential search algorithms [31, 32], fuzzy [33], and nature-inspired algorithms [34–37] are widely used for optimal DG placement in distribution network with different objectives. The objectives of existing methods for optimal DG planning are of broadly classified into three groups: loss minimization [14–17, 19, 27], voltage stability improvement [3, 13, 18], and hybrid objectives [4, 8, 11, 24, 26]. Hybrid functions are combination of two different objectives that may be either reduction of loss through nominal voltage deviation in the network or some other cost functions.

The existing DG resources can be categorized into four clusters [1] as given in Table 1.

It can be concluded from Table 1 that capacitors are type of DG, which produces reactive power only and similar to capacitor placement, DG installation at unsuitable location results into increased system loss. In [38], electrical centrality indices have been used for deciding the optimal location for static VAR compensator in transmission system and in [39], centrality indices are used for multi-point reactive power compensation in radial structured network connected with numerous renewable generations. The electrical centrality indices are derived from power flow results and are used for finding the most significant bus of the network [38, 39].

Table 1 Types of existing DG technologies

Type of DG	Active power	Reactive power	Example
A	Produces	Not available	Photovoltaic arrays, microturbine, battery storage, etc.
B	Not available	Produces	Capacitor banks and synchronous condensers
C	Produces	Produces	Diesel generators, synchronous generators, and wind turbines
D	Produces	Absorbs	Induction generators connected with wind turbines

So, this paper presents an evaluation on optimal placement of two dissimilar types of DGs (type 'B' and type 'C') based on electrical centrality indices for voltage profile improvement. In proposed method, locations of DGs are identified by electrical centrality indices and only capacities of DGs are optimized. It is evident that the optimal placement of DGs leads to loss minimization as well. The proposed method is analyzed on 34-bus radial distribution network and 69-bus radial distribution network with two types of DGs (type 'B' and type 'C') separately for comparing the voltage profile. The rest of this paper is systematized as follows: Sect. 2 explains the method and objective function formulation for optimal DG placement. The results and comparison are presented in Sect. 3, and conclusions are described in Sect. 4.

2 Objective Function and Methodology of DG Placement

2.1 Objective Function for DG Placement

In literature, it has been found that objectives for optimal DG placement are of three types: minimization of network loss, voltage profile improvement, and hybrid functions, which is combination of two different objective functions. In the presented method, the objective is minimization of bus voltage deviation, which is given as Eq. (1) and constraints are represented as Eqs. (2)–(14).

$$F = \text{Minimize} \sum |1 - V_i|^2 \tag{1}$$

Subject to

(i) Constraints of Power flow:

(a) Real power:

$$P_{gi\_\text{slack}} + P_{\text{DG}}^{i} - P_{li} - P_k = 0 \tag{2}$$

$$P_k = V_k \sum_{i=1}^{n} Y_{ki} V_i \cos(\delta_k - \delta_i - \theta_{ki}) \tag{3}$$

(b) Reactive power:

$$Q_{gi\_slack} + Q_{DG}^{i} - Q_{li} - Q_k = 0 \tag{4}$$

$$Q_k = V_k \sum_{i=1}^{n} Y_{ki} V_i \sin(\delta_k - \delta_i - \theta_{ki}) \tag{5}$$

(ii) Limits used for Control variable:

(a) Limits of real power production:

$$P_{gi\_slack}^{\min} \leq P_{gi\_slack} \leq P_{gi\_slack}^{\max} \tag{6}$$

(b) Voltage limits for PV bus:

$$V_{gi}^{\min} \leq V_{gi} \leq V_{gi}^{\max} \tag{7}$$

(c) Reactive power generation limits:

$$Q_{gi\_slack}^{\min} \leq Q_{gi\_slack} \leq Q_{gi}^{E} \tag{8}$$

(iii) State variable limits:

(a) Limits for reactive power production:

$$Q_{DG}^{i} \leq 0.75 * Q_{li} \tag{9}$$

(b) Voltage limits for PQ bus:

$$V_i^{\min} \leq V_i \leq V_i^{\max} \tag{10}$$

(c) Line flow limit:

$$LF_l \leq LF_l^{\max} \tag{11}$$

(iv) DG size limits:

(a) Real power production limits:

$$P_{DG,i}^{\min} \leq P_{DG}^{i} \leq P_{DG,i}^{\max} \tag{12}$$

(b) Reactive power production limits:

$$Q_{\mathrm{DG},i}^{\min} \leq Q_{\mathrm{DG}}^{i} \leq Q_{\mathrm{DG},i}^{\max} \tag{13}$$

(v) Active power loss:

$$P_{\mathrm{loss\_}w\mathrm{\_DG}} < P_{\mathrm{loss\_}wo\mathrm{\_DG}} \tag{14}$$

2.2 Methodology for Optimal DG Placement

The flow chart for proposed method of optimal DG placement is represented in Fig. 1. The methodology of optimal DG placement starts with power flow results and buses are ranked according to electrical centrality indices calculated from these results as in [38, 39]. Buses are ranked in accordance with the descending values of average centrality index, i.e. bus with highest value of average centrality index is assigned rank 1 and so on. The average centrality index of each bus is the average of its electrical closeness centrality index and electrical betweenness centrality index. The method for calculating these indices and ranking of buses is extensively discussed in [39, 40] and signifies the most effective bus in the network.

As presented in Table 1, available DG technologies are categorized into four clusters conditional to their capacity of generating active and reactive power. In this paper, two categories of DGs (type 'B' and type 'C') are considered individually in radial distribution system for improving voltage profile. Type 'B' DGs produce reactive power only, while type 'C' DGs produce both active and reactive power. If system voltage profiles do not reach within specified voltage limits after installation of single DG, then installation of multiple DGs can be considered. According to present paper, multiple DGs will be of same type; all DGs will be either of type 'B' or type 'C'.

3 Results and Discussions

The method of optimal DG placement is demonstrated in 34-bus radial distribution configuration and 69-bus radial distribution configuration for comparing then system performance after incorporating two dissimilar forms of DGs individually. Results of system voltage profiles and real power loss are compared after placement of DG type 'B' and DG type 'C' for both the considered radial system.

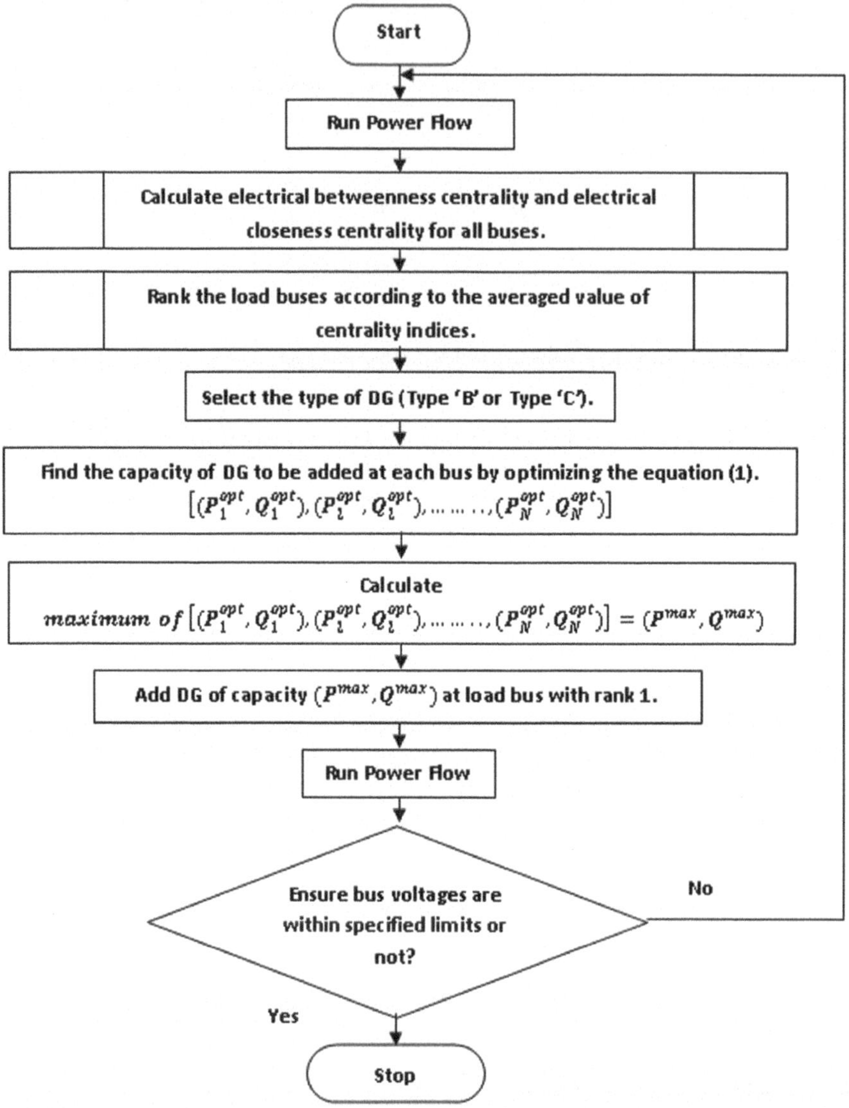

Fig. 1 Flow chart for optimal DG placement

3.1 34-Bus Radial Distribution Network

All buses and their centrality indices based ranking for 34-bus radial distribution network are listed in Table 2. Initial bus voltages are also given in Table 2.

Table 2 Ranking and bus voltages (prior and subsequent to DG placement) in 34-bus radial distribution system

Bus	Ranking	Initial bus voltages (in pu)	Bus voltages after DG placement (Type 'B') (in pu)	Bus voltages after DG placement (Type 'C') (in pu)
DG size (capacity @ location)			1300 kVAR@bus 33	(1200 kW/280 kVAR)@bus 33
1	Gen	1.0000	1.0000	1.0000
2	2	1.0000	1.0000	1.0000
3	4	0.9932	0.9950	0.9956
4	27	0.9876	0.9912	0.9922
5	25	0.9796	0.9853	0.9874
6	29	0.9726	0.9803	0.9833
7	31	0.9703	0.9791	0.9831
8	32	0.9676	0.9765	0.9804
9	30	0.9657	0.9747	0.9786
10	28	0.9636	0.9728	0.9765
11	14	0.9632	0.9723	0.9760
12	16	0.9629	0.9721	0.9758
13	26	0.9625	0.9717	0.9754
14	22	0.9929	0.9947	0.9953
15	5	0.9926	0.9945	0.9950
16	12	0.9925	0.9944	0.9949
17	21	0.9911	0.9933	0.9938
18	24	0.9673	0.9763	0.9792
19	23	0.9630	0.9730	0.9758
20	17	0.9583	0.9694	0.9721
21	18	0.9546	0.9665	0.9691
22	20	0.9515	0.9641	0.9665
23	19	0.9481	0.9614	0.9637
24	15	0.9458	0.9592	0.9614
25	13	0.9437	0.9573	0.9593
26	11	0.9428	0.9564	0.9584
27	10	0.9425	0.9561	0.9581
28	9	0.9419	0.9561	0.9589
29	8	0.9701	0.9792	0.9846
30	7	0.9700	0.9791	0.9844
31	6	0.9697	0.9788	0.9841
32	3	0.9634	0.9726	0.9762
33	**1**	0.9631	0.9724	0.9760
34	33	0.9630	0.9723	0.9759
Loss (in kW)		221.7345	177.7657	157.7861

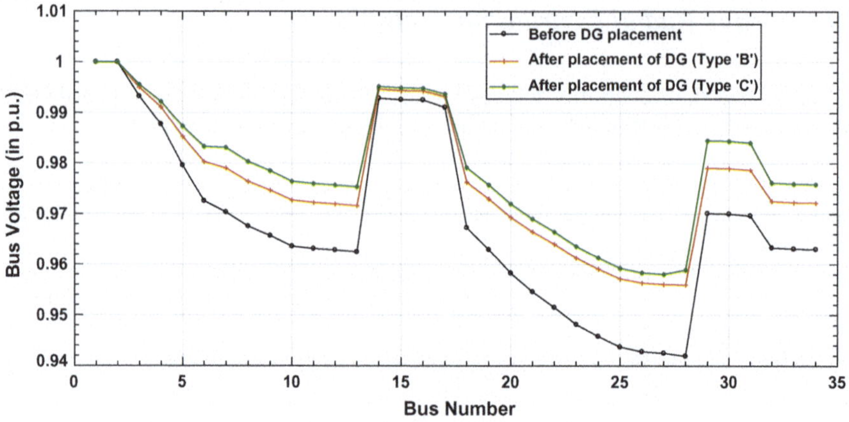

Fig. 2 Comparison of voltage profile in 34-bus radial distribution network

From results given in Table 2, it can be noticed that load bus 33 is ranked as 1. So, bus 33 is the preferred location for DG placement according to proposed method. Specified limits of bus voltages are $V_{\text{min}} = 0.95$ pu and $V_{\text{max}} = 1.05$ pu. Bus voltages after DG placement (type 'B' and type 'C') are presented in Table 2. The capacities of DGs are also put into Table 2 which shows that capacity of type 'B' DG as 1300 kVAR@bus 33, while capacity of type 'C' DG as (1200 kW/280 kVAR)@bus 33.

Bus voltages subsequent to placement of DG type 'B' and DG type 'C' are compared in Fig. 2.

From Fig. 2, it can be observed that selection of DG of type 'C' gives better results if both voltage profile improvement and reduction of active power loss are considered. Real power loss of system reduces to 177.7657 kW from initial value of 221.7345 kW after placement of type 'B' DG in the network. The installation of type 'C' DG reduces active power loss to 157.7861 kW, which is less in comparison of system loss after placement of type 'B' DG. From results of DG placement in 34-bus radial distribution network, it can be concluded that placement of type 'C' DG is advantageous over type 'B' DG placement for up-gradation of voltage profile and reduction of active power loss.

3.2 *69-Bus Radial Distribution Network*

In case of 69-bus system, specified limits of bus voltages are $V_{\text{min}} = 0.95$ pu and $V_{\text{max}} = 1.05$ pu. The under-voltage buses of network are listed in Table 3 with their voltage. Bus voltages after DG placement (type 'B' and type 'C') in 69-bus network are also presented in Table 3. According to ranking based on centrality indices, load bus 49 is ranked as 1 and first DG installation will be at bus 49. Figure 3 compares

Table 3 Bus voltages (prior and subsequent to DG placement) in 69-bus radial distribution system

Bus number	Initial bus voltages (in pu)	Type 'B' (capacity @ location)	Bus voltages after placement of DG (Type 'B') (in pu)	Type 'C' (capacity @ location)	Bus voltages after placement of DG (Type 'C') (in pu)
46	0.9401	1300kVAR@bus 49	0.9589	(800kW/250kVAR) @bus 49	0.9621
47	0.9291		0.9515		0.9547
48	0.9248		0.9486		0.9518
49	0.9198		0.9452		0.9585
50	0.9124		0.9389		0.9521
51	0.9121		0.9386		0.9519
52	0.9117		0.9383		0.9515
53	0.9098		0.9367		0.9506
54	0.9092		0.9362		0.9499
Loss (in kW)	225.9090		127.5482		95.0145

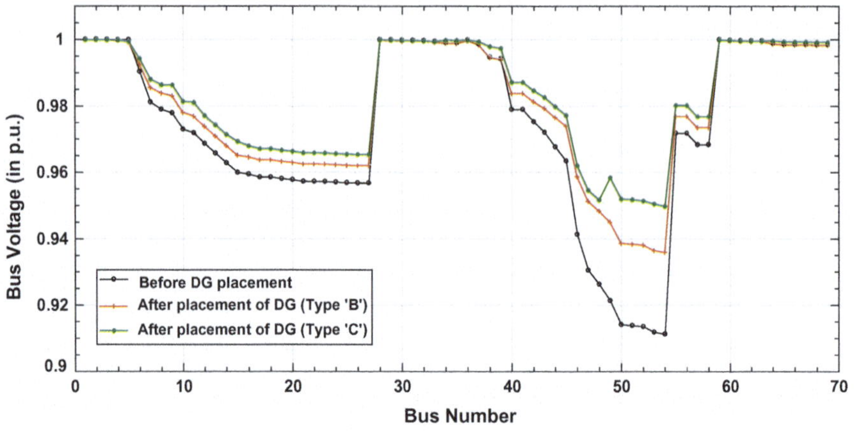

Fig. 3 Comparison of voltage profile in 69-bus radial distribution system

the bus voltages of 69-bus radial network after installation of DG type 'B' and DG type 'C'.

From Table 3, it can be examined that initially voltage at bus 46, 47, 48, 49, 50, 51, 52, 53, and 54 are not in the specified voltage limits. Initially, real power loss in network is 225.9090 kW. Type 'B' DG of capacity 1300 kVAR is installed at bus 49 by which system loss is reduced to 127.5482 kW but minimum bus voltage is 0.9362 pu which is not within specified voltage limits. Results of Table 3 indicate that specified limits of voltages are not attained by all buses of the system after placement of single DG of type 'B'.

DG of type 'C' is installed at bus 49 of with capacity of 800 kW/250 kVAR. From Table 3, it can be concluded that after placement of type 'C' DG of optimal amount, all the bus voltages reach within specified limits in 69-bus radial network and system loss is reduced to 95.0145 kW.

After placement of type 'C' DG, minimum bus voltage is 0.9499 pu at bus 54 which is approximately equal to the minimum specified voltage limit (0.95 pu).

From Fig. 3, it is clear that placement of type 'C' DG gives better voltage profile in comparison of type 'B' DG placement in 69-bus radial network if up-gradation in voltage profile and loss minimization is implicated.

3.3 Comparison of Results with Existing Method

Results of DG placement are compared with the method of paper [41] for 69-bus radial network. Results are compared for analyzing the minimization of voltage deviation, which is listed in Table 4.

Table 4 Comparison of DG placement with existing method in 69-bus radial network

Method	Location	Size (MW)	Method	Location	Size	
					Type 'B'	Type 'C'
[41]	10	1.0054	Proposed	49	1300 kVAR	(800 kW/250 kVAR)
	20	0.4185				
	61	2.2051				

It can be seen from Table 4 that method of [41] suggests placement of three DGs of capacity 1.0054, 0.4185, and 2.2051 MW at buses 10, 20, and 61, respectively. Proposed technique suggests placement of DG of capacity 1300 kVAR and (800 kW/250 kVAR) for type 'B' and type 'C,' respectively. From results of Table 4, it can be concluded that DG size by method of [41] is higher than that of proposed method.

4 Conclusions

A comparison on placement of two dissimilar types of DGs (type 'B' and type 'C') individually in radial distribution system is presented in this paper. In proposed method, suitable positions of DGs are predicted by electrical centrality indices, which can be easily computed by using power flow results. The proposed method of optimal DG placement intends to improve voltage profile, which consequently reduces system loss. System performances are being compared in 34-bus radial distribution network and 69-bus radial distribution network after DG placement of type 'B' and type 'C'. From results, it can be concluded that selection of type 'C' DG gives improved voltage profile that of type 'B' DG.

References

1. El-Fregany A (2015) Optimal allocation of multi-type distributed generators using backtracking search optimization algorithm. Int J Electr Power Energy Syst 64:1197–1205
2. Rueda-Medina AC, Franco JF, Rider MJ, Padilha-Feltrin A, Romero R (2013) A mixed integer linear programming approach for optimal type, size and allocation of distributed generation in radial distribution systems. Electr Power Syst Res 97:133–143
3. Gözel T, Eminoglu U, Hocaoglu MH (2008) A tool for voltage stability and optimization (VS&OP) in radial distribution systems using MatLab graphical user interface (GUI). Simul Model Pract Theor 16(5):505–518
4. Sultana U, Khairuddin AB, Aman MM, Mokhtar AS, Zareen N (2016) A review of optimum DG placement based on minimization of power losses and voltage stability enhancement of distribution system. Renew Sustain Energy Rev 63:363–378
5. Hung DQ, Mithulananthan N, Bansal RC (2013) Multiple distributed generators placement in primary distribution networks for loss reduction. IEEE Trans Ind Electron 60(4):1700–1708
6. Elmitwally A (2013) A new algorithm for allocating multiple distributed generation units based on load centroid concept. Alexandria Eng J 52(4):655–663

7. Kumar D, Samantaraya SR, Kamwab ICN (2014) Reliability-constrained based optimal placement and sizing of multiple distributed generators in power distribution network using cat swarm optimization. Electr Power Compon Syst 42(2):149–164
8. Juanuwattanakul P, Masoum MAS (2012) Increasing distributed generation penetration in multi phase distribution networks considering grid losses, maximum loading factor and bus voltage limits. IET Gener Transm Distrib 6:1262–1271
9. Rotaru F, Chicco G, Grigoras G, Cartina G (2012) Two-stage distributed generation optimal sizing with clustering-based node selection. Int J Electr Power Energy Syst 40:120–129
10. Hung DQ, Mithulananthan N (2014) Loss reduction and loadability enhancement with DG: a dual-index analytical approach. Appl Energy 115:233–241
11. Esmaili M, Firozjaee EC, Shayanfar HA (2014) Optimal placement of distributed generations considering voltage stability and power losses with observing voltage-related constraints. Appl Energy 113:1252–1260
12. Mohan N, Ananthapadmanabha T, Kulkarni A (2015) A weighted multi-objective index based optimal distributed generation planning in distribution system. Procedia Technol 21:279–286
13. Murty V, Kumar A (2015) Optimal placement of DG in radial distribution systems based on new voltage stability index under load growth. Int J Electr Power Energy Syst 69:246–256
14. Hung DQ, Mithulananthan N, Bansal RC (2013) Analytical strategies for renewable distributed generation integration considering energy loss minimization. Appl Energy 105:75–85
15. Hung DQ, Mithulananthan N, Lee KY (2014) Optimal placement of dispatchable and nondispatchable renewable DG units in distribution networks for minimizing energy loss. Int J Electr Power Energy Syst 55:179–186
16. Viral R, Khatod D (2015) Ananalytical approach for sizing and siting of DGs in balanced radial distribution networks for loss minimization. Int J Electr Power Energy Syst 67:191–201
17. SooHyoung L, JungWook P (2013) Optimal placement and sizing of multiple DGs in a practical distribution system by considering power loss. IEEE Trans Ind Appl 49:2262–2270
18. Raja P, Selvan MP, Kumaresan N (2013) Enhancement of voltage stability margin in radial distribution system with squirrel cage induction generator based distributed generators. IET Gener Transm Distrib 7:898–906
19. Mahmoud K, Yorino N, Ahmed A (2016) Optimal distributed generation allocation in distribution systems for loss minimization. IEEE Trans Power Syst 31(2):960–969
20. Harrison GP, Piccolo A, Siano P, Wallace AR (2007) Distributed generation capacity evaluation using combined genetic algorithm and OPF. Int J Emerg Electr Power Syst 8:1–8
21. Kansal S, Kumar V, Tyagi B (2013) Optimal placement of different type of DG sources in distribution networks. Int J Electr Power Energy Syst 53:752–760
22. Karimyan P, Gharehpetian GB, Abedi M, Gavili A (2014) Long term scheduling for optimal allocation and sizing of DG unit considering load variations and DG type. Int J Electr Power Energy Syst 54:277–287
23. Ishak R, Mohamed A, Abdalla AN, CheWanik MZ (2014) Optimal placement and sizing of distributed generators based on a novel MPS Iindex. Int J Electr Power Energy Syst 60:389–398
24. Musa H, Adamu SS (2013) Enhanced PSO based multi-objective distributed generation placement and sizing for power loss reduction and voltage stability index improvement. Energy Tech, pp 1–6
25. NguyenCong H, Mithulananthan N, Bansal RC (2013) Location and sizing of distributed generation units for loadability enhancement in primary feeder. IEEE Syst J 7:797–806
26. Kayal P, Chanda CK (2013) Placement of wind and solar based DGs in distribution system for power loss minimization and voltage stability improvement. Int J Electr Power Energy Syst 53:795–809
27. Ayodele T, Ogunjuyigbe A, Akinola O (2015) Optimal location, sizing, and appropriate technology selection of distributed generators for minimizing power loss using genetic algorithm. J Renew Energy
28. Candelo-Becerra JE, Hernández-Riaño HE (2015) Distributed generation placement in radial distribution networks using a bat-inspired algorithm. Dyna Rev Fac 82(192):60–67

29. Jamian J, Mustafa M, Mokhlis H (2015) Optimal multiple distributed generation output through rank evolutionary particle swarm optimization. Neuro Comput 152:190–198
30. Cheng S, Chen MY, Wai RJ, Wang FZ (2014) Optimal placement of distributed generation units in distribution systems via an enhanced multi-objective particle swarm optimization algorithm. J Zhejiang Univ Sci C 15:300–311
31. Injeti SK (2018) A Pareto optimal approach for allocation of distributed generators in radial distribution systems using improved differential search algorithm. J Electr Syst Inf Technol 5:908–927
32. Kumar S, Mandal KK, Chakraborty N (2019) Optimal DG placement by multi-objective opposition based chaotic differential evolution for techno-economic analysis. Appl Soft Comput J. https://doi.org/10.1016/j.asoc.2019.02.013
33. Nowdeh SA et. al (2019) Fuzzy multi-objective placement of renewable energy sources in distribution system with objective of loss reduction and reliability improvement using a novel hybrid method. Appl Soft Comput J. https://doi.org/10.1016/j.asoc.2019.02.003
34. Prakash DB, Lakshminarayana C (2018) Multiple DG placements in radial distribution system for multi objectives using whale optimization algorithm. Alexandria Eng J 57:2797–2806
35. Abdelaziz M, Moradzadeh M (2019) Monte-Carlo simulation based multi-objective optimum allocation of renewable distributed generation using OpenCL. Electr Power Syst Res 170:81–91
36. HassanzadehFard H, Jalilian A (2018) Optimal sizing and location of renewable energy based DG units in distribution systems considering load growth. Int J Electr Power Energy Syst 101:356–370
37. Ali AH, Youssef AR, George T, Kamel S (2018) Optimal DG allocation in distribution systems using Ant Lion Optimizer. In: Proceedings of international conference Innovative Trends in Computer Engineering (ITCE), pp. 324–331. IEEE, Aswan University, Egypt, 2018
38. Bharti D, De M (2018) A centrality index based approach for selection of optimal location of static reactive power compensator. Electr Power Compon Syst 46(8):886–899
39. Bharti D, De M (2019) Framework for multipoint optimal reactive power compensation in radial distribution system with high distributed generation penetration. Int Trans Electr Ener Syst. https://doi.org/10.1002/2050-7038.12007
40. Bharti D, De M (2017) A generalized bus dependency matrix based centrality measures for reactive power compensation. In: Proceedings of 2nd International conference on recent developments in control, automation and power engineering (RDCAPE), IEEE, Amity University, Noida, India, pp 49–54, 2017
41. Quadri IA, Bhowmick S, Joshi D (2018) A comprehensive technique for optimal allocation of distributed energy resources in radial distribution systems. Appl Energy 211:1245–1260

Comparison of quasi-Switched Boost Inverter and Two-Stage dc/dc Boost Inverter

P. Sriramalakshmi

Abstract This article presents the comparison between a quasi-switched boost inverter (qSBI) and a two-stage dc/dc-boosted inverter for photovoltaic applications. These two inverter topologies are compared based on their working modes, steady-state response, boost factor, voltage gain, and power loss. With the input dc voltage of 36 V, modulation index of 0.5, and shoot-through duty ratio of 0.45, both the configurations are simulated in the MATLAB/SIMULINK environment to validate the theoretical concepts. The harmonic profile of both the topologies is presented. The experimental results are presented to validate the simulation results. The detailed analysis shows that one of the single-stage qSBI topologies namely Embedded type qSBI (EqSBI) possesses the superior features compared to the traditional dc/dc-boosted inverter topologies.

Keywords Voltage source inverter · *Z* source inverter · Switched boost inverter · Shoot-through · Duty ratio · Pulse width modulation · Boost factor

1 Introduction

There is an increasing demand of electrical energy and to reduce the emission of carbon dioxide gases during the power conversion imposing the power electronic researchers to opt for non-conventional energy sources. Among all renewable energy sources (RES), the solar photovoltaic (PV) is gaining more attention due to its clean and unlimited availability in nature. The application of PV in power system acquires more importance with the vast developments in the power electronics technology [1–3]. Since it needs low maintenance cost, it plays an important role in the electric power generation. The voltage harvested from the PV system is direct voltage (dc) [4–6]. The conversion of availed dc voltage to ac voltage before connected to the grid is performed by two different configurations, namely conventional two-stage boost inversion system and single-stage boost inversion system.

P. Sriramalakshmi (✉)
School of Electrical Engineering, Vellore Institute of Technology, Vandalur–Kelambakkam Road, Chennai, Tamilnadu, India
e-mail: sriramalakshmi.p@vit.ac.in

S. N. Singh et al. (eds.), *Advances in Power and Control Engineering*, Lecture Notes in Electrical Engineering 609, https://doi.org/10.1007/978-981-15-0313-9_10

But the voltage availed from the solar panel is not sufficient to meet the grid voltage. To meet the requirement, a dc/dc boost converter is connected in cascade with the voltage source inverter (VSI). In case of grid-connected PV systems, the most attention should be given to the design and control of the inverter to attain good efficiency for the entire power systems. It forms a two-stage of power conversion. The two-stage power conversion makes the circuit more complex. To overcome the drawbacks associated with the two-stage system, impedance source inverter (ZSI) is proposed and it finds wide applications [7–9]. But it has a bulky impedance network which increases the cost as well as size of the systems. To eliminate the issue, switched boost inverter (SBI) is suggested [10–13]. SBI has discontinuous source current and the high voltage stress across the capacitor. The qSBI-based topologies are proposed to alleviate the problems [14, 15].

This article presents a comparison between two-stage dc/dc-boosted inverter topology and single-stage EqSBI topology based on the working principle and performance parameters. Both configurations are analyzed by simulation and experimentation. The simulation and hardware results are presented in detail.

2 Two-Stage dc/dc-Boosted Inverter Topology

Two-stage power conversion system from PV is depicted in Fig. 1. In the first stage, the dc source voltage availed from the solar panel is boosted and it appears across the inverter input terminals. The second stage is the inversion stage in which the boosted voltage at the inverter input terminal is inverted by the VSI and it appears across the load. But the two-stage boost inversion reduces the efficiency and limits the dynamic response in case of perturbations in the available PV output voltage due to the ambient conditions as well as sudden changes that occur in the load (Fig. 2).

Figure 3a shows the boost converter when switch S is on at $t = 0$. Inductor L starts to store energy, and the current through the inductor starts to rise. Diode D will be switched off. During this state, the load voltage is equal to the capacitor voltage. Figure 3b shows the boost converter when the switch S is in off state, to maintain

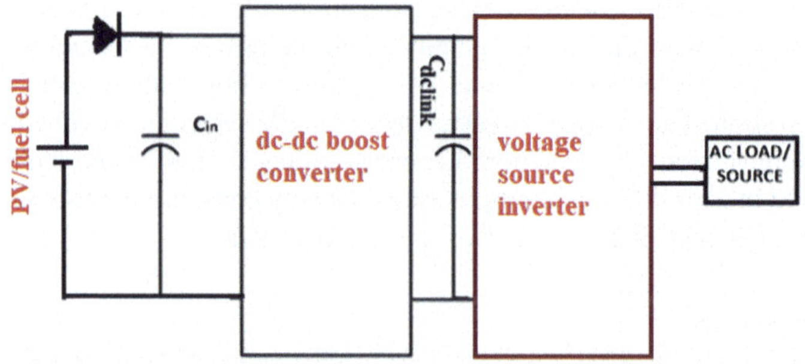

Fig. 1 Two-stage power conversion system [7]

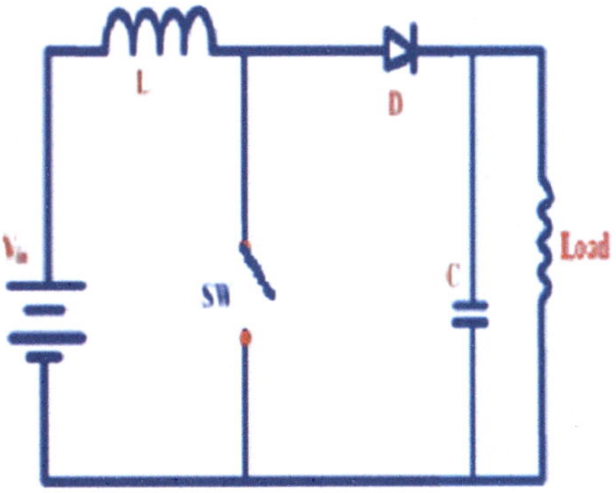

Fig. 2 Conventional boost converter

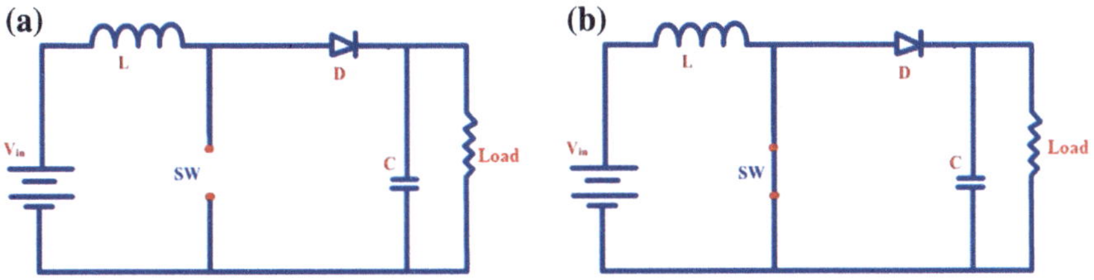

Fig. 3 **a** Switch (SW) is ON. **b** Switch (SW) is OFF

inductor current in the same direction, boost rectifier diode D is on. The current through the inductor (L) is discharged and the capacitor (C) is charged.

The VSI uses sinusoidal PWM strategies, which has active state and zero states. During active state, the power is delivered to the load.

The dc-link voltage of dc/dc-boosted VSI is given by,

$$V_{\mathrm{dclink}} = \frac{1}{1-d} V_{\mathrm{in}} \tag{1}$$

The rms ac output voltage is given by,

$$V_{\mathrm{rms}} = M \frac{1}{\sqrt{2}(1-d)} V_{\mathrm{in}} \tag{2}$$

The maximum value of M is obtained as 1.15 when sinusoidal PWM technique is applied with third harmonic injected signals.

2.1 Drawbacks of Two-Stage dc/dc-Boosted Inverter Topology

The VSI can perform only buck operation. It provides the inverted ac voltage lower than the dc input voltage. To obtain the boosted ac voltage from solar panel, a dc/dc boost converter needs to be connected in cascade with the inverter. To obtain higher boost, duty ratio must be kept at extreme value. It causes reverse recovery issues and high power loss across the diode. Several high-gain dc/dc converter topologies are available in the literature to attain higher boost. There are certain drawbacks of the two-stage power conversion such as high conversion ratio, extreme duty ratio operation, issues with switch stress, separate PWM generation and driver circuit implementation, dead time requirement, EMI immunity, and reliability issues. Moreover, two-stage configurations create a complicated circuitry and hence, the control system becomes more complex [16].

3 Single-Stage qSBI Topology

To overcome the shortcomings of the two-stage dc/dc-boosted inverter [17], single-stage boost inverters such as impedance source inverter (ZSI) and its derived topologies are introduced [6]. If ZSI has the asymmetrical impedance network that leads to unstable operation of the system. To avoid all this problems, switched boost inverter (SBI) topology is proposed [10]. This topology can provide boost as well as inversion in single stage. And it uses single inductor and capacitor which minimize the size and cost of the system. In basic SBI, a diode is directly connected to the dc source, which results in discontinuous source current. Moreover, the single capacitor needs to withstand the entire dc-link voltage. To overcome these drawbacks, a family of qSBI topologies are introduced [14]. The qSBI can provide continuous input current and less capacitor voltage stress, and hence the rating of the capacitor can be minimized. The qSBI can be classified into embedded type qSBI (EqSBI) and dc-linked type qSBI. Further, EqSBI can be categorized into two different configurations, which are depicted in Fig. 4a and b.

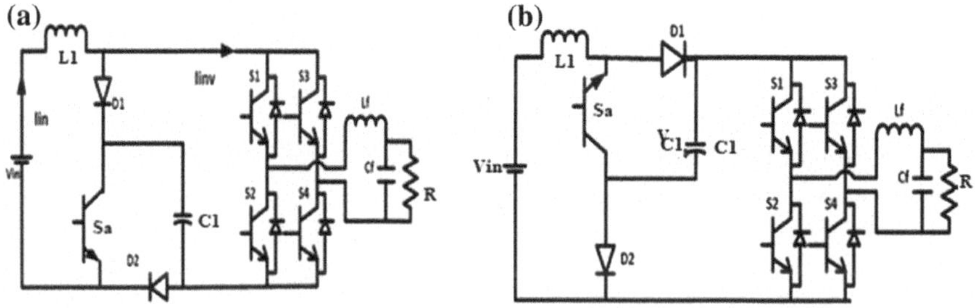

Fig. 4 **a** Embedded type-1. **b** Embedded type-2

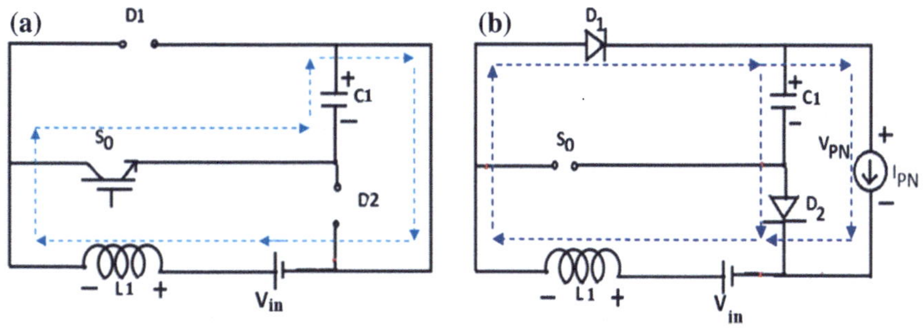

Fig. 5 **a** Shoot-through condition. **b** Nonshoot-through condition

The qSBI has shoot-through state in addition to active state and zero state. The shoot-through state during which both top and bottom switches of single phase are switched on, which results in boost action.

Among these qSBI topologies, embedded type qSBI topology provides higher boost than the dc-linked type [14] topologies. In this article, embedded type-1 qSBI is considered for the analysis and its performance is compared with two-stage dc/dc boost inversion configuration. All these qSBI topologies use shoot-through state to boost, as well as invert the available dc voltage into ac voltage.

3.1 Working Modes of EqSBI Topology [14]

The EqSBI operates in two working modes, namely shoot-through and nonshoot-through. During shoot-through state, the boost network switch is fired, and diodes D1 and D2 are turned off. The capacitor C1 charges the inductor L1 and the inverter terminals are shorted. During nonshoot-through state, the inductor charges the capacitor and the power flows through the load (Fig. 5).

3.2 Steady-State Analysis [14]

Steady-state analysis waveforms of EqSBI topology are depicted in Fig. 6a and b.

During nonshoot-through state, the inductor voltage (V_{L1}) and capacitor current (I_{C1}) are given by,

$$V_{L1} = V_{PN} - V_{C1}; \ I_{C1} = I_{C1} - I_{PN} \tag{3}$$

where I_{PN} is the inverter bridge current

During shoot-through state,

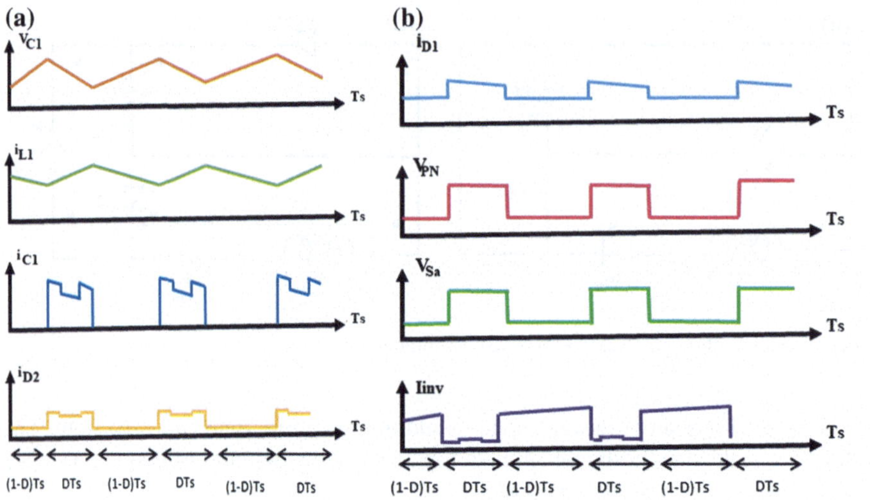

Fig. 6 **a** Capacitor voltage and current, inductor and diode current. **b** Peak dc-link voltage, switch stress, inverter current, and diode current

$$V_{L1} = V_{in} + V_{C1};\ \ I_{C1} = -I_{L1} \tag{4}$$

Applying volt-second balance to the voltage across the inductor,

$$\mathrm{d}_{ST}(V_{in} + V_{C1}) + (1 - \mathrm{d}_{ST})(V_{in} - V_{C1}) = 0 \tag{5}$$

Applying charge-second balance to the current through the capacitor,

$$\mathrm{d}_{ST}(-I_L) + (1 - \mathrm{d}_{ST})(I_L - I_{PN}) = 0 \tag{6}$$

The voltage across the dc-link of EqSBI topology is derived as

$$V_{dclink} = \frac{1}{1 - 2\mathrm{d}_{ST}} V_{in} \tag{7}$$

The rms ac voltage across the load is expressed as

$$V_{rms} = M \frac{1}{\sqrt{2}(1 - 2\mathrm{d}_{ST})} V_{in} \tag{8}$$

where d_{ST} is the shoot-through duty ratio and M is the modulation index.

The d_{ST} is limited to $(1 - M)$. The modulation techniques implemented for ZSI topologies such as simple boost control (SBC), constant boost control, and maximum boost control can be adapted to all qSBI and its derived topologies as well.

4 Simulation Results

The simulation is performed in the MATLAB/SIMULINK environment by considering the input voltage as 36 V, shoot-through duty ratio as 0.45, modulation index as 0.5, switching frequency as 10 kHz, and the resistive load as 70 Ω. The pulses are produced using SBC technique [14]. The topology is able to produce the boosted voltage of 360 V by keeping 30% ripple inductor current and 3% capacitor voltage ripple contents. The simulated waveforms of dc-link voltage, inverter input current and inductor currents are depicted in Fig. 7a. The capacitor voltage and voltage stress across the boost network switch are shown in Fig. 7b. The filtered load voltage (V_{load}) and load current (I_{load}) waveforms are presented in Fig. 7c, respectively.

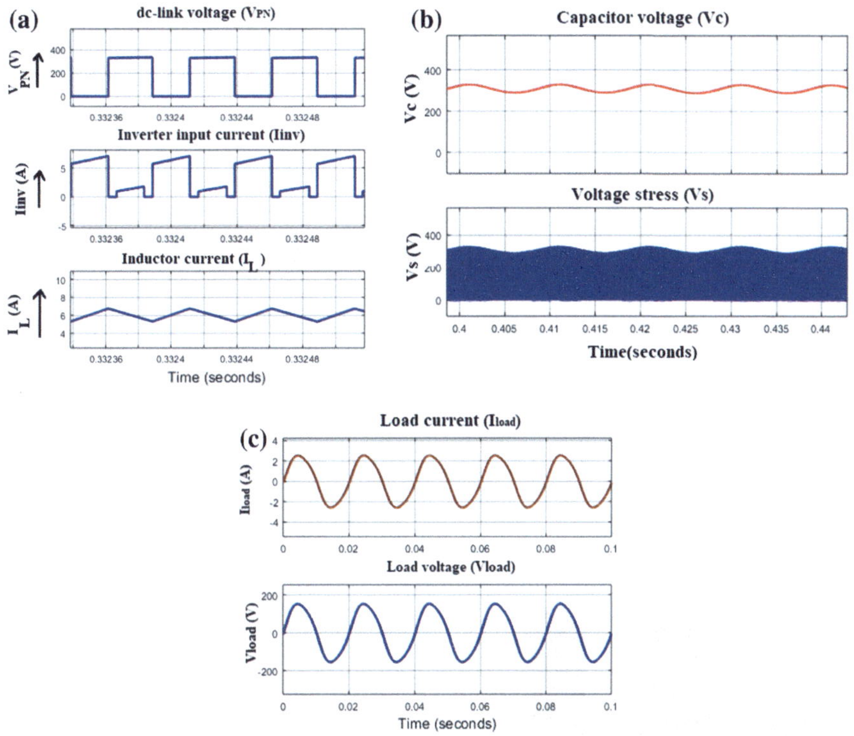

Fig. 7 Simulation results of EqSBI: **a** Waveforms of dc-link voltage, dc-link current, and current through inductor. **b** Waveforms of capacitor voltage and boost network switch stress. **c** Waveforms of filtered load current and load voltage

5 Experimental Evaluation

The EqSBI topology is implemented in hardware and the pulses are generated using FPGA platform. The pulses produced using SBC strategy are depicted in Fig. 8a and b.

Figure 8a and b show the pulses produced for the switches present in the boost network and inverter switches. The peak inductor current of 6.5 A is depicted in Fig. 8c and the dc-link voltage of 314 V is obtained as depicted in Fig. 8d. The

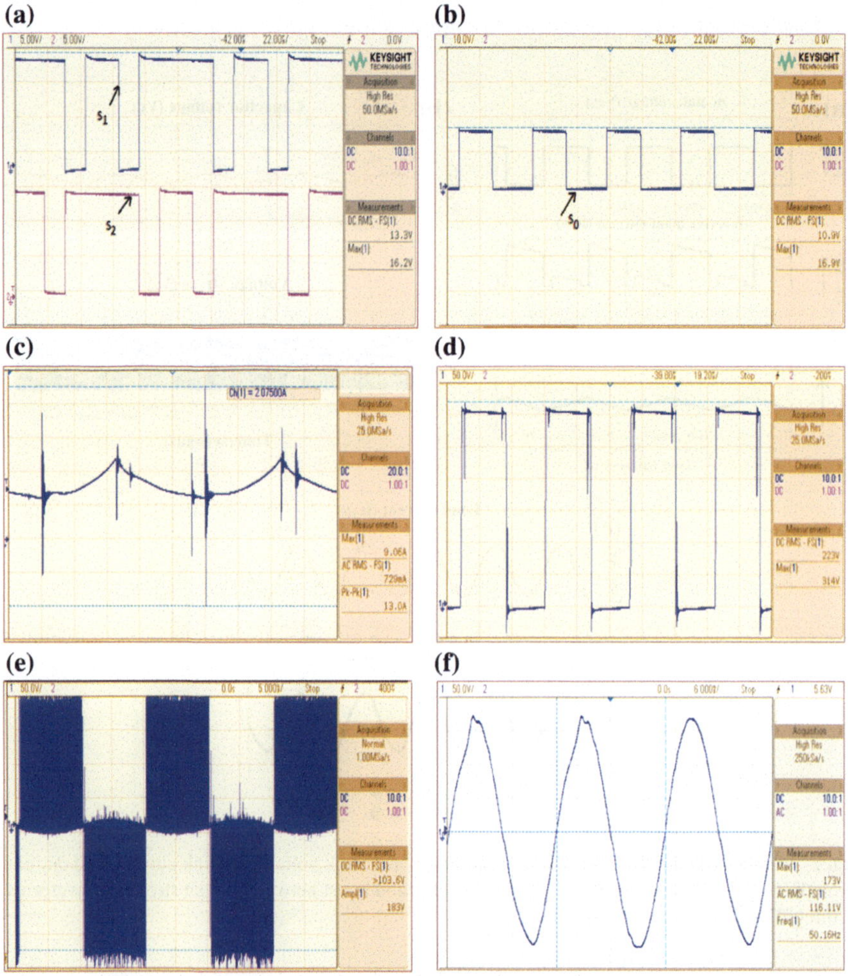

Fig. 8 **a** Pulse for phase A. **b** Pulse for boost network switch. **c** Waveform of inductor current. **d** Waveform of dc-link voltage. **e** Waveform of unfiltered load voltage. **f** Waveform of filtered load voltage. Experimental observations

unfiltered peak ac voltage of 183 V is obtained as depicted in Fig. 8e and filtered ac voltage across the load is obtained as 116.11 V as presented in Fig. 8f. There is a significant difference in the simulated and experimental values due to the non-idealities present in the passive components.

6 Gain and Voltage Stress Across Power Devices

The performance comparison is made with the same operating conditions. The modulation index is kept high which in turn reduce the dc-link voltage. Hence, the voltage stress across the capacitor is reduced. A voltage gain (G) is defined as the ratio between the peak output ac voltage $\left(V_{\text{acpk}}\right)$ and the dc input voltage (V_{in}). The expressions for voltage gain of two-stage boost inverter and qSBI topologies are given in (9) and (11), respectively.

The voltage gain (G) and boost factor (B) of dc/dc boost VSI are given as

$$G = \frac{V_{\text{acpk}}}{V_{\text{in}}} = M \cdot \frac{1}{1 - \text{d}_{\text{ST}}} \tag{9}$$

$$B = \frac{1}{1 - \text{d}_{\text{ST}}} \tag{10}$$

The voltage gain and boost factor of embedded type qSBI [14] is obtained as

$$G = \frac{V_{\text{acpk}}}{V_{\text{in}}} = M \cdot \frac{1}{1 - 2\text{d}_{\text{ST}}} \tag{11}$$

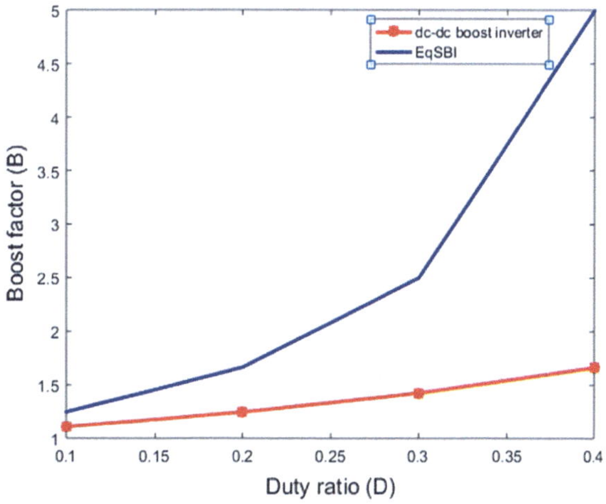

Fig. 9 Duty ratio (D) versus Boost factor (B)

Table 1 Components comparison

Components	dc/dc-boosted inverter	EqSBI [14]
Diodes (D)	1	2
Inductors (L)	1	1
Capacitors (C)	1	1
Total switches (S)	5	5

Figure 9 depicts the graph between shoot-through duty ratio (D) and the boost factor (B) of the dc-dc boost inverter and EqSBI. The EqSBI can provide the higher boost. At the shoot-through duty ratio of 0.4, two-stage boost inverter provides the boost factor of 1.66 whereas one stage boost inverter provides the boost factor of 5. The components comparison between dc/dc-boosted inverter and EqSBI topologies are presented in Table 1.

To obtain the same voltage boost, the dc/dc-boosted inverter needs one less diode compared to EqSBI topology at same operating conditions.

7 Conclusion

This article presents a comparison between a dc/dc-boosted VSI and EqSBI topologies.

With the same input voltage, duty ratio, and modulation index, the EqSBI topology can provide a high boost compared to dc/dc-boosted VSI. The voltage stress proportional to the dc-link voltage appears across the boost network switch. The voltage gain of both the topologies is proportional to the dc-link-boosted voltage. In addition, the voltage gain is directly proportional to the boost factor. Whenever a low dc voltage needs to be converted into a high ac voltage, single-stage EqSBI can be used. It is very well suitable for RES-based applications compared to two-stage power conversion system.

References

1. Tripathi L, Mishra AK, Dubey AK, Tripathi CB, Baredar P (2016) Renewable energy: an overview on its contribution in current energy scenario of India. Renew Sustain Energy Rev 60:226–233
2. Kuang Y, Zhang Y, Zhou B, Li C, Cao Y, Zeng L (2016) A review of renewable energy utilization in islands. Renew Sustain Energy Rev 59:504–513
3. Ellabban O, Abu-Rub H, Blaabjerg F (2014) Renewable energy resources: current status, future prospects and their enabling technology. Renew Sustain Energy Rev 39:748–764
4. Schonberger J, Duke R, Round SD (2006) DC-bus signalling: a distributed control strategy for a hybrid renewable nanogrid. IEEE Trans Ind Electron 53:1453–1460
5. Kakigano H, Miura Y, Ise T (2010) Low-voltage bipolar-type DC microgrid for super high quality distribution. IEEE Trans Power Electron 25:3066–3075

6. Peng FZ (2002) Z-source inverter. Proc Ind Appl Conf 2:775–781
7. Peng FZ (2003) Z-source inverter. IEEE Trans Ind Appl 39:504–510
8. Liu VY, Abu-Rub H, Ge B (2014) Z-source/quasi-z-source inverters: derived networks, modulations, controls, and emerging applications to photovoltaic conversion. IEEE Ind Electron Mag 8:32–44
9. Ge B et al (2013) An energy-stored quasi-z-source inverter for application to photovoltaic power system. IEEE Trans Ind Electron 60:4468–4481
10. Upadhyay S, Ravindranath A, Mishra S, Joshi A (2010) A switched-boost topology for renewable power application. IEEE IPEC 10:758–762
11. Adda R, Ray O, Mishra SK, Joshi A (2013) Synchronous-reference-frame-based control of switched boost inverter for standalone DC nanogrid applications. IEEE Trans Power Electron, vol 28
12. Ravindranath A, Mishra S, Joshi S (2013) Analysis and PWM control of switched boost inverter. IEEE Trans Ind Electron 60:5593–5602
13. Mishra S, Adda R, Joshi A (2012) Inverse Watkins-Johnson topology-based inverter. IEEE Trans Power Electron 27:1066–1070
14. Nguyen M-K, Le T-V, Park S-J, Lim Y-C (2015) A class of quasi-Switched boost inverters. IEEE Trans Ind Electron 62:1526–1536
15. Nguyen MK, Lim YC, Park SJ (2015) A comparison between single phase quasi-Z-source and quasi-switched boost inverters. IEEE Trans Ind Electron 62:6336–6344
16. Sriramalakshmi P, Sreedevi VT (2018) Single-stage boost inverter topologies for nanogrid applications. In: SenGupta S, Zobaa A, Sherpa K, Bhoi A (eds) Advances in smart grid and renewable energy. Lecture notes in electrical engineering, vol 435. Springer, Singapore
17. Rashid MH (1993) Power Electronics, 2nd Edn. Prentice Hall

Forecasting Soil Moisture Based on Evaluation of Time Series Analysis

Sukhwinder Singh, Sanmeet Kaur and Parteek Kumar

Abstract Precision agriculture is a technique that is incorporated to produce high crop yield with the best utilization of available resources. Traditional farming is adversely affected due to improper resource management. In order to overcome the efforts of a farmer, a model for the soil moisture forecasting has been proposed in this manuscript to deliver better after-effects of farming. The proposed model uses long short-term memory (LSTM) to predict soil moisture. The model is trained on a dataset acquired from IIT Kanpur agricultural site. For analyzing the performance of the model mean absolute error (MAE), mean absolute percentage error (MAPE), mean squared error (MSE), and root mean squared error (RMSE) has been used as performance metrics. This paper is paving way for the early prediction of the soil moisture that can be used with other advanced innovative irrigation techniques.

Keywords Smart irrigation · Time series · Soil moisture · LSTM · MAE · MSE

1 Introduction

Precision Agriculture is the most valuable innovation in the last two decades, from the perspectives of new developments in its area, resources management, and ecological impact in the agriculture industry. The precision agriculture consists of several technological tools such as remote sensing, Global Positioning Systems (GPS), aerial imagery or satellites along with Geographic Information Systems (GIS) are collaborated to generate the clear and accurate information about the agricultural production and its resources. It provides more support in the agriculture fields such as crop production, water requirement, maintaining the level of nutrients in the soil, and many more. The population of the world is also increasing day by day; there is 1% inclined in the population worldwide every year and only limited surface left for the agriculture as most of the land is used as commercial and residential purpose. To overcome its effect, this domain is helpful to balance the ratio between the food and the people.

S. Singh (✉) · S. Kaur · P. Kumar
CSED, Thapar Institute of Engineering and Technology, Patiala, Punjab, India
e-mail: sukhsinghturka02@gmail.com

S. N. Singh et al. (eds.), *Advances in Power and Control Engineering*, Lecture Notes in Electrical Engineering 609, https://doi.org/10.1007/978-981-15-0313-9_11

The farmer can use these advanced techniques and utilize his available resources in a particular manner to decrease human efforts and resulting in the yield of the crops.

In agriculture, soil moisture plays an important role in exploring the extent of the fertility of the crop. It is monitored using electromagnetic radiation emitted by the soil at different temperatures [1]. The mensuration of soil moisture content is crucial in agriculture domain. It acts as a food nutrient for the growth of the plant and a chemical or biological catalyst for the fertility of the soil. New technologies like wireless sensor networks (WSN) and remote sensing devices are incorporated to analyze the soil moisture of larger areas concurrently [2]. This soil moisture can be utilized in a particular way to gain more production of crop using forecasted data. The forecasted data helps the farmer to arrange the required resources. Traditionally, the forecasting of the numerical value is implemented using statistical models like automatic regression (AR), moving averages (MA), and automatic regression integrated moving average (ARIMA). These models are suitable only for limited data values and short-term forecasting. There are several other machine learning approaches such as decision tree, random forest, and neural networks that can be used to do calculations of information for forecasting longer and larger sets of data. The LSTM of neural networks is used as a time series model for computation of the respective dataset and predicts the future values of the input data with its best accuracy.

The objective of this study aims to develop a model by utilizing advanced innovation techniques to measure soil moisture, that is, an essential component to anticipate the yields is to investigate the accuracy of the soil moisture forecasting using LSTM of neural network. Several performance metrics are used to compute the validation of the dataset.

The rest of this paper is structured as follows: Sect. 2 presents the literature review of existing soil moisture approaches based on machine learning models. In Sect. 3, the detailed description of datasets is provided. Section 4 presents the proposed framework in detail. The experimental results and discussions have been depicted in Sect. 5. Finally, the conclusion and future scope of this study are presented in Sect. 6.

2 Literature Review

In this section, several existing approaches such as forecasting of soil moisture with respect to the usage of water, precipitation, and other parameters of crops have been discussed. Apart from this, several different studies on time series with different types of datasets are also described in this section latterly.

Zhang et al. [3] proposed an algorithm for an automatic valve. The valve is used to open or close automatically with respect to the moisture level within the soil. Soil sensors and other real-time devices monitor and collect the data about the growth of the crops and upload it to Shandong Agricultural University, China's data center. After mining and extraction of the data, it forecasts the water required for crops in different periods and makes an automatic irrigation system [3].

Lenka et al. [4] developed an optimization technique for the prediction of soil moisture. Gradient descent with momentum-based neural networks is used to predict the contents of the soil moisture before an hour in advance. The results of this study are evaluated using RMSE and MSE as performance metrics, and these have 0.039622 and 0.19905 resulted values, which are high on the accuracy of the model [5].

Han et al. [6] depict the estimation of the soil moisture level using the CART algorithm. A CART is a classification and regression tree algorithm. The soil moisture is estimated at a high spatial resolution of 1 km distance from earth surface using meteorological satellites. It uses precipitation, temperature, and soil properties to predict the soil moisture. The results of this study have high accuracy and high spatial resolutions [6].

Zhu et al. [7] discuss the sensor technology used for smart farming. Researchers designed an application in C programming language for the environment monitoring system based on WSN. This proposed system has consisted of the software layer and the hardware layer. Within the hardware layer, it had a data acquisition module to fetch data and a processing module to process the information. A wireless communication module is used to transmit the data from the node to the server. After acquiring the data from the real environment using sensors, there is only ± 1 °C change in temperature in actual and predicted data [7].

Priya et al. [8] use Naive Bayes classifier integration of the MapReduce paradigm. Researchers recommended a crop such as chili, rice, cotton, and maize to the farmer in the Telangana region of India. They use several parameters like rainfall, soil moisture, temperature, and atmospheric pressure. MapReduce integrated with Naive Bayes paradigm predicts the behavior of the different crop with respect to the given attributes [8].

Ponce-Guevara et al. [9] proposed an application for vegetable crops using data mining techniques in Java-based platform. Researchers of Ecuador divide the application into two phases. In the first phase, model is built using classification-based decision tree, and the data is predicted in the next phase. The soil moisture is predicted with respect to the relative humidity, temperature, luminosity, and carbon-dioxide parameters. The predicted results are 96% accurate [9].

Soni et al. [10] apply ARIMA model to simulate monthly mean Terras MODIS (Moderate Resolution Imaging Spectroradiometer) and AOD550 nm (aerosol optical depth) on a dataset of the eight sites of Indian-Gangetic plain from March 2000 to March 2012. This study analyzes the data in annual and seasonal variation, and several AODs of all the sites are compared with their respective terms [10].

Zhang et al. [11] use ARIMA model for trend analysis and forecasting of data of $PM_{2.5}$ in the air in Fuzhou, China to find the change value in the air pollution. It includes data of two years (2014–2016) for modeling and analysis and uses the next year 2017 for the validation of the model. This study has no actual increase in the trend for the year 2017 as expected. In fact, there is a decrease in trend due to new policies in the nation [11].

Qing et al. [12] predicted the hourly based weather data using the LSTM model. The study uses a dataset of *Cape Verde* in the Africa continent from March 2011 to December 2013 and has a gap of 6 months from September 2012 to December 2012 with features of temperature, humidity, visibility, wind speed, and weather type. It predicted the last six months' data with 18.34% more accuracy as compared with BPNN (back propagation neural network) [12]. A tabular representation of existing studies used in this paper is described in Table 1.

Inspired from the above studies, this paper uses an LSTM model to forecast data of soil moisture. LSTM is a time series model of machine learning. It was developed in 1997 *by Sepp Hochreiter* and *Jürgen Schmidhuber* [13]. LSTM-RNN model shows the transient changes in the information because of their intermittent design and memory units. To the best of our insight, this is the first paper that utilizes LSTM-RNN to forecast soil moisture.

Table 1 Existing technologies

Author	Dataset	Algorithm	Proposed scheme
Zhang et al. [3]	Soil moisture, water level, and video-image data	Decision model	Real-time automatic valve based on water requirement is built
Lenka et al. [5]	Temperature, humidity, soil type, air-flow rate, UV index, soil moisture, soil temperature, and carbon-dioxide	Neural networks	Prediction of the soil moisture contents before an hour
Han et al. [6]	Precipitation, land surface temperature, evapotranspiration, biophysical, elevation, topography, vegetation condition, soil texture type, and soil water content	CART decision tree	Analysis and estimation of the soil moisture
Priya et al. [8]	Rainfall, soil moisture, temperature, atmospheric pressure, and crop type	Naive bays classifier	Analysis and prediction for the best crop of the season
Guevara et al. [9]	Soil moisture, relative humidity, temperature, luminosity, and carbon-dioxide	Decision tree	Analysis of different crop data and its growth in a special green environment
Soni et al. [10]	Quarterly aerial data of eight Indian cities	ARIMA	Statistical analysis of Ganga plain site of 12 years
Zhang et al. [11]	Month, $PM_{2.5}$, SO_2, NO_2, $PM_{1.0}$. MCR, MMMT, and MMH	ARIMA	Forecasting of $PM_{2.5}$ for measuring the effects of pollution
Qing et al. [12]	Temperature, dew point, humidity, visibility, wind speed, and weather	LSTM	Prediction of hourly based data of solar irradiance

3 Description of the Dataset

To explore the problem, it is necessary to study the dataset. This section emphasized on the dataset that is used for this experiment.

3.1 Dataset

The dataset used for modeling is acquired from the IIT Kanpur [14]. The dataset has the soil moisture readings at certain depths of 10, 25, 50, and 80 cm from November 11, 2011 to November 22, 2012. Initially, this dataset is composed of the different file of each sensor installed at a particular depth level. After choosing the dataset, the UTC *date_time* is the universal time coordinated time of the day in 24-hour format. The UTC is the global time clock used by Britishers. The reading of the soil moisture of each depth such as 10, 25, 50, and 80 cm is added to the next column of the dataset, respectively. In the end, there are 5 columns in the dataset that are used in this study.

4 Proposed Methodology

In artificial neural networks, LSTM is a special kind of neural network that uses short-term and long-term memories. The structure of an LSTM cell is shown in Fig. 1. This model uses gates to flow data from the input state to the output state. There are three gates and one cell state through which the information passes and operations are performed on the model. These gates are the input gate, the output gate, and the forget gate. The new data of present time t is passed to LSTM through input gate (i). The forget gate (f) calculates the proportion of previous data to be kept. It receives some input data using a sigmoid function and calculates how much of the current data should be kept. The output gate (o) is used to obtain results using another sigmoid function [15]. The cell state preserves the data of the previous hidden layer and passes it to the next layer for the processing. In LSTM, each gate has its own activation function. There are three sigmoid functions and two *tanh* function used for the modeling of data. h_{t-1} is the output of the previous hidden layer and h_t is the output of the current hidden layer. X_t is the current input to the model, C_{t-1} is the cell state of previous time t and C_t is the current cell state of the model, O_t is the output of the model. These defined states of LSTM are described in the following equations [16, 17].

$$i_t = \sigma(W_{xi}X_t + W_{hi}h_{t-1} + W_{ci}C_{t-1} + b_i) \quad (1)$$

$$f_t = \sigma\left(W_{xf}X_t + W_{hf}h_{t-1} + W_{cf}C_{t-1} + b_f\right) \quad (2)$$

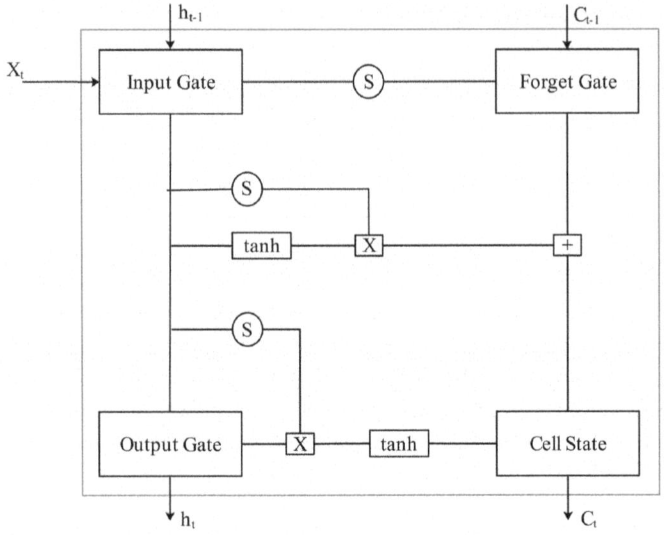

Fig. 1 Structure of an LSTM

$$c_t = f_t c_{t-1} + i_t \tan h(W_{hc} h_{t-1} + W_{xc} x_t + b_c) \quad (3)$$

These above equations are representing the function of the input gate (i_t), forget gate (f_t), and the cell state. There is a logistics sigmoid function which is represented as σ. W_{xi}, W_{hi}, and W_{ci} are the weight matrix of the input gate. W_{xf}, W_{hf}, W_{cf} are the weight matrix of the forget gate. In cell state, $f_t c_{t-1}$ is the vector multiplication of forget gate and the previous cell state, the i_t tan$h$ is the tanh function applied on input state (i_t), and $W_{hc,}$ W_{xc} are the weight matrix for hidden state and cell state.

$$O_t = \sigma\left(W_{xo} X_t + W_{ho} h_{t-1} + W_{co} C_{t-1} + b_f\right) \quad (4)$$

W_{xo}, W_{ho}, and W_{co} are the weight matrix for the output gate. There is a b_f, i.e., a bias function is used in the neural network to adjust the outputs along with the weighted sum of the inputs. This bias provides more efficiency to fit the model in its best way.

$$h_t = o_t \tan h(c_t) \quad (5)$$

In the last Eq. 5, a tan$h$ function is performed on the cell state and the resulting information is passed to the forget gate to perform an operation on next t state. The next inputs of the dataset are passed to the LSTM network at each timestamp t for the processing of the model as denoted in the equations above. Once the last input of the dataset being processed, the final output for the whole dataset is returned by the network. Hence, this process is iterative for the whole dataset and the model is built using historical dataset [18].

5 Experiment and Results

The forecasting model is implemented in Python 3.6.5, and Spyder IDE is used for the development of the model on a system with i3 processor, 4 GB RAM, and 64-bit Ubuntu operating system. Initially, the dataset is in the form of *CSV* format and imported from the local disk. The required libraries and directories which have been used for the model building such as *numpy, pandas,* and *sklearn* that are used for the arrays and dataset processing, *matplotlib* for graph plotting. *Tensorflow* and *Keras* libraries used for model building are imported. The fetched dataset is processed using *MinMaxScaler* function on the scale from 0 to 1 float values. After scaling the acquired dataset, it is further split into training (70) and testing dataset (30), where training dataset has the readings taken from November 11, 2011 to November 11, 2012, and the values are recorded from October 8, 2012 to November 22, 2012 are used for the testing of the forecasting model.

In the model, the *Keras* library uses *Tensorflow* as a backend to perform the operations. The *Keras* has the LSTM, it uses 50 units of neurons of the LSTM layer and has a dense layer with 1 neuron and an epoch value of 100 is used for model fitting. This model is also trained with different parameters by changing its number of neurons, layers, and epochs. In the end, these parameters depict better performance than others. Hence, these are used for the model building.

An LSTM model is trained using training data of the acquired dataset and the predictions from the model are processed. These predictions are inversely transformed to the normal dataset for the actual values and saved into a new variable. In the final step, the predicted values have been compared with the testing dataset to compute the error rate of the model. The whole process of this experiment is elaborated in Fig. 2. The results of the actual and the predicted datasets of all depth are graphically compared and analyzed below.

Figure 3 depicts the graphical representation of complete dataset of the soil moisture from November 22, 2011 to November 22, 2012, and four different lines are used to illustrate the different readings of the respected time. The green, blue, yellow, and brown lines represent soil moisture at different depths of 10, 25, 50, and 80 cm, respectively.

The results of this study are depicted in the different depths of the soil moisture. The comparison of the actual data and the predicted data is analyzed in the figures below. The actual data is the testing dataset and is represented through the green line. The predicted data is the outputs of the model and is represented by the yellow line.

In Figs. 4, 5, 6, and 7, it has been analyzed that the actual values and the predicted values are quite similar in the graph. These values are compiled using the validation strategy in Sect. 5.1.

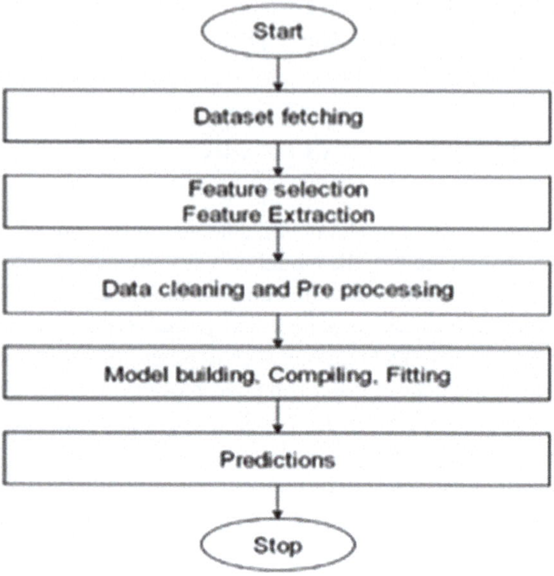

Fig. 2 Flowchart of the experiment

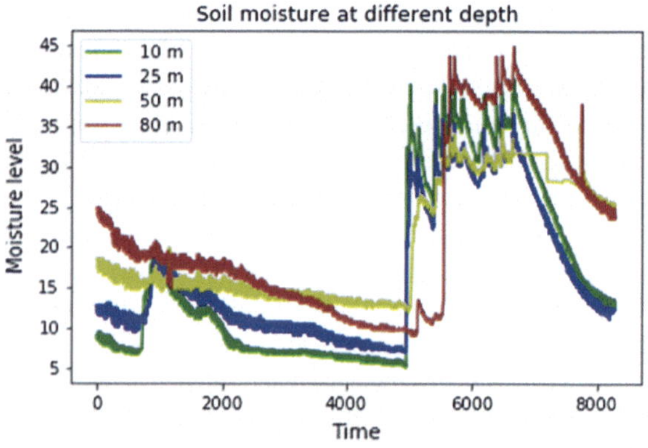

Fig. 3 Soil moisture at different depths

5.1 Validation Strategy

The validation of the model is very crucial to compute the performance of the model. The validation is done using different statistical models of the performance and evaluation metrics such as mean absolute error (MAE), mean absolute percentage error (MAPE), mean squared error (MSE), and root mean squared error (RMSE) are used [19–21]. The reason for computing several metrics is to compare this study to another forecasting model. The lower value of the error depicts the best model for the experiment. In this study, the testing dataset is used as a validated dataset and the outputs from the model are used as the predicted dataset.

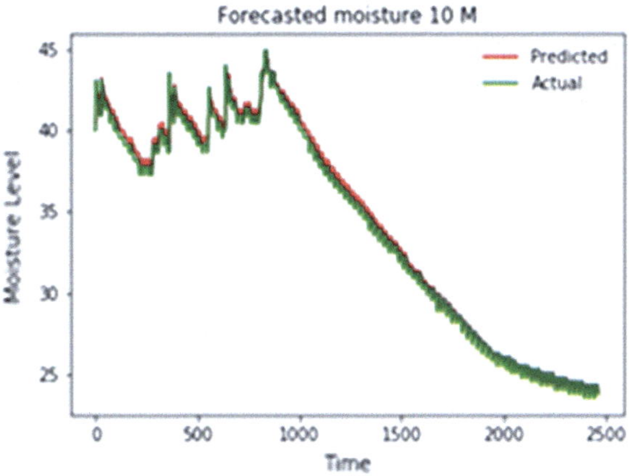

Fig. 4 Soil moisture at 10 m

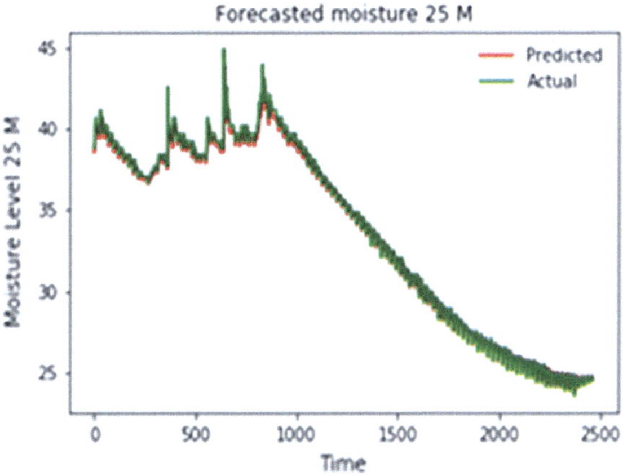

Fig. 5 Soil moisture at 25 m

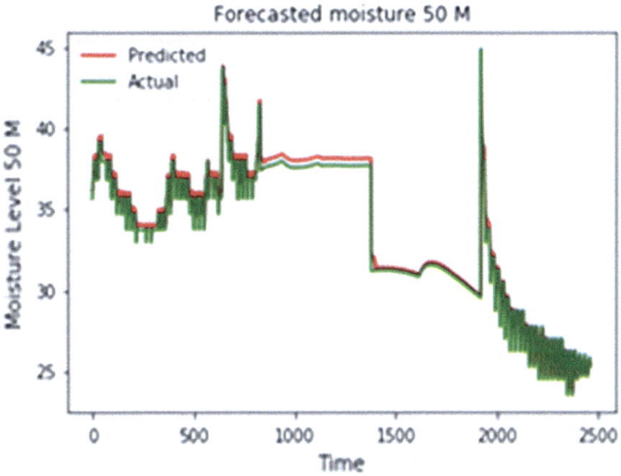

Fig. 6 Soil moisture at 50 m

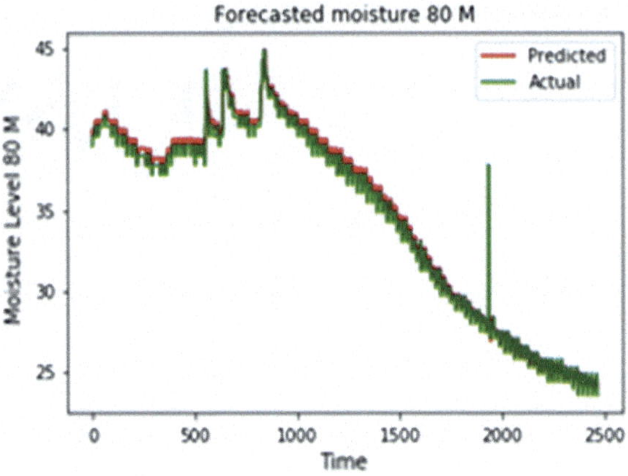

Fig. 7 Soil moisture at 80 m

$$\textbf{Absolute Error} = X_i - Y_i \tag{6}$$

$$\textbf{MAE} = \frac{\sum_{i=1}^{n} |\text{Absolute Error}|}{n} \tag{7}$$

$$\textbf{MAPE} = \frac{\sum_{i=1}^{n} \text{Absolute Error}_i}{\text{Actual}_i} * 100 \tag{8}$$

$$\textbf{MSE} = \frac{\sum_{i=1}^{n} (\text{Absolute Error})^2}{n} \tag{9}$$

$$\textbf{RMSE} = \sqrt{\frac{\sum_{i=1}^{n} (\text{Absolute Error})}{n}} \tag{10}$$

where n is the total number of the readings, i is the reading at t time. The absolute error is the difference between the actual data (X_i) and the predicted data (Y_i). Table 2 describes the performance of the model using several performance metrics.

Table 2 Error rate at several depths of soil

Depth level	MAE	MAPE	MSE	RMSE
Soil moisture at 10 cm	0.13	0.38	0.06	0.24
Soil moisture at 25 cm	0.16	0.50	0.07	0.26
Soil moisture at 50 cm	0.17	0.53	0.23	0.47
Soil moisture at 80 cm	0.16	0.46	0.13	0.36
Soil moisture of all depths[a]	0.16	0.46	0.12	0.33

[a]The soil moisture of all depth is calculated by taking the mean of all the soil moisture

MAE is calculated as the average of the absolute error. There are pipes on both ends of the error that makes the nature of this result always positive.
MAPE of each time *t* is calculated as an absolute error and divided by the time *t*.
MSE is the square mean of the absolute error.
RMSE is the mean of the square root of differences between actual readings and predicted readings.

6 Conclusion

The proposed study presented in this manuscript discusses soil moisture forecasting model. The LSTM model is implemented on the dataset of the soil moisture that acquired from the IIT Kanpur's agricultural site. It forecasts the values of soil moisture at different depths of the soil. The performance analysis of this study is computed using MAE, MAPE, MSE, and RMSE as performance metrics to compute the accuracy of the model. From the error rate of all depth levels, the results show that the proposed model provides substantial results with a high degree of accuracy. On the other hand, this study performs better in comparison with a dataset of tourism forecasting of eight cities [22]. The proposed study performs better from the existed study and the error rate of performance metrics is in lesser values. Keeping in views of the existing studies, this study can be hyperextended to develop an automatic valve for water saving. Forecasted readings of the soil moisture can depict the level of the water absorbed by the soil. The automatic valve pretends as the water requirement for the crop and saves an unutilized and excessive amount of water.

Acknowledgements The author would like to thanks Dr. Rajib Maity of IIT Kanpur, India for providing dataset based on the different depths of the soil moisture and also acknowledge to Council of Scientific and Industrial Research (CSIR) for funding grants vide No. 38(1464)/18/EMIR-II for carrying out research work.

References

1. Shinde D, Siddiqui N (2018) IOT based environment change monitoring and controlling in greenhouse using WSN. In: 2018 international conference on information, communication, engineering and technology (ICICET), IEEE, Aug 2018, pp 1–5
2. Calla OPN et al (2008) Measurement of soil moisture using microwave radiometer. In: Proceedings of the IEEE international conference on microwave '08, 2008, pp 621–624
3. Zhang P, Zhang Q, Liu F, Li J, Cao N, Song C (2017) The construction of the integration of water and fertilizer smart water saving irrigation system based on big data. In: 2017 IEEE international conference on computational science and engineering (CSE) and IEEE international conference on embedded and ubiquitous computing (EUC), vol 2, pp 392–397. IEEE, July 2018
4. Lenka SK, Mohapatra AG (2015 December). Gradient descent with momentum based neural network pattern classification for the prediction of soil moisture content in precision agriculture.

In: 2015 IEEE International Symposium on Nanoelectronic and Information Systems (pp. 63–66). IEEE
5. Ip RH, Ang LM, Seng KP, Broster JC, Pratley JE (2018) Big data and machine learning for crop protection. Comput Electron Agric 151:376–383
6. Han J, Mao K, Xu T, Guo J, Zuo Z, Gao C (2018) A soil moisture estimation framework based on the CART algorithm and its application in China. J Hydrol 563:65–75
7. Zhu Y, Song J, Dong F (2011) Applications of wireless sensor network in the agriculture environment monitoring. Procedia Eng 16:608–614
8. Priya R, Ramesh D, Khosla E (2018) Crop prediction on the region belts of India: a Naïve Bayes MapReduce precision agricultural model. In: 2018 international conference on advances in computing, communications and informatics (ICACCI), pp 99–104. IEEE, Sept 2018
9. Ponce-Guevara KL, Palacios-Echeverría JA, Maya-Olalla E, Domínguez-Limaico HM, Suárez-Zambrano LE, Rosero-Montalvo PD, Peluffo-Ordóñez DH, Alvarado-Pérez JC (2017) GreenFarm-DM: a tool for analyzing vegetable crops data from a greenhouse using data mining techniques (First trial). In: 2017 IEEE second ecuador technical chapters meeting (ETCM), pp 1–6. IEEE, Oct 2017
10. Soni K, Kapoor S, Parmar KS, Kaskaoutis DG (2014) Statistical analysis of aerosols over the Gangetic-Himalayan region using ARIMA model based on long-term MODIS observations. Atmos Res 149:174–192
11. Zhang L, Lin J, Qiu R, Hu X, Zhang H, Chen Q, Tan H, Lin D, Wang J (2018) Trend analysis and forecast of $PM_{2.5}$ in Fuzhou, China using the ARIMA model. Ecol Ind 95:702–710
12. Qing X, Niu Y (2018) Hourly day-ahead solar irradiance prediction using weather forecasts by LSTM. Energy 148:461–468
13. Hochreiter S, Schmidhuber J (1997) Long short-term memory. Neural Comput 9(8):1735–1780
14. Pal M, Maity R, Dey S (2016) Statistical modelling of vertical soil moisture profile: coupling of memory and forcing. Water Resour Manage 30(6):1973–1986
15. Li Y, Cao H (2018) Prediction for tourism flow based on LSTM neural network. Procedia Comput Sci 129:277–283
16. https://www.cse.iitb.ac.in/~cs725/notes/lecture-slides/2016b/lecture-23-annotated.pdf
17. Bouktif S, Fiaz A, Ouni A, Serhani M (2018) Optimal deep learning LSTM model for electric load forecasting using feature selection and genetic algorithm: Comparison with machine learning approaches. Energies 11(7):1636
18. Sharma A, Rani R (2017) An optimized framework for cancer classification using deep learning and genetic algorithm. J Med Imag Health Inform 7(8):1851–1856
19. Fischer T, Krauss C (2018) Deep learning with long short-term memory networks for financial market predictions. Eur J Oper Res 270(2):654–669
20. https://en.wikipedia.org/wiki/Mean_absolute_percentage_error
21. https://www.dataquest.io/blog/understanding-regression-error-metrics
22. Haiges R, Wang YD, Ghoshray A, Roskilly AP (2017) Forecasting electricity generation capacity in Malaysia: an auto regressive integrated moving average approach. Energy Procedia 105:3471–3478

Modeling and Estimation of Lithium-ion Battery State of Charge Using Intelligent Techniques

S. Hemavathi

Abstract The Li-ion battery is an energy storage system in consumer and industrial applications. Because of their cell and pack level protection, the Li-ion battery requires a battery management system. The important function of the battery management system is to monitor the Li-ion battery state of charge (SOC), to indicate the charge level of the battery. In this research article, efficient intelligent techniques-based SOC estimation algorithm is presented. The proposed techniques are feedforward neural network and layer recurrent neural network with a Scaled Conjugate Gradient (SCG) and Levenberg-Marquardt (LM) training methods. The proposed estimators are applied on 18650 single-cell Li-ion battery to test the performance of the neural networks to estimate the SOC. A real-time experiment carried out on 18650 single-cell Li-ion battery, and experimental results and characteristics are analyzed. The battery cell voltage and current obtained from experimental results are used as the input parameter to proposed networks and battery SOC as the output. The proposed estimation is carried out in the MATLAB. The simulation results show that layer recurrent neural network with LM training method has the best performance to estimate the Li-ion battery SOC in terms of accurate measurement with actual SOC and highest convergence speed.

Keywords Feedforward neural network · Levenberg-Marquardt · Li-ion battery · Recurrent neural network · Scaled conjugate gradient · State of charge

1 Introduction

Battery is an energy storage device which can be widely used in portable, industrial, telecommunication, electric vehicle, and microgrids. Recently, the rechargeable Li-ion battery plays a vital role in automotive and grid-connected systems due to low weight, high energy density, long lifetime, high capacity and fast charge capability, low level of self-discharge and eco-friendly. Li-ion battery cell level and pack-level

S. Hemavathi (✉)
Battery Division, Central Electrochemical Research Institute, CSIR Madras Complex, Chennai, India
e-mail: hemavathi@cecri.res.in

S. N. Singh et al. (eds.), *Advances in Power and Control Engineering*, Lecture Notes in Electrical Engineering 609, https://doi.org/10.1007/978-981-15-0313-9_12

control variables are needed to be maintained accurately for safe operation. These control variables are monitored and protected by battery management system (BMS) [1–4]. The BMS is an electronic device which protects Li-ion battery pack from over current, over voltage during charging, under voltage during discharging, over temperature and under temperature. The most important function of BMS is to monitor the battery state of charge (SOC) which indicates the actual charge level of the battery. The estimation of battery SOC is a challenging phenomenon due to its non-linear time-varying characteristics and function of battery voltage and current. In real time, the BMS used various methods to estimate SOC based on applications, cost and accuracy [5–8].

One of the most common methods is to estimate the battery SOC by coulomb counting, but the accuracy of this method is affected by battery voltage, current and temperature after few cycles. Other direct methods are open-circuit voltage method, terminal voltage method and impedance spectroscopy method. These direct methods are depended on the battery physical parameters and characteristics which are affected by many chemical factors, and hence these methods are not practically giving accurate results in real time.

On the other hand, the adaptive methods are introduced to overcome difficulties in the direct methods and very accurate estimation of SOC and reliable in various applications. The developed techniques are fuzzy logic method, neural networks, Kalman filter, slide-mode observer and support vector machine [9–12].

The neural network is one of the intelligent techniques which is based on the machine learning algorithm and uses data-driven technology [13, 14]. This technique implementation will not depend on battery electrochemical models, hence the execution is not too complex and few computational resources are needed compared to the Extended Kalman Filter [15, 16]. In this article, feedforward neural network and layer recurrent neural network were proposed to accurately estimate the Li-ion battery SOC with the help of SCG and LM training functions [17–20]. The battery cell voltage and battery current are taken as input vector to train the network and the battery SOC as an output signal. Also, the testing data estimated SOC results are compared with actual SOC based on the execution time, convergence speed, gradient and mean square error to find the best suitable neural network and training function for efficient operation of Li-ion battery under different operating conditions [21–24].

The research article is organized as follows: Section II explains the real-time experimental studies on Li-ion battery. In Section III, the intelligent techniques are proposed for battery state of charge estimation. The Section IV provides the simulation and comparison results of feedforward and recurrent neural networks which is trained by SCG and LM methods.

Table 1 Specifications of single-cell Li-ion battery

Parameters	Value
Nominal capacity	2600 mAh
Charging voltage	4.2 ± 0.05 V
Nominal voltage	3.7 V
Discharge cut-off voltage	2.75 V
Charging method	Constant current-constant voltage
Maximum charge current	2600 mA
Maximum discharge current	5200 mA
Operating temperature	Charge: 0–45 °C Discharge: −20–60 °C
Weight of cell	47 g
Dimension of cell	Diameter: 18.4 mm Height: 65 mm
Initial internal impedance	≤100 mΩ

2 Real-Time Experimental Studies on Li-ion Battery

2.1 Modeling of Li-ion Battery

Li-ion battery is a type of rechargeable battery which is classified by different aspects. They are chemistry, performance, cost and safety characteristics. According to the classification, Lithium Nickel Manganese Cobalt Oxide (NMC) chemistry-based 18,650 cylindrical Li-ion battery cell is taken in this paper, and the parameters used for modeling are listed in Table 1.

2.2 Testing and Characteristics Analysis of Single Cell Li-Ion Battery

The indigenous Li-ion battery real-time experimental test setup in laboratory is shown in Fig. 1. The battery test system consists of 100 channels for testing battery cycle life; performance monitoring includes voltage, current, capacity, temperature and internal impedance. Also, the system supported to testing of different sizes and chemistries of Li-ion battery cells at the standard and rapid charging and discharging C-rates.

From Fig. 1, single-cell 18,650 Li-ion battery is tested in battery test system by performing continuous charge and discharge test at C/2 rate for eleven cycle numbers. The testing results and characteristics are stored in personal computer with the help of control software which is interfaced with the battery test system.

Fig. 1 Li-ion battery experimental test setup model

Voltage and Current Characteristics: The battery cell voltage and battery current characteristics are shown in Figs. 2 and 3 with respect to time. This test is performed at C/2 rate for eleven cycle numbers and constant voltage constant current method. As per the datasheet, the cell voltage is varied from 4.2 to 2.8 V for

Fig. 2 Battery voltage-time characteristics

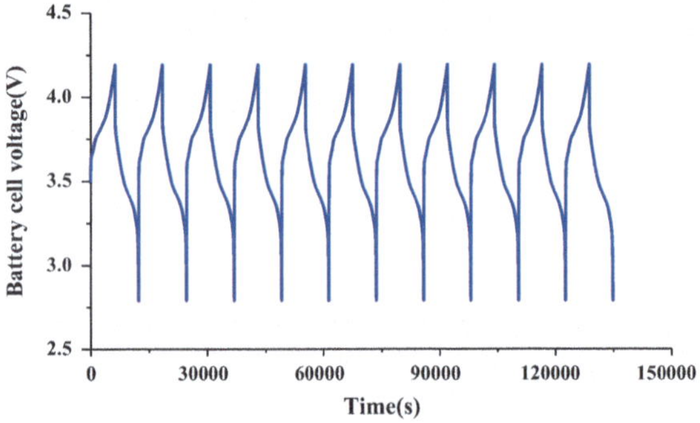

Fig. 3 Battery current-time characteristics

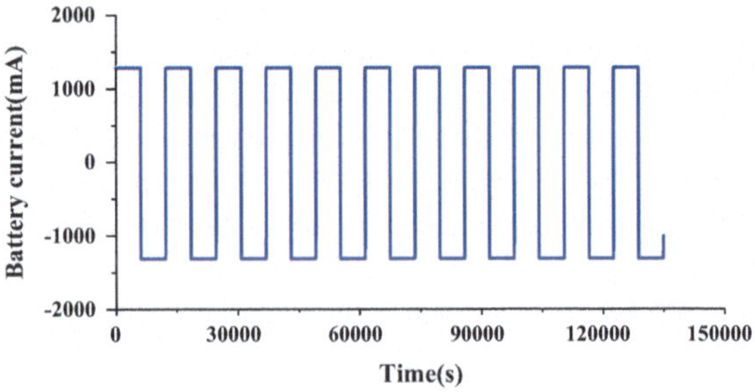

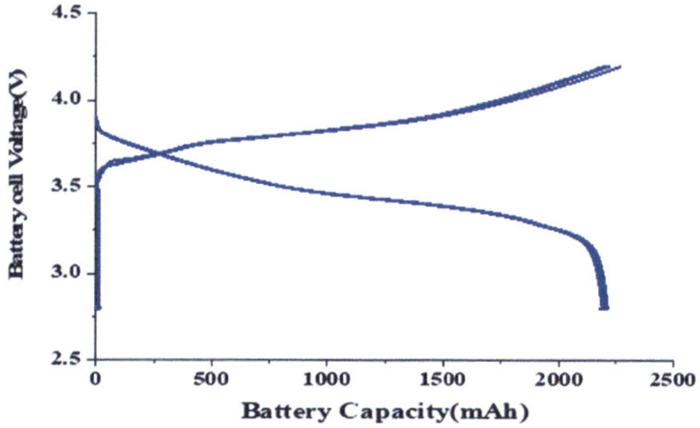

Fig. 4 Battery voltage-battery capacity characteristics

every charge and discharge cycle. The battery current is maintained constant that is 1300 mA during charge and -1300 mA during discharge.

Voltage-Capacity Characteristics. Figure 4 shows that battery voltage is varied according to the battery capacity for eleven charge and discharge cycles.

3 State of Charge Estimation Based on Intelligent Techniques

State of charge estimation is an important factor in battery technology and cannot be measured directly due to non-linear characteristics. There are different techniques to estimate the Li-ion battery state of charge. The traditional mathematical technique is coulomb counting method as shown in (1).

$$\%\mathrm{SOC}_{t+1} = \%\mathrm{SOC}_{t_0} + \left(\frac{\int_{t_0}^{t+1} I \, \mathrm{d}t}{\mathrm{C}_{\mathrm{rated}}} \times 100 \right) \quad (1)$$

where SOC_{t_0} is initial state of charge at time t_0, I is charge (or) discharge current and assumed positive sign for charge current and negative sign for discharge current, c_{rated} is Nominal capacity of the Li-ion battery.

This technique is very sensitivity to find the initial SOC value. At present, the most popular intelligent technique is neural network which overcomes difficulties faced in other estimation methods. The neural network-based SOC estimation is progressively appropriate for different working conditions of batteries and lacking accuracy of battery test system in actual operation.

The block diagram of the proposed neural network-based Li-ion battery SOC estimation is shown in Fig. 5. In this technique, the battery voltage and battery current are chosen as input vector to train the neural network model and SOC as output vector.

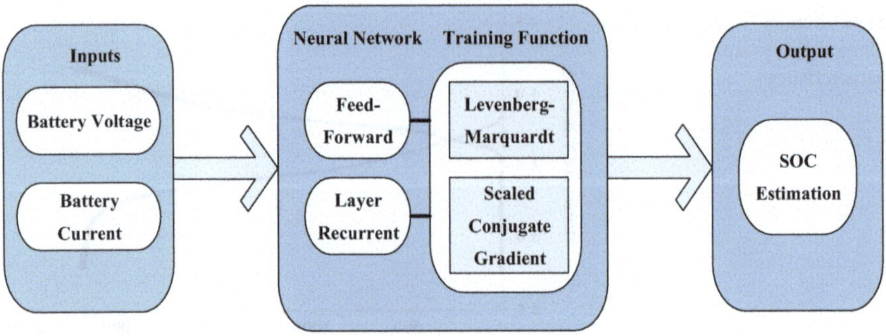

Fig. 5 Battery block diagram of proposed model

3.1 Feedforward Neural Network

A feedforward neural network is a single or multilayer perception network wherein connections from input node to output node do not frame a loop. It is consists of one or more nonlinear hidden layers which is activated by hyperbolic tangent sigmoid transfer functions. This function is a good trade-off for neural networks to get the nonlinear relationship between inputs and outputs. In this architecture, battery voltage (V) and battery current (I) are inputs, and output is state of charge. These two input parameters can be measured easily and significant relationship with the battery SOC.

3.2 Layer Recurrent Neural Network

Recurrent neural network is contrast from feedforward neural network where the output node from previous stage is fed as input node to the current stage. It is also known as long short-term memory due to remembering each previous output by giving each output as input to the following hidden layer which is activated by hyperbolic tangent sigmoid transfer function based on same weights and bias to all hidden layers. In this network, battery voltage (V) and battery current (I) are input parameters, and output is battery state of charge.

3.3 Scaled Conjugate Gradient (SCG) Training Method

The conjugate gradient training method is a learning algorithm in neural networks. The Scaled Conjugate Gradient (SCG) is a variation of conjugate gradient method which ignores the line search per iteration so as to scale the step size, and the amount of training time is shortened by reduced calculation of gradient.

The algorithm of SCG training method is as shown below.

Select weight vector $\overline{w_1}$ and scalars $0 < \sigma \leq 10^{-4}$, $0 < \lambda \leq 10^{-6}$, $\overline{\lambda_1} = 0$. Set $\overline{p_1} = \overline{r_1} = -E'(\overline{w_1})$, $i = 1$ and success = true.

If success = true, then calculate second order information

$$\sigma_i = \frac{\sigma}{|\bar{p}_i|} \tag{2}$$

$$\bar{s}_i = \big(E'(\overline{w}_i + \sigma_i \bar{p}_i) - E'(\overline{w}_i)\big)/\sigma_i \tag{3}$$

$$\delta_i = \bar{p}_i^{\mathrm{T}} \bar{s}_{\mathrm{i}} \tag{4}$$

Scale δ_{i}

$$\delta_i = \delta_i + \big(\lambda_i - \overline{\lambda}_i\big)|\bar{p}_i|^2 \tag{5}$$

If $\delta_i \leq 0$ then make the Hessian matrix positive definite

$$\overline{\lambda}_i = 2(\lambda_i - \delta_i)|\bar{p}_i|^2 \tag{6}$$

$$\delta_i = -\delta_i + \lambda_i|\bar{p}_i|^2 \tag{7}$$

$$\lambda_i = \overline{\lambda}_i \tag{8}$$

Calculate step size

$$\mu_i = \bar{p}_i^{\mathrm{T}} \bar{r}_i \tag{9}$$

$$\alpha_i = \mu_i/\delta_i \tag{10}$$

Calculate the comparison parameter

$$\Delta_i = 2\delta_i[E(\overline{w}_i) - E(\overline{w}_i + \alpha_i \bar{p}_i)]/\mu_i^2 \tag{11}$$

If $\Delta_i \geq 0$, then a successful reduction in error can be made

$$\overline{w}_{i+1} = \overline{w}_i + \alpha_i \bar{p}_i \tag{12}$$

$$\bar{r}_{i+1} = -E'(\overline{w}_{i+1}) \tag{13}$$

$$\overline{\lambda}_i = 0, \text{ success} = \text{true}$$

If i mod $N = 0$ then restart algorithm

$$\bar{p}_{i+1} = \bar{r}_{\text{i}+1} \tag{14}$$

Else create new conjugate direction

$$\beta_i = \left(|\bar{r}_{i+1}|^2 - \bar{r}_{i+1}^{\text{T}}\bar{r}_i\right)/\mu_i \tag{15}$$

$$\bar{p}_{i+1} = \bar{r}_{i+1} + \beta_i \bar{p}_i \tag{16}$$

If $\Delta_i \geq 0.75$, then reduce the scale parameter

$$\lambda_i = \frac{1}{4}\lambda_i \tag{17}$$

Else a reduction in error is not possible

$$\overline{\lambda}_i = \lambda_i, \text{ success} = \text{false}$$

If $\Delta_i < 0.25$, then increase the scale parameter

$$\lambda_i = \lambda_i + \frac{\delta_i(1 - \Delta_i)}{|\bar{p}_i|^2} \tag{18}$$

If the steepest descent direction $\bar{r}_i \neq 0$, then set $i = i + 1$ and go to step 2.

Else terminate and return $\overline{w}_{i+1}$ as desired minimum.

Where $\bar{s}_i$ is the approximation of Hessian matrix, E is the total error function, E' is the gradient of E, σ_i and λ_i are parameters to estimate the Hessian matrix.

3.4 Levenberg-Marquardt (LM) Training Method

The Levenberg-Marquardt (LM) training method is a combination of Gradient-descent (GD) and Gauss-Newton (GN) method. The LM method detects the minimum of a function that is stated as a second-order sum of square of nonlinear function which gives the faster training speed than GD and GN method, less number of iterations, more accurate near a minimum error and no need to calculate Hessian matrix.

The damping parameter λ is important parameter to decide the effectiveness of LM algorithm in neural networks. When λ is too small (nearly zero), the LM algorithm approximates to Gauss-Newton method, and λ is very large. The LM algorithm approximates to Gradient-descent method. As solution improves to faster convergence, the λ is decreased by increasing step size.

The algorithm of LM training method is as shown below.

Initialize input vector x, weight vector w and parameter λ. Set iteration $i = 1$

Compute total error function

$$E_i(x, w) = \frac{1}{2} \sum_{m=1}^{M} \sum_{n=1}^{N} e_{mn}^2 \tag{19}$$

Compute Jacobian matrix

$$J = \begin{bmatrix} \frac{\partial e_{11}}{\partial w_1} & \frac{\partial e_{11}}{\partial w_2} & \cdots & \frac{\partial e_{11}}{\partial w_K} \\ \frac{\partial e_{12}}{\partial w_1} & \frac{\partial e_{12}}{\partial w_2} & \cdots & \frac{\partial e_{12}}{\partial w_K} \\ \vdots & \vdots & \vdots & \vdots \\ \frac{\partial e_{MN}}{\partial w_1} & \frac{\partial e_{MN}}{\partial w_2} & \cdots & \frac{\partial e_{MN}}{\partial w_K} \end{bmatrix} \tag{20}$$

Compute rule

$$w_{i+1} = w_i - \left(J_i^{\mathrm{T}} J_i + \lambda I\right)^{-1} J_i e_i \tag{21}$$

Compute total error for new weight vector $E_{i+1}(x, w_{i+1})$ using (18).
If $E_{i+1} \leq E_{\max}$ terminate
If $E_{i+1} \leq E_i$, then set $\lambda = \frac{\lambda}{\alpha}$ and $w_i = w_{i+1}$, and go to step 1.
If $E_{i+1} > E_i$, then set $\lambda = \lambda \cdot \alpha$ and restore $w_i = w_i$, and go to step 4.
Where e_{mn} is error vector at output N when pattern M and it is defined as difference between desired output vector and actual output vector, K is the number of weights.

4 Simulation Results

4.1 Design of Neural Network Model

The feedforward and layer recurrent neural networks are developed, trained and tested in MATLAB using the neural network tool. The properties used to create neural networks are listed in Table 2.

4.2 Training and Testing Data

The set of data were generated using the experimental results and characteristics to train and test the neural network. The actual SOC is obtained from experimental setup compared with proposed methods for estimation of the Li-ion battery SOC. From the total data set, the 70% of data used for training and remaining 30% of data used for testing purpose. The training of neural network is an important factor to accurately estimate the battery state of charge with the help of training functions.

Table 2 Neural network properties

Parameters	Feedforward neural network	Recurrent neural network
Input data	Battery cell voltage and current	Battery cell voltage and current
Target data	Battery state of charge	Battery state of charge
Training function	Scaled conjugate gradient	Scaled conjugate gradient
	Levenberg-Marquardt	Levenberg-Marquardt
Performance function	Mean Squared Error(MSE)	Mean Squared Error(MSE)
Number of layers	Hidden layer and output layer	Hidden layer and output layer
Number of neurons in hidden layer	10	10
Hidden layer transfer function	Hyperbolic tangent sigmoid function	Hyperbolic tangent sigmoid function
Selection of initial weights	Random	Random

The default initial training parameters are used to train the neural network by SCG training method are maximum number of epochs to train, performance goal, Maximum time to train in seconds, Minimum performance gradient, Maximum validation failures, Determine change in weight for second derivative approximation and parameter for regulating the indefiniteness of the Hessian.

The default initial training parameters are used to train the neural network by LM training method are maximum number of epochs to train, performance goal, Maximum validation failures, Minimum performance gradient, initial λ, λ decrease factor, λ increase factor, maximum λ, and Maximum time to train in seconds.

4.3 Battery State of Charge Estimation Results

First, the feedforward neural network model was trained by applying SCG and LM training methods, and the testing data estimation results of battery state of charge are shown in Figs. 6 and 7. The SCG method passed 122 epochs to converge during training, and the Mean Square Error (MSE) was 0.871. The dotted line curve in Fig. 6a is regarded as actual SOC obtained from experimental results, and the solid line curve is regarded as an estimated SOC obtained from SCG training method

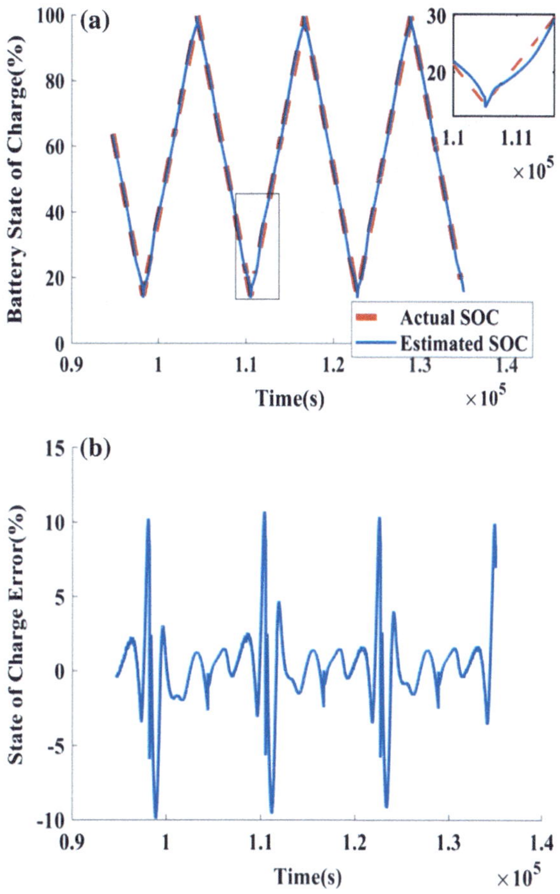

Fig. 6 SOC estimation using SCG method of feedforward neural network **a** Battery state of charge **b** State of charge error

of feedforward neural network. Figure 6b shows that battery state of charge error between actual quantity and estimated SOC.

The LM training method recorded 75 epochs, and the MSE was 0.292, which was better than that of SCG training of feedforward neural network. From Fig. 7a, the dotted line curve is observed as actual SOC, and the solid line curve is observed as an estimated SOC obtained from LM training method of feedforward neural network. Figure 7b shows that the battery state of charge error between actual SOC and estimated SOC, which was very lower than that of SCG training of feedforward neural network.

Second, the SOC estimation was performed with layer recurrent neural network using the same two training methods, and the results are shown in Figs. 8 and 9. The SCG method converged to 80 epochs during training, and the test MSE was 0.57, which was better than that of SCG training method of feedforward neural network and inferior than that of LM training of feedforward neural network. The dotted line curve in Fig. 8a is viewed as actual SOC obtained from experimental results, and the

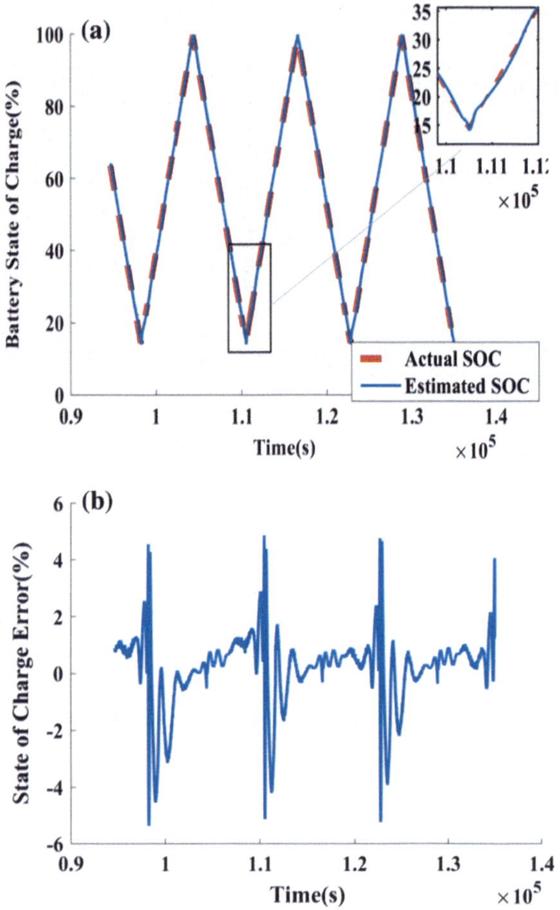

Fig. 7 SOC estimation using LM method of feedforward neural network
a Battery state of charge
b State of charge error

solid line curve is viewed as an estimated SOC obtained from SCG training method of recurrent neural network.

The LM training method completed to six epochs and the test data MSE was 0.241, which was the best method in all the proposed networks. Figure 9a shows that the dotted line curve is observed as actual SOC, and the solid line curve is observed as an estimated SOC obtained from LM training method of Recurrent Neural Network, which gives accurate results and very less marginal error between actual SOC and estimated SOC as shown in Fig. 9b.

All the estimated results are comprehensively compared in the Table 3. From these results, the Levenberg-Marquardt (LM) training of recurrent neural network is demonstrated excellent performance among all the networks. Moreover, the highest convergence speed and most accurate with minimum MSE among all the methods. Thus, the layer recurrent neural network with LM training method is suitable for estimating the Li-ion battery state of charge.

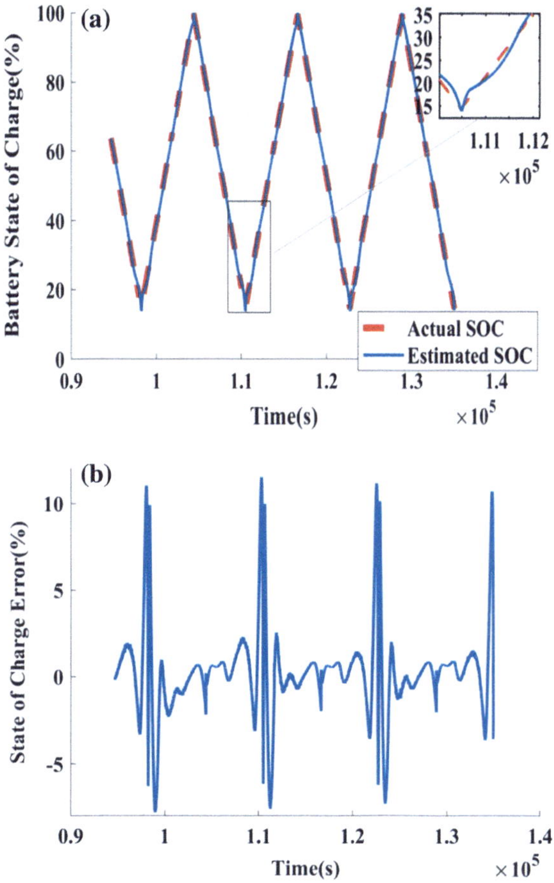

Fig. 8 SOC estimation using SCG method of recurrent neural network **a** Battery state of charge **b** State of charge error

5 Conclusion

In this paper, an efficient SOC estimation algorithm is developed based on intelligent techniques for 18,650 single-cell Li-ion battery. Two intelligent networks, feedforward and layer recurrent neural networks, are presented, and training and testing data are generated from real-time experiment carried out on 18,650 single-cell Li-ion battery. The experimental results and characteristics are analyzed and the battery cell voltage and current applied as input data and SOC as output for these two networks which are trained and tested by SCG and LM training methods. The comparison results show that the layer recurrent neural network with LM training method that has the best intelligent technique among all the methods for estimating the Li-ion battery state of charge due to fast convergence speed and most accurate. Also, this technique does not depend on the material science of batteries, since a neural network is a data-driven methodology. Therefore, the developed technique can be promptly applied to batteries with different chemistries.

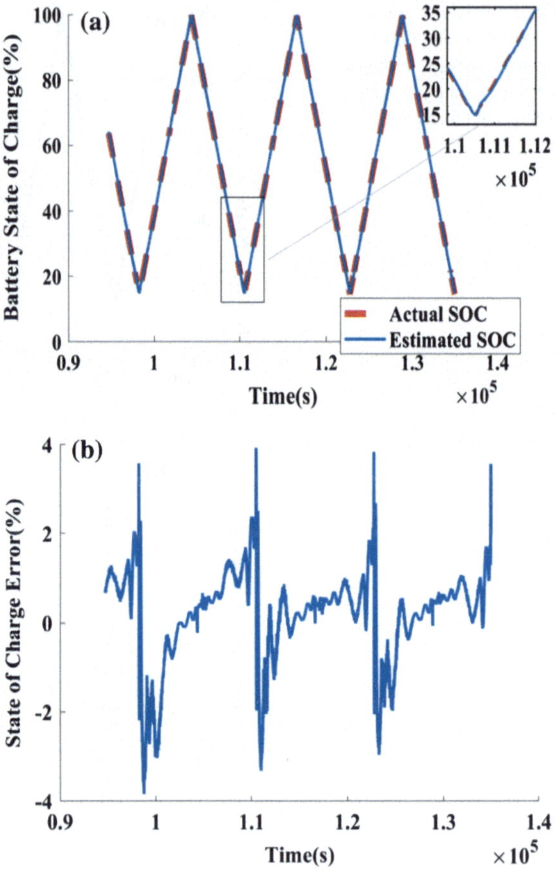

Fig. 9 SOC estimation using LM method of recurrent neural network **a** Battery state of charge **b** State of charge error

Table 3 Comparison of neural network training parameters

Training parameters	Neural network			
	Feedforward		Recurrent	
	SCG	LM	SCG	LM
Epoch	122	75	80	6
Performance (MSE)	0.871	0.292	0.57	0.241
Gradient	1.22	0.0998	1.14	0.0447
λ	–	0.001	–	0.001

Acknowledgements Author would like to thank the Research and development in battery division of Central Electrochemical Research Institute, Council of Scientific and Industrial Research in India, for the financial assistance. The work presented in this paper is a part of the In House Project (IHP) "Design and Development of 1 kWh Solar Energy Storage System using Li-ion Battery cells," funded by Central Electrochemical Research Institute, India.

References

1. Corno M, Bhatt N, Savaresi SM, Verhaegen M (2015) Electrochemical model-based state of charge estimation for Li-ion cells. IEEE Trans Control Syst Technol 23(1):117–127
2. Chang WY (2013) The state of charge estimating methods for battery: a review. ISRN Appl Math
3. Pang S, Farrell J, Du J, Barth M (2001) Battery state of charge estimation. In: Proceedings of the 2001 American control conference, pp 1644–1649. June 2001
4. Xiong R, Cao J, Yu Q, He H, Sun F (2018) Critical review on the battery state of charge estimation methods for electric vehicles. IEEE Access 6:1832–1843
5. Berrueta A, San Martin I, Sanchis P, Ursua A (2016) Comparison of state-of-charge estimation methods for stationary Lithium-ion batteries. In: 42nd annual conference of the IEEE industrial electronics society, pp 2010–2015. Dec 2016
6. Kribsky P, Krivka J, Valda L, Zahour J (2014) Li-ion state of charge estimation methods. In: 22nd telecommunications forum, pp 649–651. Nov 2014
7. Chiasson J, Vairamohan B (2005) Estimating the state of charge of a battery. IEEE Trans Control Syst Technol 13(3):465–470
8. Parthiban T, Ravi R, Kalaiselvi N (2007) Exploration of artificial neural network [ANN] to predict the electrochemical characteristics of Lithium-ion cells. Electrochimica Acta 53:1877–1882
9. Jeon S, Yun JJ, Bae S (Oct 2015) Comparative study on the battery state-of-charge estimation method. Indian J Sci Technol 8(26)
10. Chaoui H, Mandalapu S (2017) Comparative study of online open circuit voltage estimation techniques for state of charge estimation of Lithium-ion batteries. Batteries 3(2):12
11. Lee S, Kim L, Lee J, Cho BH (2008) State-of-charge and capacity estimation of Lithium-ion battery using a new open-circuitvoltage versus state-of-charge. J Power Sour 185(2):1367–1373
12. Ma Yan, Li Bingsi, Xie Y, Chen H (2016) Estimating the state of charge of Lithium-ion battery based on sliding mode observer. IFAC-Papers OnLine 49(11):54–61
13. Yadaiah N, Sowmya G (2006) Neural network based state estimation of dynamical systems. In: International joint conference on neural networks, pp 1042–1049. July 2006
14. Qazi A, Fayaz H, Wadi A, Raj RG, Rahim NA, Khan WA (2015) The artificial neural network for solar radiation prediction and designing solar systems: a systematic literature review. J Clean Prod 104:1–12
15. Yu Z, Huai R, Xiao L (2015) State-of-charge estimation for Lithium-ion batteries using a Kalman filter based on local linearization. Energies 8:7854–7873
16. Hussein AA (2014) Kalman filters versus neural networks in battery state-of-charge estimation: a comparative study. Int J Mod Nonlinear Theory Appl 3:199–209
17. Youssef C, Omar D, Ahmed G, Fatima E, Najia ES (2017) Designand simulation of an accurate neural network state-of-charge estimator for Lithium ion battery pack. Int Rev Autom Control 10(2):186–192
18. Jimenez-Bermejo D, Fraile-Ardanuy J, Castano-Solis S, Merino J, Alvaro-Hermana R (2018) Using dynamic neural networks for battery state of charge estimation in electric vehicles. Procedia Comput Sci 130:533–540

19. Narang SB, Singh M, Pubby K (2015) Comparison of feed forward neural network training methods for visual character recognition. Adv Comput Sci Inf Technol 2(4):366–369
20. Prerana ParveenSehgal (2015) Comparative study of GD, LM and SCG method of neural network for thyroid disease diagnosis. Int J Appl Res 1(10):34–39
21. Cetisli B, Barkana A (Mar 2009) Speeding up the scaled conjugate gradient algorithm and its application in neuro-fuzzy classifier training. Soft Comput, pp 365–378
22. Moller MF (1993) A scaled conjugate gradient algorithm for fast supervised learning. Neural Netw 6:523–533
23. Lv, Xing Y, Zhang J, Na X, Li Y, Liu T, Cao D, Wang FY (2017) Levenberg-marquardt backpropagation training of multilayer neural networks for state estimation of a safety critical cyber-physical system. IEEE Trans Ind Inf
24. Ismail M, Dlyma R, Elrakaybi A, Ahmed R, Habibi S (2017) Battery state of charge estimation using an artifical neural network. IEEE Transp Electrif Conf Expo, pp 342–349

Improved DTC-Drive-Based Photovoltaic Water Pumping System with INC-MPPT

Priyanka Singh and Prerna Gaur

Abstract Induction motors are horse powers and widely in demand. Photovoltaic (PV) system has been now widely accepted as renewable energy source. In this paper, an INC-MPPT-based DTC drive system is designed for PV-based induction motor drive to be used in water pumping. A modified space vector modulation technique based on imaginary switching times is used to overcome the disadvantage of hysteresis-based conventional DTC for induction motor. The conventional DTC is compared with improved SVM-based DTC, and performance of motor parameters is analyzed for PV parameter variations in the system using MATLAB/SIMULINK.

Keywords Photovoltaic · MPPT · Variable frequency drive · Conventional DTC · Imaginary switching times · Improved SVM-DTC · Induction motor · PV water pump

1 Introduction

Tremendous fast growth in the energy field inspires researchers to adopt the alternate source of power generation like from wind or photovoltaic power systems [1]. Among all the different types of renewable energy sources feasible to acquire electricity, photovoltaic energy is becoming most popular source of energy, since it is inexhaustible, clean, and requires very less maintenance [2]. It is widely used in many applications, among which power source for the water pumping is regarded as one of the most optimistic. Mainly in the remote areas where the adequate amount of electricity is not attainable, the photovoltaic water pumping systems are taken in use for water supply. Low maintenance, reliability, easy installation, and the emulating between the water usage needs and the power generated are the major benefits of water pumps energized by photovoltaic systems. Mainly, the PV water pumping

P. Singh (✉) · P. Gaur
Netaji Subhas Institute of Technology, Azad Hind Fauj Marg, Sector-3, Dwarka, New Delhi 110075, India
e-mail: siradhna.p@gmail.com

P. Gaur
e-mail: prerna.gaur@nsit.net.in

S. N. Singh et al. (eds.), *Advances in Power and Control Engineering*, Lecture Notes in Electrical Engineering 609, https://doi.org/10.1007/978-981-15-0313-9_13

system is composed various components such as PV array, a dc-dc converter, controller, inverter, motor, and the pump. And it becomes mandatory to acquire the best performance from each component of the system for a vast range of input power. The PV array current and power characteristics are non-linear and are highly pretentious by the temperature and irradiance variation. Thus, a maximum power point tracker (MPPT) control is essential for the maximum power extraction from the PV arrays and handle problems such as temperature and irradiance variation. Many different types of techniques for MPPT have been proposed [3], and these techniques vary from each other in accessibility, simple in implementation, cost, time response, accuracy, easiness, and many other technical features. Papers [2–5] present various classical MPPT techniques like incremental conductance (INC-MPPT) method, perturb and observe method, etc., used for water pumping system, among which incremental conductance is considered to be the best method. By the advancement of the power converters with high switching frequency operated with the digital signal processors, the torque and speed of AC motors can be smoothly controlled. Today, the AC motors are widely used in industrial applications as they are low in cost and weight, have better reliability, and less maintenance requirement. This paper presents a water pumping system where the pump is steered by three-phase induction motor powered by a three-phase voltage source inverter. For some applications like for agricultural irrigation , control of water flow is needed, so therefore, it becomes mandatory to establish a system for tracking and monitoring the speed of the drive motor. Several methods for control of motor drive have been developed [6–8]. Performance of conventional DTC and a simplified SVPWM-DTC with the idea of imaginary switching time is compared and introduced in our work, where lookup table, identifying angle and sector are not required by which the burden on the processor is reduced. Also, the ripples are reduced by a substantial amount with the switching patterns and the results are alike to SVM-DTC.

2 Design of the Proposed System

Figure 1 is the basic structure of the smart water pumping system proposed which consists of three-phase induction motor drive with improved DTC control in which PWM ac voltage is provided by voltage source inverter and a boost converter with INC-MPPT control for SPV system.

2.1 PV System Characteristics and Model

When the sunlight strikes on the semiconducting material in the PV cell, the electric field gets initiated which then leads to the momentum and direction to light stimulated-electrons that get accelerated, generating a drift of electricity. Modules are generated for high demand of power which are created with numerous PV cells

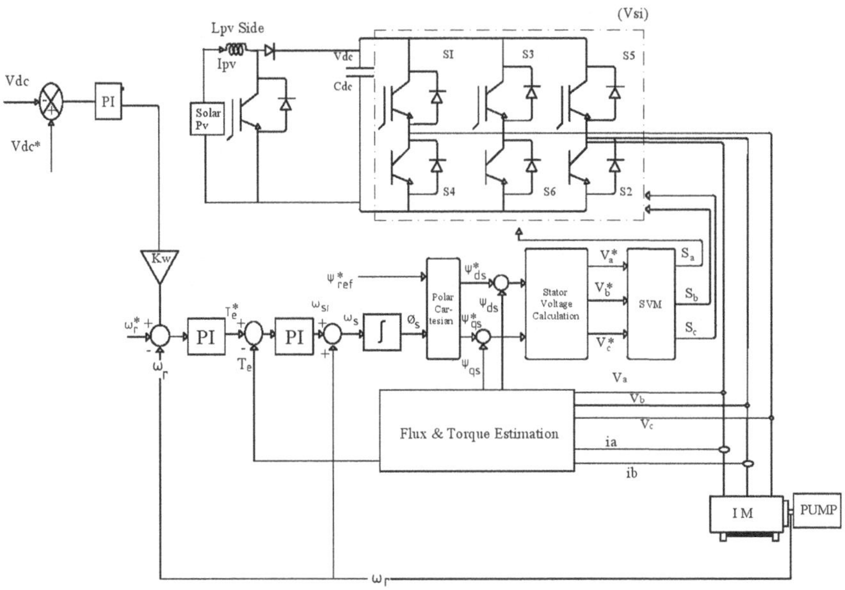

Fig. 1 Block diagram of proposed solar water pumping system

linked in parallel and series circuit, and these several modules are then electrically linked in parallel and series combination for generating the required voltage and current.

Using the solar cells equivalent circuit as in Fig. 2, the I–V characteristics of PV module, by pretermitting internal series resistance is represented as (1) [5].

$$I = I_{\text{ph}} N_{\text{p}} - I_{\text{s}} N_{\text{p}} \left\{ \mathrm{e}^{\frac{V + I R_S}{n V_t C N_S}} + 1 \right\} - I_{\text{sh}} \quad (1)$$

where V and I are the output voltage and current of the PV module, respectively, I_{ph} is the photoelectric current, n is the diode ideality factor, V_{t} is thermal Voltage, C is number of cells in PV module, N_{p} is the actual number of cells connected in parallel, and I_{sh} is the shunt current of the PV module.

Fig. 2 Equivalent circuit diagram of a PV cell

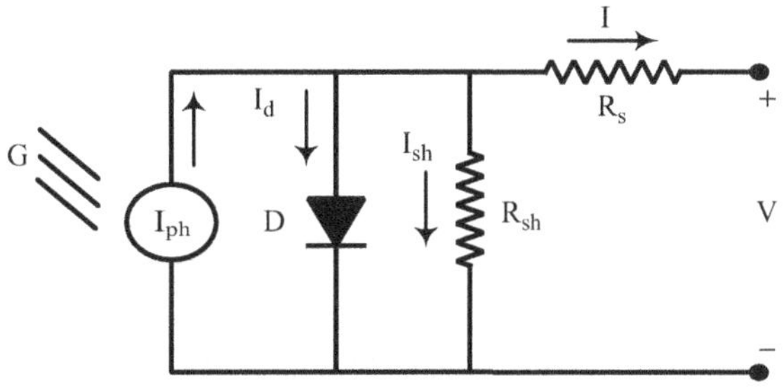

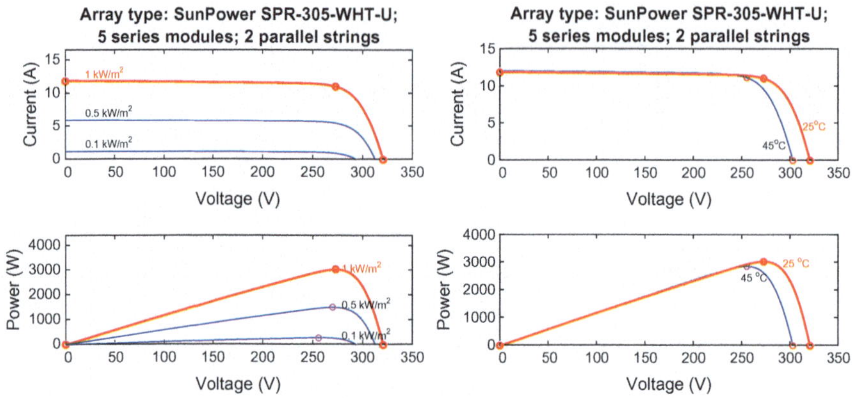

Fig. 3 V–I and P–V characteristics of a PV module for different irradiation levels

The saturation current I_s and reverse saturation current I_{rs} of the PV module which vary with temperature are as given in Eqs. (2) and (3) [4].

$$I_s = I_{rs} - \left[\frac{T_{on}}{T_{ref}}\right]^3 e^{\left[\frac{qE_g}{nK}\left(\frac{1}{T_{on}} - \frac{1}{T_{ref}}\right)\right]} \tag{2}$$

$$I_{rs} = \frac{I_{rs}}{\left[e^{\left(\frac{V_{oc}q}{KCT_{on}n}\right)} - 1\right]} \tag{3}$$

where, E_g is the band gap energy, T_{ref} is reference temperature, K is the Boltzmann's constant, q is the charge of one electron, and V_{oc} is open-circuit voltage of the PV module.

From Fig. 3, it is observed that the increment in temperature leads to decrement in open-circuit voltage depriving of any considerably alter in the short-circuit current, and from Fig. 3, it is observed that for small values of solar irradiations the change in open circuit is very less, whereas the short-circuit current is reducing considerably, which proves that the maximum power from the module is descending.

It is always desired that the PV sources should always deliver the maximum available energy, for this purpose, the MPPT methods are considered to automatically search the current I_{mpp}, or voltage V_{mpp} at which maximum power is delivered by PV array under a given irradiance and temperature.

3 Description of the Tracking Algorithm

According to the results presented [9], water pumping system tested without MPPT is not efficient and gives prolonged response. To solve this situation, tracking device is used at the PV generator to extract the maximum available power. Numerous

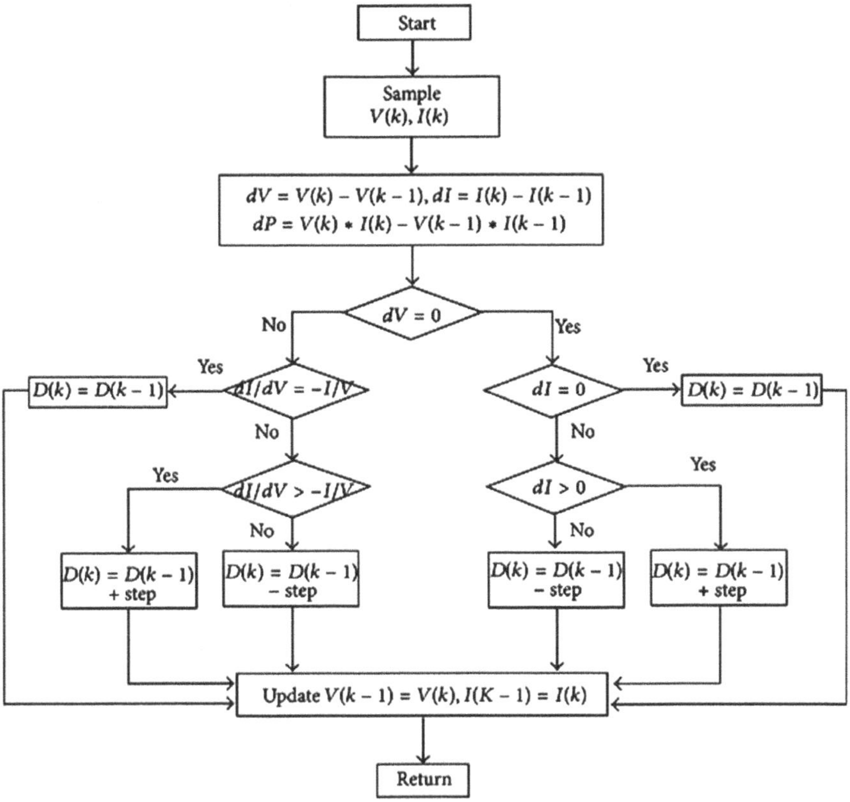

Fig. 4 Flowchart of INC algorithm

techniques for tracking the maximum power point (MPP) have been proposed over the years [2–5]. Among which perturb and observe (P&O) and incremental conductance (INC) are considered most favoured and taken in use for many applications. INC-MPPT is used in this paper for system. And Fig. 4 shows flow chart of INC-MPPT.

4 Control of Induction Machines

The "voltage/frequency" (V/f) of induction motor is widely taken in use in many industrial applications as it is considered to be the simplest controller, although torque and stator fluxes are not directly controlled in this scheme which results in not very good accuracy in both torque and speed responses. Another very popular method is field-oriented control (FOC) which is also known as vector control. FOC method deals with the coordinate system as motor equations are transformed. This control method operates with fast responses and is excellent for handling transients, but due to its costly driver circuits and complexity, direct torque control (DTC) was introduced [9–11].

In direct torque control (DTC), the torque and speed of three-phase AC electric motors are directly controlled. From the measured values of motor current and voltage, the magnetic flux and torque are estimated. In the later stages, DTC has been improvised and SVM-based DTC has been implemented [12]. Thus, DTC based on hysteresis controllers is often termed as "classical DTC."

4.1 Torque Control Scheme in DTC of Induction Motor Drive

Flux equations from both the d and q axes are as below [13]

$$\psi_{ds} = L_{\mathrm{s}} i_{ds} + L_{\mathrm{m}} i'_{d\mathrm{r}} \tag{4}$$

$$\psi'_{d\mathrm{r}} = L_{\mathrm{r}} i'_{d\mathrm{r}} + L_{\mathrm{m}} i_{ds} \tag{5}$$

The d-axis current and q-axis current can be derived as

$$i_{qs} = \frac{1}{\vartheta L_{\mathrm{s}}}\left(\psi_{qs} - \frac{L_{\mathrm{m}}}{L_{\mathrm{r}}}\psi'_{q\mathrm{r}}\right) \tag{6}$$

$$i_{ds} = \frac{1}{\vartheta L_{\mathrm{s}}}\left(\psi_{ds} - \frac{L_{\mathrm{m}}}{L_{\mathrm{r}}}\psi'_{d\mathrm{r}}\right) \tag{7}$$

where $\vartheta = 1 - \frac{L_{\mathrm{m}}^2}{L_{\mathrm{s}} L_{\mathrm{r}}}$ is the leakage factor, ψ_{ds} and $\psi'_{d\mathrm{r}}$ are d-axis stator and rotor flux, ψ_{qs} and $\psi'_{q\mathrm{r}}$ are q-axis stator and rotor flux, i_{ds} and $i'_{d\mathrm{r}}$ are d-axis stator and rotor current, i_{qs} is q-axis stator current, L_{s}, L_{r}, and L_{m} are stator, rotor and mutual inductances, respectively.

By applying above equations, the expression for torque can be simplified as

$$T_e = \left(\frac{3}{2}\right)\left(\frac{P}{2}\right)\frac{L_{\mathrm{m}}}{\vartheta L_{\mathrm{s}} L_{\mathrm{r}}}|\psi_{\mathrm{s}}||\psi_{\mathrm{r}}|(\sin(\theta)) \tag{8}$$

Here, stator flux vector and rotor flux vector has angle θ in between them. Thus, from Eq. (8), it is concluded that when the rotor and stator flux are in constant magnitude then the rate of change in torque is directly proportional to the rate in change of θ.

4.2 Flux Control Strategy of Induction Motor

The stator flux for the stationary reference frame can be written as [13]

$$\frac{d\psi_s}{dt} = V_s - [r_s]i_s \tag{9}$$

$$\psi_s = \int V_s - [r_s]i_s \tag{10}$$

One voltage vector is applied in specific sampling period of time interval among the six voltage vectors, and each voltage vector is constant which is applied during the pre-defined sampling period of time interval. Thus, the above Eq. (10) can be written as

$$\psi_s = \int_0^t V_s dt - [r_s]\int_0^t i_s dt \tag{11}$$

$$\psi_s = V_s \cdot \Delta t + \psi_s/_{t=0} \tag{12}$$

where $\psi_{s/t=0}$ is the primary stator flux linkage at the immediate time of switching; V_s and i_s are the measured stator voltage and current. Also r_s is estimated stator resistance. By neglecting the resistance drop we get:

$$\Delta\psi_s = V_s\Delta t \tag{13}$$

By applying an appropriate required voltage vector to the inverter, the stator flux linkage amplitude can be controlled.

4.3 Flux Regulator

The flux regulator is a two-level hysteresis comparator. The output of the regulator is a two-valued function which gives either increase or decrease in the flux. When the flux developed in the machine is less than the reference value by a value greater than the hysteresis band of the regulator, then the output of the flux regulator gives a flux increase command ($\psi^* = 1$). When the flux developed in the machine is greater than the reference value by a value greater than the hysteresis band of the regulator, then, the output of the flux regulator gives a flux decrease command ($\psi^* = 0$) (Fig. 5).

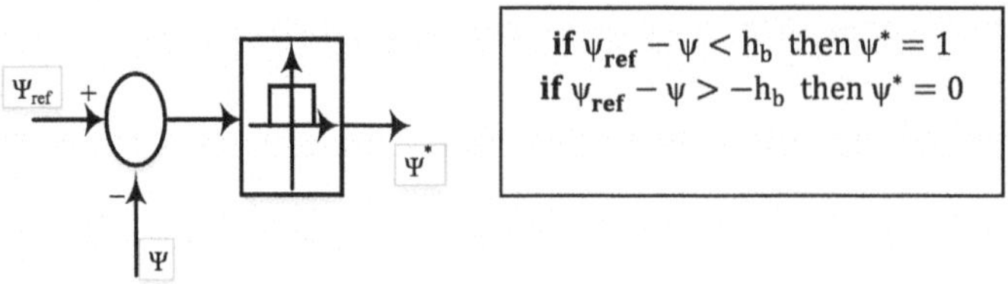

Fig. 5 Flux regulator

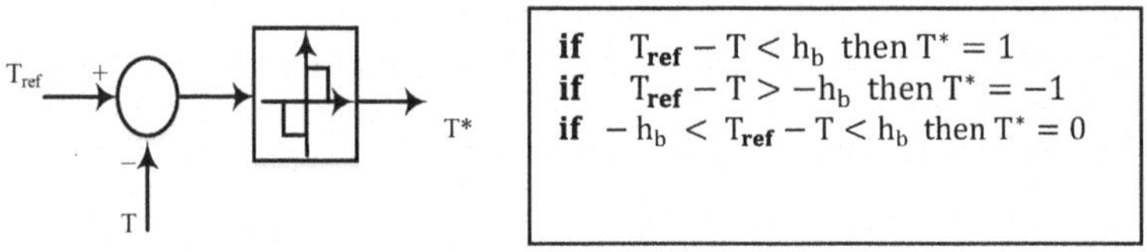

Fig. 6 Torque regulator

4.4 Torque Regulator

The torque regulator is a three-level hysteresis regulator whose output states are 1, 0, and -1. When the torque developed in the machine is less than the reference value by a value greater than the hysteresis band of the regulator, then the output of the torque regulator gives torque increase command ($T^* = 1$). When the torque developed in the machine is greater than the reference value by a value greater than the hysteresis band of the regulator, then the output of the torque regulator gives a Torque decrease command ($T^* = -1$). When the torque error is within the hysteresis band then the torque regulator output will be $0(T^* = 0)$ (Fig. 6).

4.5 Voltage Vector Selection

Table 1 shows the switching table which governs both the rotating direction and amplitude of the stator flux. The switching table consists of six sectors of voltage vectors. Every sector has four non-zero voltage vectors out of six. Flux sectors θ, T^*, and ψ^* are the outputs of the torque and flux hysteresis comparators. In Table 1, for $\psi^* = 1$, the real flux is less than the reference value, and for $\psi^* = 0$, the real flux is higher than the reference value. This also implies for the torque.

Table 1 Switching table

Sector		$\theta(1)$	$\theta(2)$	$\theta(3)$	$\theta(4)$	$\theta(5)$	$\theta(6)$
$\psi^* = 1$	$T' = 1$	V_2'	V_3'	V_4'	V_5'	V_6'	V_1'
	$T' = 0$	V_7'	V_0'	V_7'	V_0'	V_7'	V_0'
	$T' = -1$	V_6'	V_1'	V_2'	V_3'	V_4'	V_5'
$\psi^* = 0$	$T' = 1$	V_3'	V_4'	V_5'	V_6'	V_7'	V_8'
	$T' = 0$	V_0'	V_7'	V_0'	V_7'	V_0'	V_7'
	$T' = -1$	V_5'	V_6'	V_1'	V_2'	V_3'	V_4'

5 SVM-DTC Based on Imaginary Switching Concept

Considerably, high ripples in current, flux, and torque are the foremost drawback of classical DTC. Over a period of time, this drawback is overcome by many different techniques proposed in the literature. Like, in [14] reduced ripples in torque, flux and current with constant switching frequency is obtained by space vector modulation-based DTC introduced by T. G. Habelter. Dithering technique is employed in [12] which ripples are reduced by a carrier wave with high frequency superimposed with torque and flux error. It is shown in [13] that by suitably applying the voltage vector and zero vector for a specific period of time in the sampling period, the torque ripples are reduced considerably. But the prime features, i.e. facile and ease in implementation of DTC are demolished by these techniques as they are very complex to implement. A simple scheme is proposed in this paper where lookup table, identifying angle, and sector are not required by which the burden on the processor is reduced. Also, the ripples are reduced by a substantial amount with the switching patterns and the results are alike to SVM-DTC. This scheme is simulated in MATLAB and results for the same are compared with the classical DTC. From the results, it is seen that this technique posses much superiority than the classical one.

By measuring motor parameters either by two-phase currents and voltages or by dc-link voltage with any two-phase current the electromagnetic torque and the real stator flux vectors in both d-axis and q-axis components can be derived. As given in [11], when the synchronous speed which is attained by adding motor speed to the slip speed that is directly proportional the torque error is integrated, the flux vector's phase angle is calculated. And, also the reference flux space vector in both *d*-axis and *q*-axis components can be calculated by the motor speed and torque error.

Error in between these two flux vectors then generates imaginary switching time reference and its *d*–*q* components are determined by the following methodology.

Stator voltage equation in stationary reference frame in *d*–*q* is as given:

$$V_s = R_s i_s + \frac{d\psi}{dt} \tag{14}$$

After neglecting the stator resistance, the equation is simplified as:

$$\Delta\psi_s = V_s \cdot \Delta t \tag{15}$$

$$\Delta\psi_{sd} + j \cdot \Delta\psi_{sq} = \left(V_{sd} + j \cdot \Delta V_{sq}\right)\Delta t \tag{16}$$

By analogizing the imaginary and real components of Eq. (3), results:

$$V_{sd} = \frac{\Delta\psi_{sd}}{\Delta t} = \frac{\psi_{sd}^{*} - \psi_{sd}}{\Delta t} \tag{17}$$

$$V_{sq} = \frac{\Delta\psi_{sq}}{\Delta t} = \frac{\psi_{sq}^{*} - \psi_{E}}{\Delta t} \tag{18}$$

where Δt is sampling time T_{s*}.

Thus, the imaginary switching instants in *d–q* stationary reference frame are derived as following:

$$T_{sd} = \frac{V_{sd}}{V_{DC}}.T_{s*} = \frac{\Delta\psi_{sd}}{\Delta t(= T_{s*}) * V_{DC}} = \frac{\psi_{sd}^{*} - \psi_{sd}}{V_{DC}} \tag{19}$$

$$T_{sd} = \frac{V_{sd}}{V_{DC}} \cdot T_{s*} = \frac{\Delta\psi_{sd}}{\Delta t(= T_{s*}) * V_{DC}} = \frac{\psi_{sd}^{*} - \psi_{sd}}{V_{DC}} \tag{20}$$

Also,

$$T_s = T_{sd} + j \cdot T_{sq} = \frac{V_{sd}}{V_{DC}} \cdot T_{s*} + j \cdot \frac{V_{sq}}{V_{DC}}.T_{s*} \tag{21}$$

Hence,

$$T_s = \frac{V_s}{V_{DC}} \cdot T_{s*} = \frac{\psi_s^{*} - \psi_s}{V_{DC}} \tag{22}$$

From Eq. (22), the real switching times of the inverter are calculated from the imaginary switching time vectors. However, the sampling time should always be greater than the imaginary time vector's magnitude. The imaginary switching times T_{sa}, T_{sb}, and T_{sc} obtained from two-to-three-phase conversion of imaginary time vector components, and is resolute and resolved in every sampling period, respectively. Therefore, the switching times are generated for the PWM inverter. Following every sampling interval, real stator flux vector Ψ_s is rectified by the error and then it strives to procure reference flux space vector Ψ_s^{*}. Moreover, the torque demand as discussed above is taken care by the reference stator flux as it is the summation of slip speed calculated from the real rotor speed and torque error. Hence, the torque and flux waveforms are improved as both the torque and flux errors are indemnified in every sampling interval. The d–q components of imaginary switching instant are derived

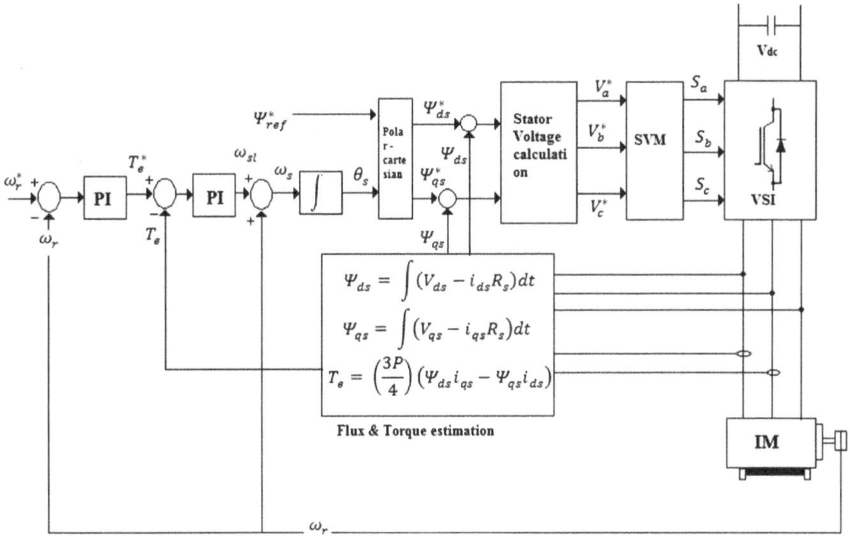

Fig. 7 SVM-DTC scheme

by (21) and (22). In this method, the complexity of the system is reduced as identifying sector and angle is not required. Block diagram for SVM-DTC is as given in (Fig. 7).

6 Simulation Results

The classical DTC and SVM-based DTC both are simulated in the MATLAB for a 3 hp and 2.7 kw PV array system. The flux locus of classical DTC is shown below. It can be observed that the locus is circular. When compared to the classical DTC, the flux locus is simply a pure circle without any ripples and also the torque ripple is also reduced in SVM-based DTC as shown in Figs. 8 and 9.

The torque response, current, and speed response of the SVM-DTC and classical DTC for motor is as shown below. The response is plotted for a step change in the load torque from 0 to 8 Nm at 1.5 s (Figs. 10 and 11).

Flux and torque analysis and the ripples comparison can be easily verified from Figs. 12 and 13 for both the schemes.

Comparison between the two schemes is shown in Table 2, which clearly shows that the settling time and the ripples in the SVM-DTC are much less as compared to the classical DTC.

The steady-state performance of the system is shown in Fig. 14 for radiation 800 wb/m^2, here the DC-link voltage V_{DC}, V_{mpp} , and I_{mpp} are maintained at their respective values.

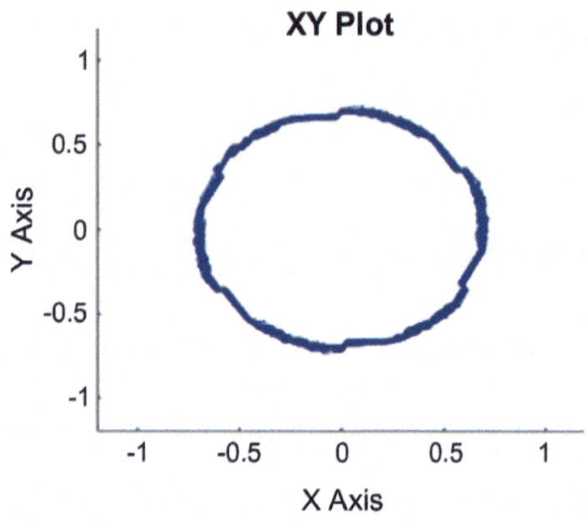

Fig. 8 Flux locus of classical DTC

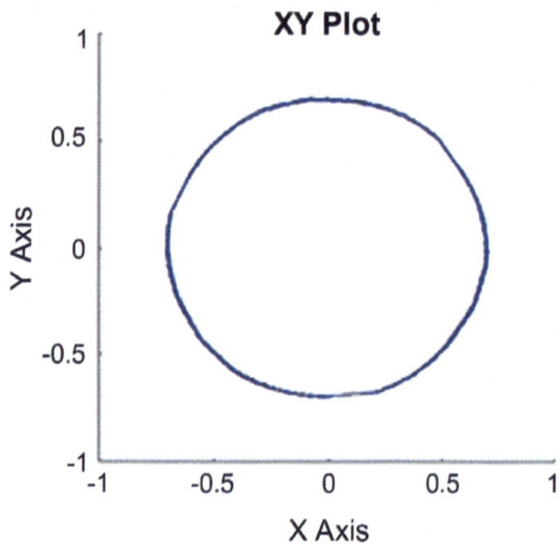

Fig. 9 Flux locus of SVM-DTC

Effect of change in radiation is seen in current and power of PV array and motor parameters, i.e. when the radiation is decreased from 800 to 500 wb/m^2 the *I*pv and *P*pv are also decreased which also lead to decrease in torque, speed, and current of motor drive and when the radiation is increased from 500 to 800 wb/m^2, current *I*pv and power P_{pv} of PV array is increased and same is seen in torque, speed, and current of motor drive parameters Figs. 15 and 16, whereas change in V_{pv} and DC-link voltage, V_{DC} is almost negligible in both the cases.

7 Conclusion

In this paper, simplified SVM-DTC with new switching scheme that reduces the complexity of identifying the sector and the angle of the voltage vector for switching

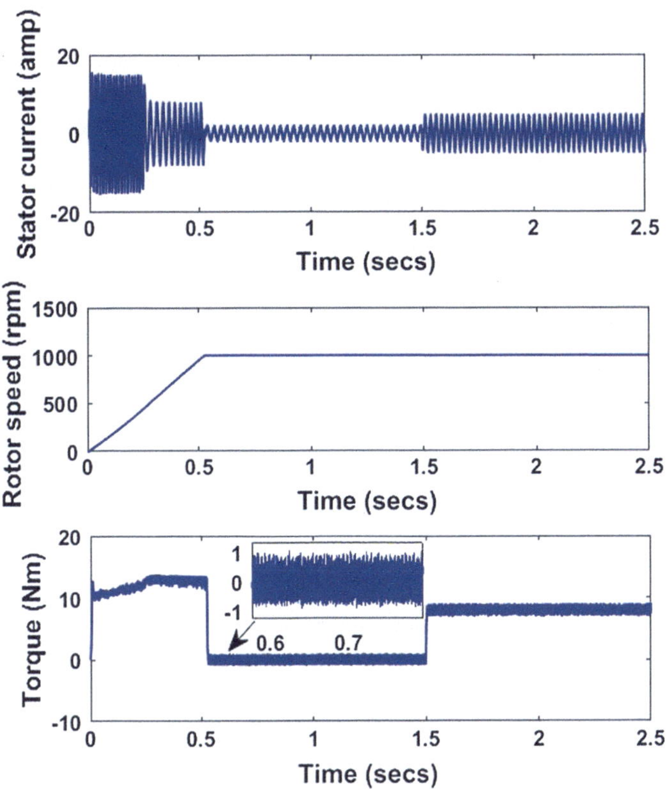

Fig. 10 Torque, speed, and current response of classical DTC

the inverter is proposed. Simplified SVM-DTC in combination with INC-MPPT-based PV-powered induction motor for water pumping system has reduced harmonics distortion, that helps in power saving and improves the efficiency further with less complexity and less burden on processor. The faster settling time and less torque ripples are the added advantage of the proposed SVM-DTC method as applied to induction motor control.

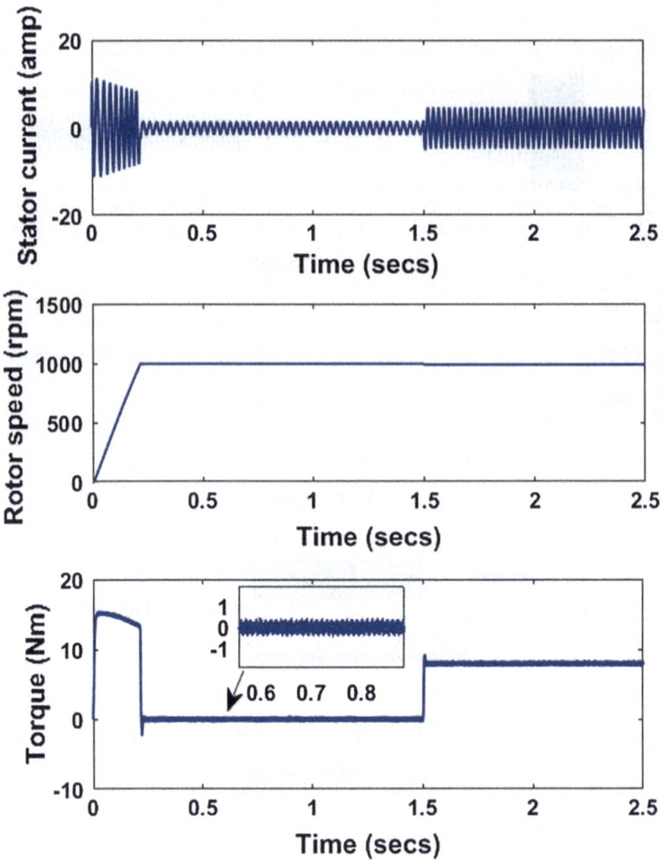

Fig. 11 Torque, speed, and current response of SVM-DTC

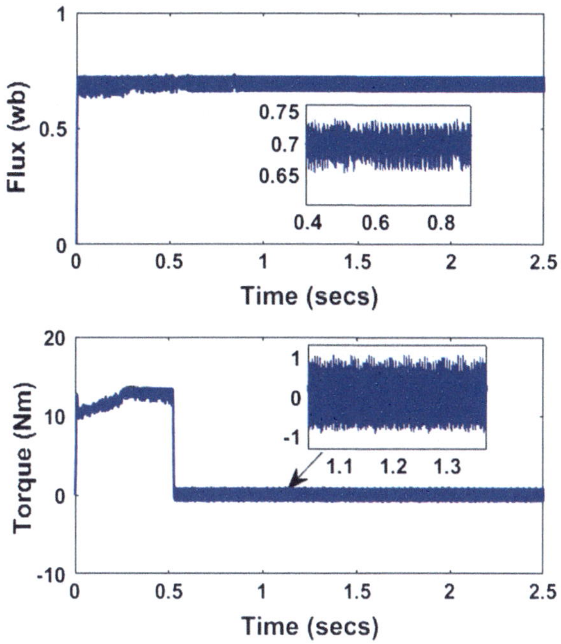

Fig. 12 Estimated torque and flux for classical DTC

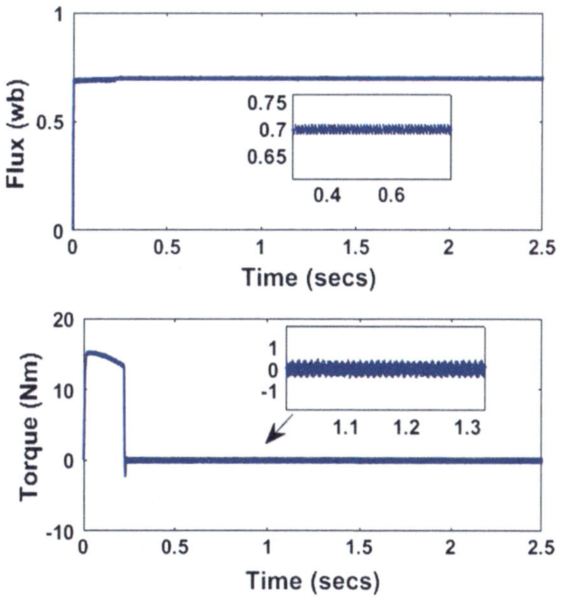

Fig. 13 Estimated torque and flux for SVM-DTC

Table 2 Comparison table

Method	Parameter	Settling time (s)	Ripple
Classical DTC	Torque	0.5232	1.8 Nm
	Flux	0.5236	0.08 wb
	Speed	0.5244	
Improved SVM-DTC	Torque	0.2250	0.8 Nm
	Flux	0.2247	0.01 wb
	Speed	0.2238	

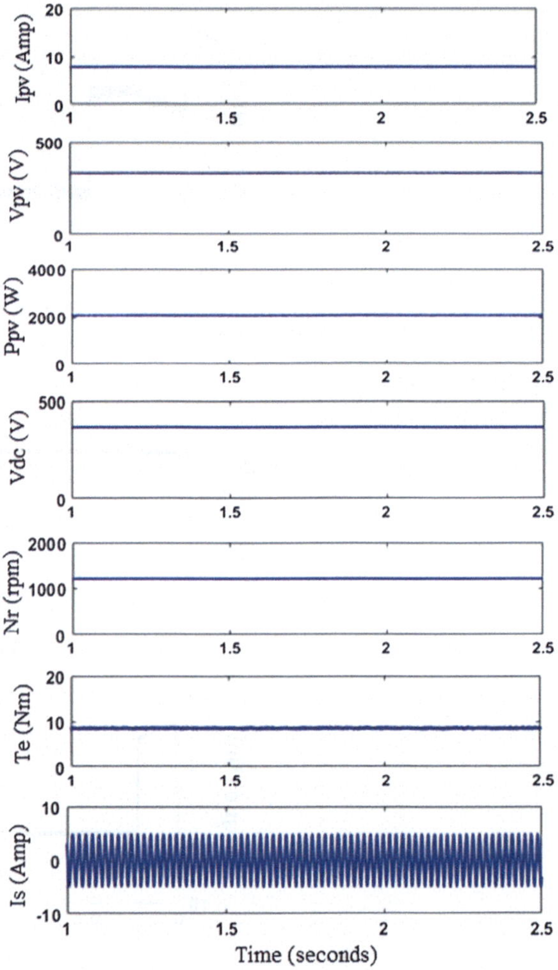

Fig. 14 Steady state performance insolation performance

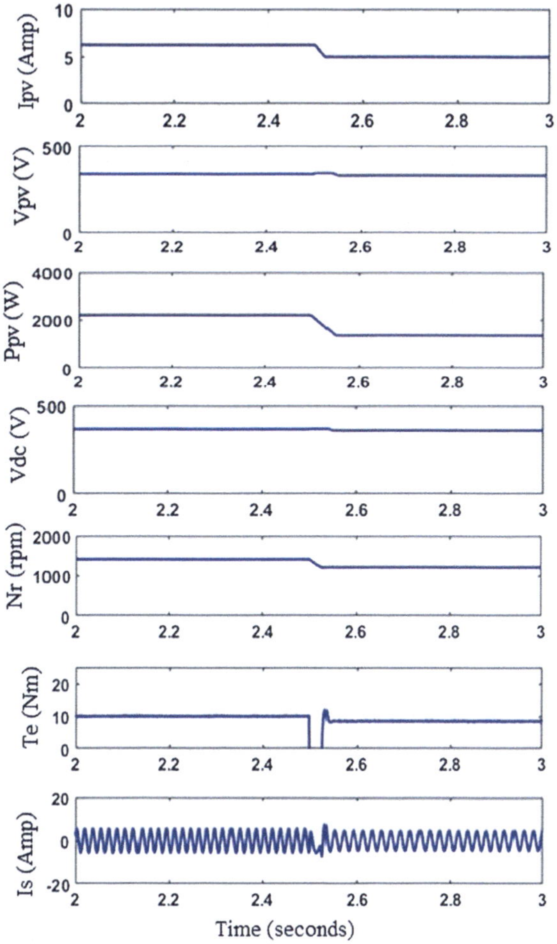

Fig. 15 Decrease in 800–500 wb/m^2

Fig. 16 Increase in insolation performance 500–800 wb/m^2

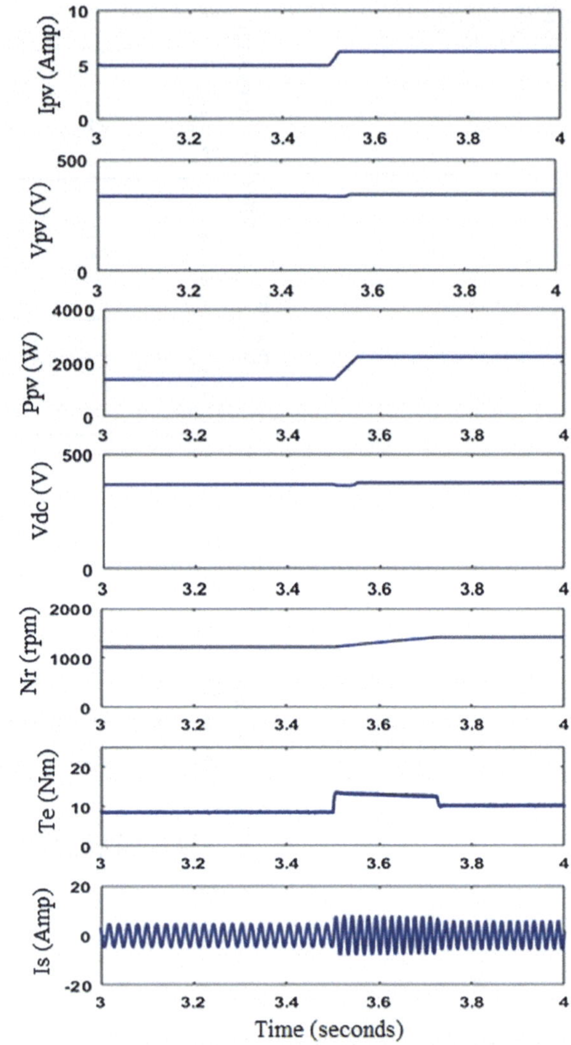

References

1. Blaabjerg F, Chen Z, Kjaer SB (2004) Power electronics as efficient interface in dispersed power generation systems. IEEE Trans Power Electron 19(5):1184–1194
2. Petrone G, Spangnuolo G, Teodorescu R, Veerachary M, Vitelli M (2008) Reliability issues in photovoltaic power processing systems. IEEE Trans Ind Electron 55(7):2569–2580
3. Esram T, Chapman PL (2007) Comparison of photovoltaic array maximum power point tracking techniques. IEEE Trans Energy Convers 22(2):439–449
4. Xiao W, Ozog N, Dunford WG (2007) Topology study of photovoltaic interface for maximum power point tracking. IEEE Trans Ind Electron 54(3):1696–1704
5. Mutoh N, Ohno M, Inoue T (2006) A Method for MPPT control while searching for parameters corresponding to weather conditions for PV generation systems. IEEE Trans Ind Electron

53(4):1055–1065
6. Mokeddem A, Midoun A, Kadri D, Hiadsi S, Raja IA (2011) Performance of a directly- coupled PV water pumping system. Energy Convers Manag 52(10):3089–3095
7. Kolhe M, Joshi JC, Kothari DP (2014) Performance analysis of a directly coupled photovoltaic water-pumping systems. IEEE Trans Energy Convers 19(3):813–818
8. Ramya K, Reddy SR (2005) Design and simulation of a Photovoltaic induction motor water pumping system. In: Proceedings of the international conference on computing, electronics and electrical technologies, Kumaracoil, India, 21–22 Mar 2005, pp 32–39
9. Kumar A, Fernandez BG, Chatterjee K (2004) Simplified SVPWM-DTC of three phase induction motor using the concept of imaginary switching times. In: Proceedings of the 30th annual conference of IEEE industrial electronics society, Bussan, Korea, 2–6 Nov 2004, pp 341–346
10. Takahasi I, Noguchi T (1986) A new quick response and high efficiency control of an induction motor. IEEE Trans Ind Appl 1A-22(5):820–827
11. Zhang J, Rahman MF (2005) Analysis and design of a novel direct flux control scheme for induction machine. In: Proceedings of the IEMDC 2005-IEEE international electric machines and drives conference, San Antonio, USA, 15–18 May 2005, pp 426–430
12. Behera RK, Das SP (2008) Improved direct torque control of induction motor with dither injection. Sadhana 33(5):551–564
13. Depenbrock M (1988) Direct self control of inverter-fed induction machines. IEEE Trans Power Electron 3(4):420–429
14. Chung DW, Kim JS, Sul SK (1998) Unified voltage modulation technique for real time three phase power conversion. IEEE Trans Ind Appl 34(2):374–380

Artificial Neural Network-Based Battery Energy Storage System for Electrical Vehicle

Neha Kumari and Vani Bhargava

Abstract In this review paper, we studied and implementing techniques of HESS for EV. It is very important to perform EV. This is well performed in matlab simulation on EV For HESS consists Li-ion batteries, sup-cap, to charging and discharging the EV. Main controlling system is designed using ANN to performing to get their result by using PI controller. It reduces the calculation complexity of system by reducing the values of proportional constant and integral constant calculations. The neural network promotes self-learning of system and reduces fluctuation if any. An innovation for down to energy is the capacity to store any energy for short time and recover the energy.

Keywords Energy storage system · PI controller · Artificial neural network

1 Introduction

By petroleum product, energy sources by various products, new sources will be ceaselessly developed. The snappy direct of each section was surveyed. A short framework of the motor execution at different atmosphere and weight conditions was outfitted together with a portrayal of the effect of intensity instabilities on the battery current and on the power exchange with the system [1]. Two sorts of HESS are looked in this examination; one using vitality-thick lithium particles batteries joined with ultracapacitors and the other using vitality-thick lithium particle batteries coordinated with ultraground-breaking batteries. These two frameworks are taken a gender at against a control framework using just batteries. Three sizes of each framework are resolved with identical volume in each size. They are taken a gender at for vitality stockpiling, vitality capability, vehicle range, and mass and relative intrigue change when reproduced for controlling a model Pacifica through each of five distinctive drive cycles [2]. The outcomes demonstrated that the current drawn from the battery current in the scope of rates has been diminished over it. When utilizing the proposed system for indicated driving cycles, in spite of the fact that this decrease in little it very well may be noteworthy if thinking about the transport is working entire day

N. Kumari (✉) · V. Bhargava
Ajay Kumar Garg Engineering College, Ghaziabad, UP, India

S. N. Singh et al. (eds.), *Advances in Power and Control Engineering*, Lecture Notes in Electrical Engineering 609, https://doi.org/10.1007/978-981-15-0313-9_14

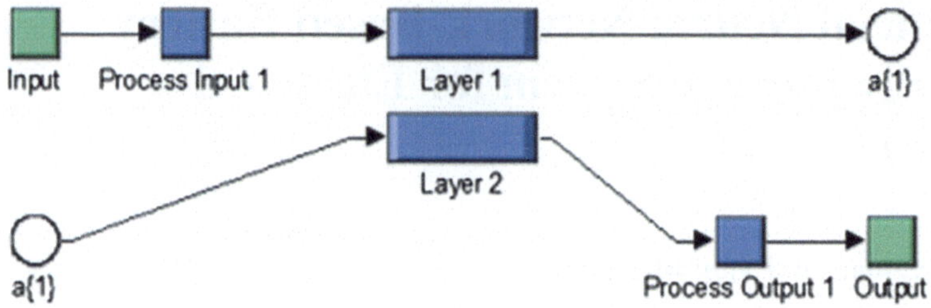

Fig. 1 Artificial neural network layer

[3]. In solicitation to improve the headway capability, a vectorized fluffy gathering framework which licensed enormous size of fluffy rationale controllers working in parallel is planned. At long last, the Pareto perfect game plans of different HESS are obtained and diverged from exhibiting the accomplished enhancements of the bi-level perfect evaluating and vitality the official structure [4]. The ultracapacitor module will ingest the unforeseen weight assortments and purpose of containment the battery management necessities by maintaining an ideal power balance between the two embedded capability units. The execution of the agreement vitality based mostly controller is differentiated and also the customary PI controller and its ability to accomplish good transient execution is thoroughbred [5].

In Fig. 1, it shows that in human mind, data is stored so as to be circulated, and we can extract more than one bit of this data when essential from our memory in parallel. We are not mixed up when we state that a human mind is comprised of thousands of powerful parallel processors. In multi-layer artificial neural networks, there are likewise neurons placed along these lines to the human mind. Each neuron is associated with different neurons with specific coefficients. Amid preparing, data is disseminated to these association focuses with the goal that the network is found out.

2 ANN Subsystem

The network design has an info layer, concealed layer (there can be multiple), and the yield layer. It is likewise called multi-layer perceptron (MLP) in light of the different layers. The shrouded layer can be viewed as a "refining layer" that distills a portion of the significant examples from the sources of info and passes it onto the following layer to see. It makes the network quicker and productive by recognizing just the significant data from the sources of info forgetting the repetitive data (Fig. 2).

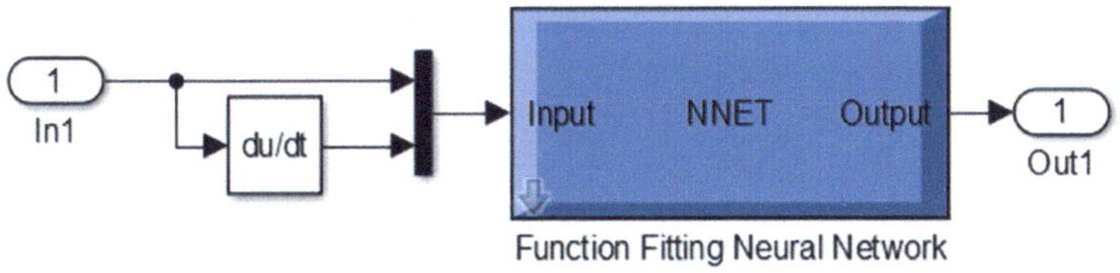

Fig. 2 ANN subsystem

2.1 *Advantages of ANN*

- Storing data on the whole network: Information, for example, in customary writing computer programs is stored on the whole network, not on a database.
- Ability to work with inadequate learning: After ANN preparing, the information may deliver yield even with deficient data.
- This paper gives an example where a multilayered feed-forward back-propagation neural network is trained offline to perform as a controller for a temperature control system with no a priori knowledge regard.

Table 1 shows the comparison table of THD of PI controller and an ANN. Here we can see the different parameter values in THD in PI techniques and THD in ANN techniques. And the parameters are battery current, ultracapacitive current, and load voltage DC motor values which are shown in table. The low value we have got in *v* is compared to THD in PI techniques.

3 Implementation and Results

Figure 3 shows the hybrid energy storage system applied to electric vehicles in PI model, it consists of a bridge rectifier super-capacitor battery and sub system, and here we are adding the proportional integral controller.

Figure 4 shows that where V_{dc} and $V_{\text{dc-sen}}$ are, respectively, the actual voltage and rated voltage of DC motor, i_{UC}^{*} and $i_{\text{UC}-\text{sen}}^{*}$ are, respectively, the per unit of super-capacitor actual current and rated current; f_{s} is the switching frequency; $G_{1,2}$ are the switching signal of T_1 and T_2.

Figure 5 shows the proposed hybrid energy storage system applied to electric vehicles in ANN; here we are using the artificial neural network; the ANN has many

Table 1 Comparison of THD of PI controller and ANN

Signal name	THD PI value	THD ANN value
IB	109.85	80.59
IUC	76.48	66
VDC	80.31	77.13

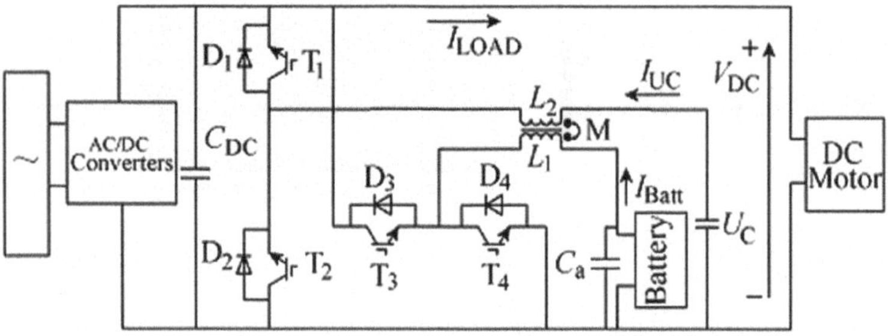

Fig. 3 Hybrid energy storage system applied to electric vehicles in PI model

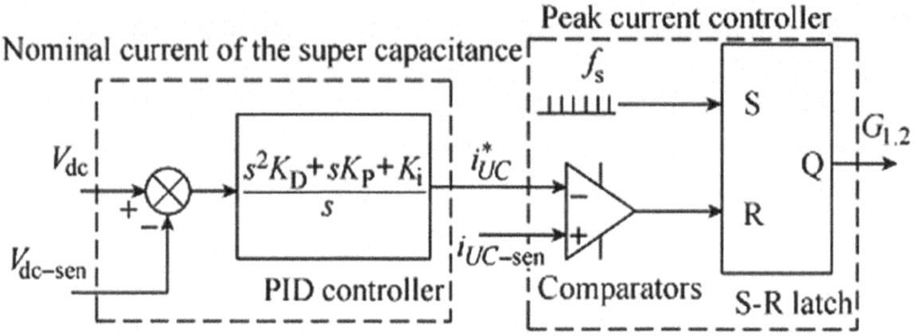

Fig. 4 Block diagram of controllers

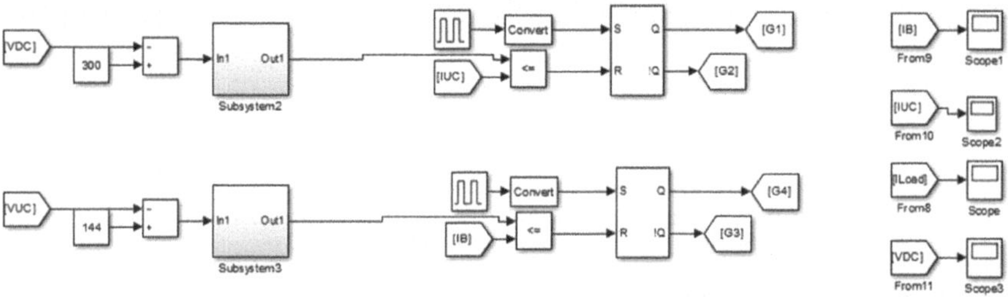

Fig. 5 Proposed hybrid energy storage system applied to electric vehicles in ANN

advantages compared to PI as we already discussed above. The main advantage of artificial neural network compare to proportional integral can handle large amount data sets and it has the ability to the implicitly detect nonlinear relationship between the dependent and independent values.

3.1 Results

Figure 6 shows the battery current of proposed system; the battery current changes are smoother with no instantaneous perturbations. We can see in this graph the output current of the battery pack is smooth and has minimal ripple content.

Figure 7 shows the super-capacitor current of proposed system. Compared to battery current, the super-capacitor current is having high ripple current.

Figure 8 shows the load voltage of proposed system, the load voltage having small amount fluctuation.

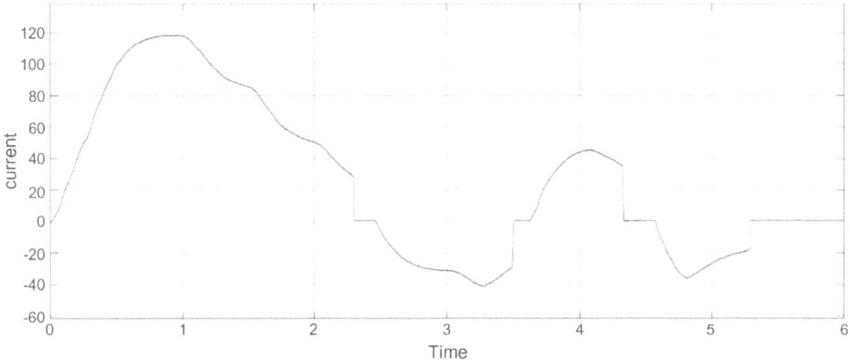

Fig. 6 Battery current proposed system

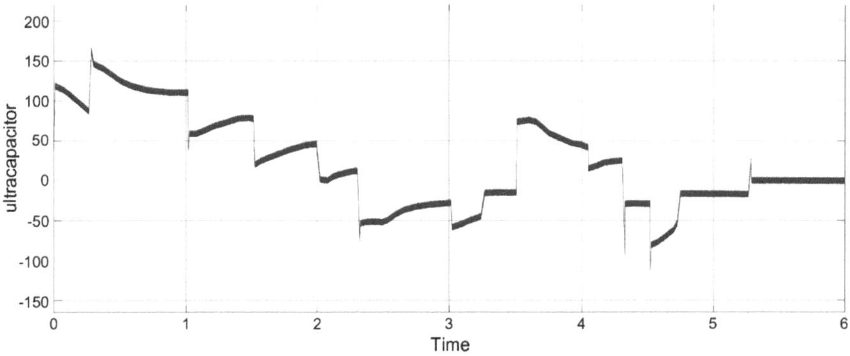

Fig. 7 Super-capacitor current of HESS applied on electric vehicles

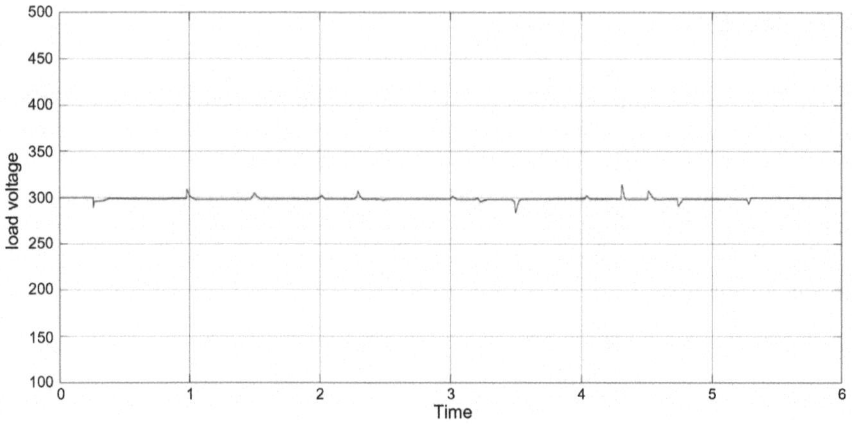

Fig. 8 Load voltage of HESS applied on electric vehicles

4 Conclusion

Hence, we are studied and implemented HESS in EV using ANN. Hybrid energy storage system is plan depending on lithium-ion battery and DC/DC converter with artificial neural network (ANN). In the system is contracted with HESS, it have crucial standpoint of counts and give Voltage, Current to applied Electric Vehicle, and thusly the battery life is improved. It is easily seen in this, location of thyristors and capacitors, battery and converter mostly suitable management, the system of HESS and ability selection must be considering for a long life cycle, instead the controlling terminology are created like topology when a HESS and once the electric battery cycle is known as the HESS to be picked up to controlling techniques and give equipment to product it. HESS an investigated one in that work to locate the most appropriate answers for their application.

References

1. Barelli L, Bidini G (2018) Dynamic analysis of a hybrid energy storage system coupled to a photovoltaic plant. Energies 11:396. https://doi.org/10.3390/en11020396
2. Zhang X (2010) A multi hybrid energy storage system for hybrid electric vehicle. World Electr Veh J 4:1–6. EVS25 Shenzhen, China, Nov 5-9, 2010. ISSN 2032-6653
3. Yu H et al (2009) Multi-objective optimal sizing and energy management of hybrid energy storage system for electric vehicles. Cao, Member IEEE, Wang F-Y, Fellow IEEE, Mar 2009
4. Karl BA et al (2010) Design and evaluation of hybrid energy storage systems for electric power train Waterloo, Ontario, Canada
5. Anderssona T (1992) Alternative energy storage system for hybrid electric vehicles. IEEE AES Mag:14–19

Data Communication Between DC Microgrids for Real-Time Converter Control

Victor George, Pradipkumar Dixit, Ayushi Priya, Deeksha Singh, Tanya Swaroop and D. Nikita

Abstract The scope of DC microgrids using renewable resources includes electric vehicle charging stations because of the increased conversion efficiencies caused from the reduced power converter stages between the DC source and DC load. Reduction in carbon emission is the common goal between DC microgrid and the electric vehicle DC fast-charging stations. Proper control of DC/DC converters is essential to transfer power between autonomous microgrids in order to exploit maximum available renewable power. An IoT-based real-time data acquisition is proposed to generate the necessary reference voltages for the respective converters for the efficient and smooth transfer of excess solar power. ThingSpeak cloud platform and MATLAB/SIMULINK software are used to design the proposed control mechanism for the energy transfer.

Keywords DC/DC converter · DC microgrid · Fast-charging stations · IoT · Real-time data acquisition

1 Introduction

Energy market is shifting its focus toward the efficient generation and utilization of renewable energy resources. Power conversion losses can be minimized by feeding DC loads with DC source, thereby increasing the overall efficiency [1]. Maximum utilization of renewable sources with minimum use of battery storage is a challenging task. The intermittent nature of solar energy and the dynamic loads like electric vehicles (EV) at the DC charging stations necessitates the real-time energy transfer between microgrids. This can reduce the dependence of the AC grid even when the available solar power at a DC microgrid is insufficient to meet its local demand. To have satisfactory stability and efficiency, a proper control system is required in a microgrid. There are several centralized, decentralized, distributed, and hierarchical methods which have been proposed to obtain prolonged and reliable operation of DC

V. George (✉) · P. Dixit · A. Priya · D. Singh · T. Swaroop · D. Nikita
Department of Electrical and Electronics Engineering, M S Ramiah Institute of Technology, Bangalore, India
e-mail: victorgeorge@msrit.edu

S. N. Singh et al. (eds.), *Advances in Power and Control Engineering*, Lecture Notes in Electrical Engineering 609, https://doi.org/10.1007/978-981-15-0313-9_15

microgrid [2–4]. The hierarchy of the control system of microgrid consists of primary, secondary, and tertiary levels [5]. Although the centralized controller provides optimal operation of the microgrids by managing the data centrally, system reliability is the main concern due to the single point failure probability. Droop control is the widely used decentralized control method which can ensure proportional current sharing between multiple sources of microgrid [6]. Different voltage and current control modes based on DC voltage deviation have been proposed in [7] for regulating the DC microgrid under variable voltage generation and control. Autonomous power control and management techniques are proposed in [4, 8] to synchronize energy distribution and management between DC microgrids without any dedicated communication infrastructure. Cloud-based data collection for generating a reference voltage to control the energy transfer is yet to be analyzed. Also such a setup is suitable for the electric vehicle charging stations. DC fast-charging stations can be powered from solar pv generators with suitable DC/DC converters. Grid support will increase the reliability of such an arrangement. Hence an attempt is made to generate the reference voltages for various digitally controllable DC/DC converters and communicate those values to the respective converters through a cloud platform. The proposed structure of DC microgrids and the interconnection with local AC grids through AC/DC converters is well suited for controlled solar energy transfer between DC microgrids based on the available solar energy and the possible loads at the local level.

2 Structure of the Proposed System

2.1 General Structure of Interconnected Microgrids

The proposed model consists of three DC microgrids interconnected through controllable switches. The common DC bus onto which different renewable sources and DC loads connected through power converters. This setup is shown in Fig. 1. The system under consideration consists of DC loads powered by solar pv generators and backed up by the grid. The loads can be electric vehicles, and the DC microgrids are part of the DC fast-charging stations.

The microgrids are normally not connected to each other until an external signal initiates the bidirectional switch. Since the load is dynamic and the solar is intermittent in nature, it is important to have a proper mechanism that can effectively transfer the solar power between microgrids. The various hierarchical control methods can be implemented more efficiently by establishing a communication link between all the converters. Internet of things (IoT) can be used for this purpose.

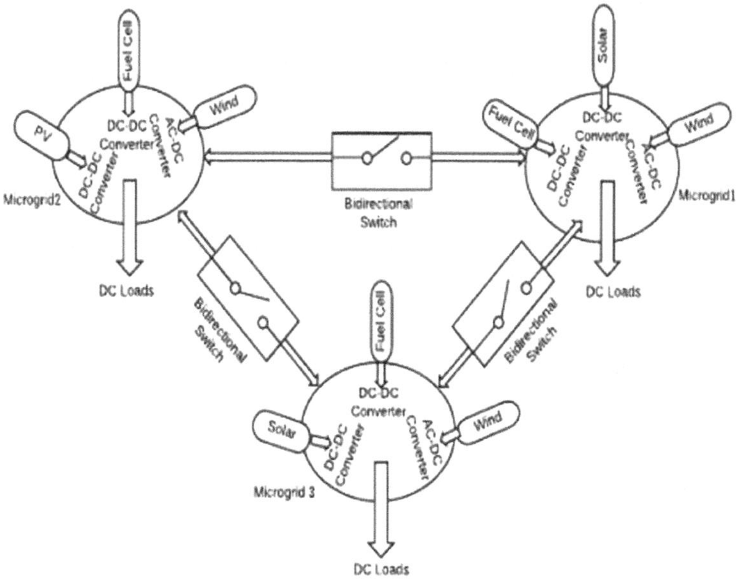

Fig. 1 Interconnection of microgrids through bidirectional switch

2.2 *Specifications of the Experimental Setup*

Each microgrid of the proposed model consists of a battery representing a DC source and a regulated power supply, powered by AC mains, representing the AC grid. The DC loads connected to the microgrids comprise mainly of variable resistances and DC motors. A lead acid battery of rating 12 V, 26 Ah is used to represent the solar generator. The voltage range of the regulated power supply used is 0–30 V. The rating of the rheostat used in the proposed work is 50 Ω, 5 A. Due to the practical difficulties in getting controllable DC/DC converters , SIMULINK model of boost converters is used to test the effect of calculated values of V_{refn} in controlling the output voltage of the converter.

3 Proposed Control Strategies for Converters

IoT can be incorporated in electrical power system to constantly monitor the power requirement of microgrids, thereby identifying the microgrids having excess and deficit power at any point of time.

Different IoT platforms can be used to have real-time data monitoring and transfer. ThingSpeak is chosen for the proposed work as it is an open IoT platform with MATLAB analytics. The real-time data acquired by different sensors will be sent to an IoT platform (ThingSpeak) via Wi-Fi computer module. The data from the cloud can then be sent to the local controller for analysis and control.

The situations where few microgrids are having excess power and few are having deficit power can be handled efficiently by transferring power from the microgrid having excess power to the one having deficit power by controlling bidirectional switches. The deciding criteria that are being considered in this proposed work for selecting microgrids for transferring power between them are the cable losses.

The current through each source is measured by connecting a voltage sensor across a 1 Ω resistor, which is connected in series with the source. After identifying the proper microgrids to transfer energy, it is proposed to increase the voltage reference of the converter having excess power in steps depending on the minimum possible step size of the converter.

4 Data Sensing and Cloud Communication

The current values from each source of the microgrid as well as the load voltage are sensed with the help of voltage sensor. These values are sent to the controller (Node MCU). The sensed data that is available on the controller is sent to the IoT application ThingSpeak where three channels have been created and each channel represents one microgrid. There are three fields in each channel. Field 1 and 2 represents source currents and field 3 represents load voltage. The acquired data is shown in the form of plots with date and time and it keeps on updating at the interval of 15 s each. Same notation has been followed for other microgrids as well. The current and voltage values from the cloud are retrieved at the MATLAB software where it is further analyzed to calculate the reference voltage for converter control. Circuit connections used for data sensing is shown in Fig. 2. The algorithm for calculation of reference voltage of the converter is given below:

- Retrieve the individual source current (I_s) and load voltage (V_L) values.
- The consecutive fields of a particular channel of ThingSpeak get updated every 15 s. Hence, there is a need for synchronization so that continuous current and voltage values can be obtained. This is done by interpolating the values between two consecutive readings.
- Total current (I_t) from each source of a particular microgrid is calculated as $I_{t1} = I_{s1} + I_{s2}$.
- Load power is calculated as $P_{L1} = V_{L1} * I_{t1}$.
- Excess/deficit power is calculated as the difference between available power and load power in which available power is assumed to be present at the cloud.
- The number of microgrids having excess and deficit power is found out.
- Cable losses are readily obtained and considered for effective power transfer.
- Switching between the microgrid is done by considering four cases:

  i. *Three microgrids have excess power*

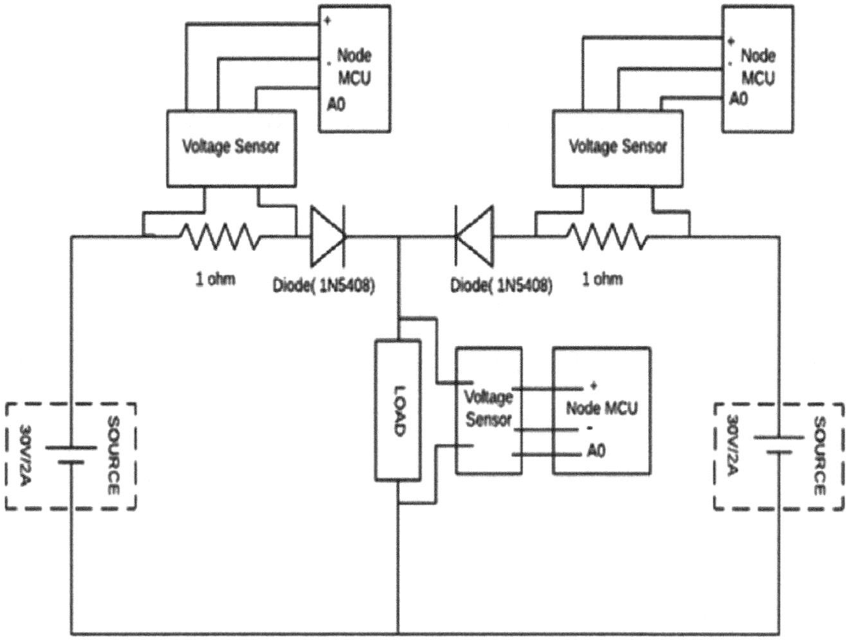

Fig. 2 Circuit diagram for data sensing in an individual microgrid

ii. *Two microgrids have excess power and one has deficit power*
iii. *One microgrid has excess power and two have deficit power*
iv. *All the three microgrids have deficit power.*

- The microgrid pair with lesser cable losses between them are selected for power transfer.
- Now, the reference voltage (V_{ref}) is calculated by increasing the voltage of the excess microgrid in steps till it exceeds that of deficit microgrid ($V_{\text{ref}} = V_{\text{excess}} +$ step value). The step size can be decided from the resolution of the converter. The level of voltage and current is ensured all the time well within the allowable range before increasing the step size of the V_{ref}.
- The calculated reference voltage is sent to the SIMULINK model of boost converter where V_{out} (output voltage of the converter) is calculated and sent back to MATLAB. The details of excess and deficit powered microgrid and V_{out} will then be sent to a channel of the cloud (ThingSpeak). The data is then retrieved back from that channel of the cloud to the Wi-Fi controller for switching the relay between the selected microgrids.

5 Experiment Studies

The values of individual source current and load voltage of each microgrid is sensed and validated using appropriate meters. The sensed data are then sent to a channel on ThingSpeak. The output is obtained in the form of plots on ThingSpeak. The current and voltage data of each microgrid from its respective channel has been retrieved at MATLAB. The retrieved values are synchronized by the method of interpolation. Fig. 3 explains the idea for the generation of duty cycle. The table obtained after synchronization of current values is shown in Fig. 4. The calculated value of V_{ref} is used for simulation of boost converter model in SIMULINK to calculate the output voltage. The output voltage of the converter is compared with the reference voltage calculated in the MATLAB code and an error signal is generated, given as $V_{error} = V_{ref} - V_{out}$. This error signal is fed to a PI controller for stabilization and later compared with the saw tooth waveform to generate the control pulse. This generated duty cycle waveform is fed to the switch (MOSFET) of the converter for generating the desired output voltage. The obtained output voltage is retrieved in MATLAB. The screenshots of the same has been taken and shown in Fig. 5. The results obtained in the command window of MATLAB are shown in Fig. 6.

The complete experimental setup used to test the proposed algorithm to generate the reference voltages for individual converters is shown in Fig. 7. The data flow diagram of the developed model is shown in Fig. 8.

The index number of excess and deficit microgrid, their respective voltages and the output voltage value will be sent to ThingSpeak. The same values are retrieved at a local controller using Wi-Fi controller module. Arduino IDE is used

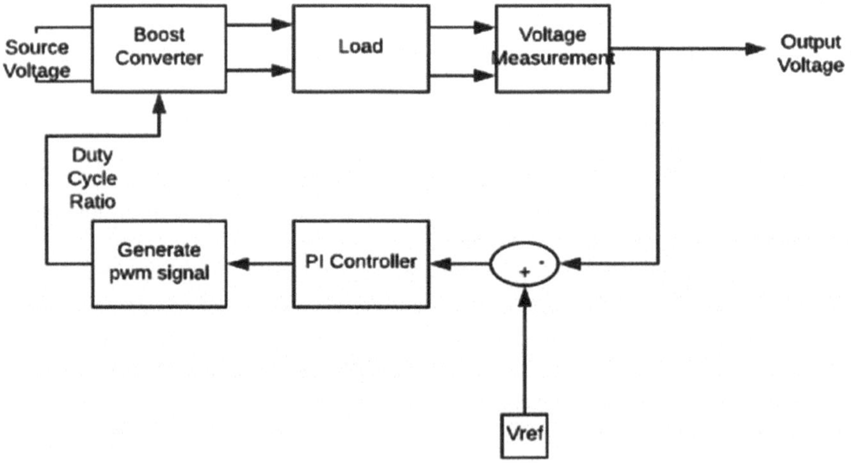

Fig. 3 Block diagram for generation of duty cycle

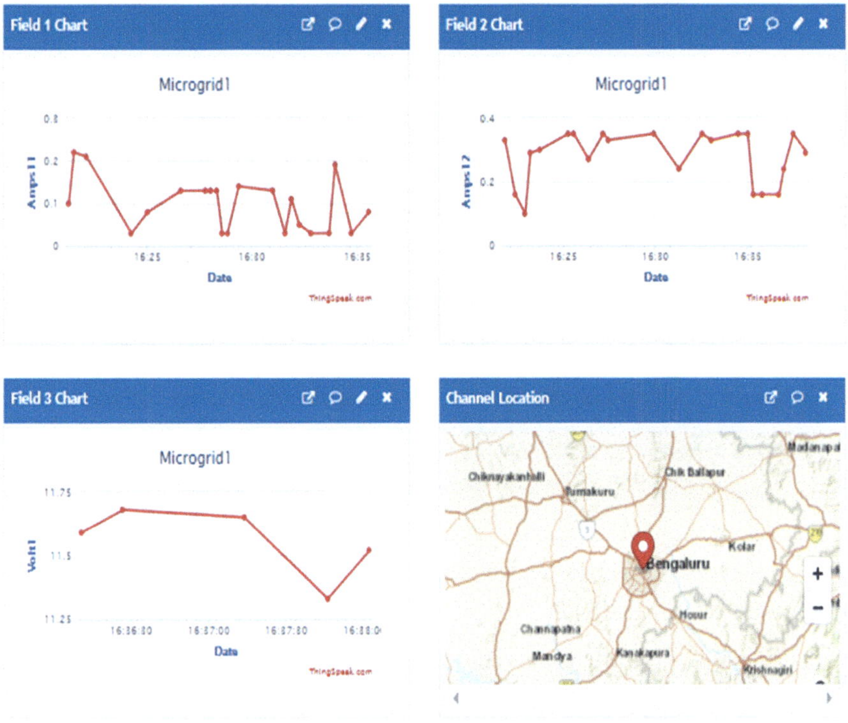

Fig. 4 Plots of source currents and load voltage of microgrid 1

for checking the condition for transfer of power. The power transfer cases obtained after the implementation of the proposed control methodology is summarized in Tables 1 and 2.

The real-time voltage control circuit developed for the interconnected DC-DC converters can be useful in controlled energy transfer between selected microgrids. It is still possible to have a proportional current sharing within each microgrid. The control circuit is successfully implemented with the help of wireless controllers and IoT-based data communication. The developed data acquisition system is found to be efficiently transferring the various voltage and current values to the cloud platform 'ThingSpeak.' The reference voltage generated with the help of MATLAB software is applied to the SIMULINK model of the DC-DC boost converter. The reference voltage is then increased step by step till it reaches the pre-calculated voltage level ensuring the energy transfer from the microgrid where excess power is available to that of deficit in power. The reference voltage can be applied directly to a digitally controlled DC/DC converter without the MATLAB interference. The developed mechanism is successfully operated the relay switches after setting the desired voltage level at the sending and receiving microgrids.

Timestamps	Amps31	Amps32
18-May-2019 16:53:25	0.46	0.67788
18-May-2019 16:53:43	0.4576	0.67242
18-May-2019 16:54:03	0.45493	0.66636
18-May-2019 16:54:20	0.45267	0.66121
18-May-2019 16:54:40	0.45	0.65515
18-May-2019 16:54:57	0.45	0.65
18-May-2019 16:55:15	0.45	0.64455
18-May-2019 16:55:30	0.45	0.64
18-May-2019 16:55:48	0.45	0.64
18-May-2019 16:56:06	0.45	0.64
18-May-2019 16:56:22	0.45	0.64
18-May-2019 16:56:37	0.45	0.64227
18-May-2019 16:56:54	0.45	0.64485
18-May-2019 16:57:12	0.45	0.64758
18-May-2019 16:57:28	0.45	0.65

Fig. 5 Synchronized current values

```
Output_Voltage
  timeseries with properties:

            Events: []
              Name: ''
          UserData: []
              Data: 13.0971
          DataInfo: [1×1 tsdata.datametadata]
              Time: 1
          TimeInfo: [1×1 tsdata.timemetadata]
           Quality: []
       QualityInfo: [1×1 tsdata.qualmetadata]
       IsTimeFirst: 1
  TreatNaNasMissing: 1
            Length: 1
```

Fig. 6 Output voltage of boost converter

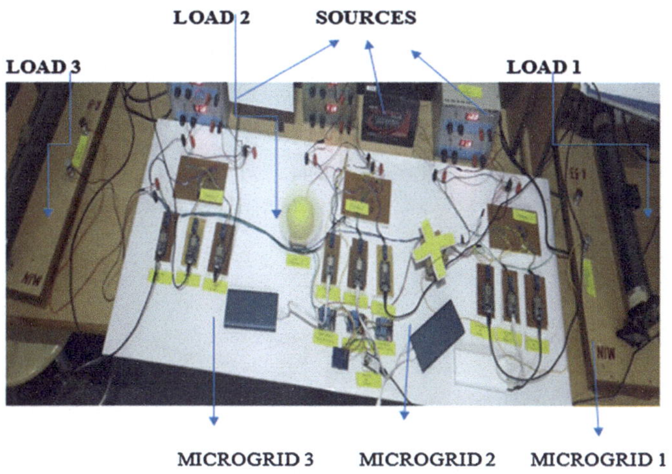

Fig. 7 Circuit connection of the setup

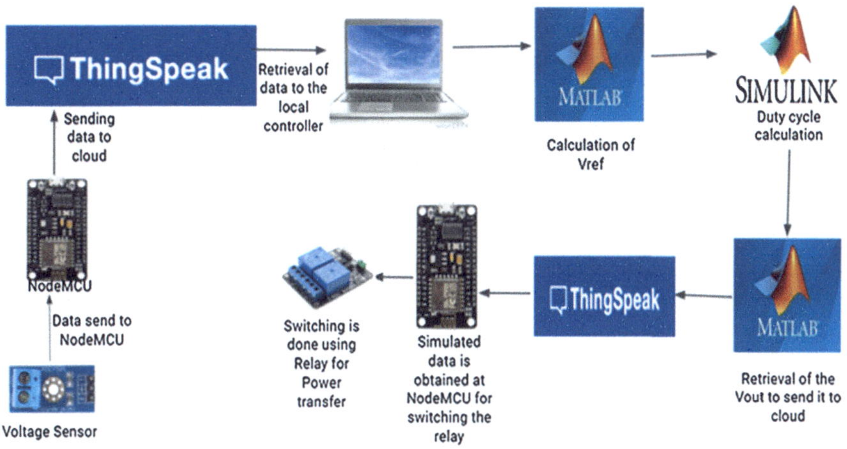

Fig. 8 Overall flow of implementation of the proposed model

Table 1 Observation of Case 2 for power transfer

Microgrid	1	2	3
Source 1 current (A)	0.21	0.23	0.22
Source 2 current (A)	0.21	0.21	0.25
Total current (A)	0.42	0.44	0.47
Load voltage (V)	11.57	11.16	10.57
Load power (P_d) (W)	4.8594	4.9104	4.9679
Available power (P_a) (Assumed) (W)	4	5.2	5.9
Difference ($P_d - P_a$) (W)	−0.8594	0.2896	0.9321
Losses (in W)	0, 0.02112, 0.0528	0.02112, 0, 0.02112	0.0528, 0.02112, 0
Status	Deficit	Excess	Excess
Case Two excess, one deficit	Receives power from second microgrid	Transfers power to first microgrid (less cable losses)	Does not transfer power to first microgrid (more cable losses)

Table 2 Observation of Case 3 for power transfer

Microgrid	1	2	3
Source 1 current (A)	0.21	0.23	0.22
Source 2 current (A)	0.21	0.21	0.25
Total current (A)	0.42	0.44	0.47
Load voltage (V)	11.57	11.16	10.57
Load power (P_d) (W)	4.8594	4.9104	4.9679
Available power (P_a) (Assumed) (W)	5	4	4.2
Difference ($P_d - P_a$) (W)	0.1406	−0.9104	−0.7679
Losses (in W)	0, 0.02112, 0.0528	0.02112, 0, 0.02112	0.0528, 0.02112, 0
Status [excess/deficit]	Excess	Deficit	Deficit
Case One excess, two deficit	Transfers power to second microgrid	Receives power from first microgrid (less cable losses)	Does not receive power from first microgrid (more cable loss)

References

1. Jadhav AB, Pawar SH (2017) Power management of DC micro-grid and AC grid. In: 2017 international conference on innovations in information, embedded and communication systems (ICIIECS), Coimbatore, pp 1–5
2. Khorsandi A, Ashourloo M, Mokhtari H (2014) A decentralized control method for a low-voltage DC microgrid. IEEE Trans Energy Convers 29(4):793–801
3. Dragičević T, Lu X, Vasquez JC, Guerrero JM (2016) DC microgrids—part I: a review of control strategies and stabilization techniques. IEEE Trans Power Electron 31(7):4876–4891
4. Chen D, Xu L, Yao L (2013) DC voltage variation based autonomous control of DC microgrids. IEEE Trans Power Delivery 28(2):637–648
5. Guerrero JM, Vasquez JC, Matas J, de Vicuna LG, Castilla M (2011) Hierarchical control of droop-controlled AC and DC microgrids—a general approach toward standardization. IEEE Trans Ind Electron 58(1):158–172
6. Liu KX, Wang P, Loh PC (2011) A hybrid AC/DC microgrid and its coordination control. IEEE Trans Smart Grid 2(2):278–286
7. Justo JJ, Mwasilu F, Lee J, Jung JW (2013) AC-microgrids versus DC-microgrids with distributed energy resources: a review. Renew Sustain Energy Rev 24:387–405
8. Sanjeev P, Padhy NP, Agarwal P (2018) Autonomous power control and management between standalone DC microgrids. IEEE Trans Ind Inform 14(7):2941–2950

Performance Analysis of Mechanical Rotary System Using Linear Quadratic Regulator Controller

Piyali Das, Ram Krishna Mehta and Om Prakash Roy

Abstract This study explains about the SIMO plant statistics of observer-based control to a rotational electro-mechanical system. A robust linear quadratic regulator controller is proposed here, in this study, to observe all the states after recognizing the linear model. The unknown states were determined by the method and for the tuning of the weighted matrices various methods were applied. This full state feedback controller was able to produce faster and smoother output stabilization, which leads to better robustness. The model is highly nonlinear, and model identification is done by MATLAB simulation procedure.

Keywords Inverted pendulum · Observer · Arithmetic Riccati equation · State feedback · Linear quadratic regulator

1 Introduction

In modern industrial system, gantry cranes as well as rotary cranes are widely used for the heavy loads transfer. Rotational electro-mechanical systems are, nowadays, used to minimize the laborious procedure and sustain a better output. For the anti-sway control of these type of mechanical system like traveling cranes, Gait systems, etc., there are several solutions, such as fuzzy control, optimal control, etc., and each of them is reported to be effective. Inverted pendulum is a classic problem of rotational system in dynamics and control theory to verify different algorithms for best response category. The pendulum zero-swing angle matched to a suspended pendulum in the vertical rest down position [1, 2]. In the used system, an input u (motor voltage), two outputs θ (pendulum angle), and x (arm position angle) are considered. There are various developments in optimization process inspiring by the nature like plants, insects and species some beautiful analogies were developed. These methods were executed in simulation by many researchers using mathematical platforms

P. Das (✉) · R. K. Mehta · O. P. Roy
Department of Electrical Engineering, NERIST, Nirjuli, Arunachal Pradesh 791109, India
e-mail: pd@nerist.ac.in

R. K. Mehta
e-mail: rkm@nerist.ac.in

S. N. Singh et al. (eds.), *Advances in Power and Control Engineering*, Lecture Notes in Electrical Engineering 609, https://doi.org/10.1007/978-981-15-0313-9_16

and became useful to the engineering applications. Among all of them, some of these approaches are based on swarm fish named as particle swarm optimization (PSO) [3], following Darwinian's survival of the fittest and natural evolution, one method has been approached in [1] named as genetic algorithm (GA). Next section describes about the model and formulation of problem and Sects. 2.1–2.4 imply the methodology of the paper and Sect. 3 is briefing about the compared results.

2 Model Description

The model here is used as inverted pendulum. The plant consists of a DC motor in a solid aluminum frame. The motor parts include a gearbox and the output of that drives external gears. The basic unit is equipped with a potentiometer to measure the output/load angular position. The model is designed by M/s FEEDBACK Instruments.

The real model of the pendulum is a nonlinear function, which means out of all the states, an argument of the model is a nonlinear function. From that nonlinear model, a linearized model has been attempted to find for the use of control engineering. The linear model identification is being provided at the result section where it has been seen that the dynamic linearized model representation follows the control law of the model. The input in this model is used as voltage and two outputs are associated with the model, i.e. position and angle movement.

The equation of motion is proposed in (1) where summing all the forces which includes acting on the pendulum as well as cart system.

$$(m + M)\ddot{x} + b\dot{x} + ml\ddot{\theta}\cos\theta - ml\dot{\theta}^2\sin\theta = F \tag{1}$$

$$\left(I + ml^2\right)\ddot{\theta} - mgl\sin\theta + ml\ddot{x}\cos\theta + d\dot{\theta} = 0 \tag{2}$$

The plant characteristics are vastly nonlinear and it is very difficult to control the plant and design a controller for the same.

This type of nonlinear models possesses very small deviation in state to become linearized. This such small deviation is named as equilibrium point. This plant of pendulum has two such points, one is when $\theta = 0$ and the other is when $\theta = \pi$. The first position is for inverted pendulum and the other is when the pendulum is hanging freely at the time of crane is controlled. The state of inverted pendulum some times may become very destructive, for such an instance to carry out simulation tests are very much appropriate before examining in real plant. To validate all the nonlinear model, Table 1 is been considered. The control voltage supply may be bounded −2.5 to +2.5 V and the force is bounded with a magnitude of −20.0 to +20.0 N. The cart position length may be adjusted within −0.5 m to +0.5 m. The model is single input multiple output (SIMO) plant where the translation between the force F and control voltage u is given in Eq. (3).

Table 1 Description of plant parameters [1]

Symbols	Description	Values
g	Gravitational force	9.81 m/s^2
l	Length of the pole	0.39 m
M	Mass of the cart	2.4 kg
m	Mass of the pole	0.23 kg
I	Moment of inertia of the pole	0.099 kg m^2
b	Coefficient of cart friction	0.05 Ns/m
d	Coefficient of pendulum damping	0.005 Nms/rad

$$F = K_{FU}\frac{\mathrm{d}u}{\mathrm{d}t} \tag{3}$$

where K_{FU} may be considered as gain between the first-time derivative of u voltage and the force F (Fig. 1).

2.1 Model Linearization

The nonlinear functions present in the model equation may be linearized as per Taylor approximation for the nonlinear functions. For minimum deviations of an equilibrium point of θ, the following magnitudes may be considered.

$$\sin\theta \cong \theta$$
$$\cos\theta \cong 1$$

$$\dot{\theta}^2 = 0 \tag{4}$$

Both the linear and nonlinear model has one input and two outputs, which are F and θ (angle) and x (displacement), respectively.

$$\begin{aligned} x &= \left[\theta_0\theta_1\dot{\theta}_0\dot{\theta}_1\right] \\ &= [0\;0\;0\;0] \end{aligned} \tag{5}$$

2.2 Model Identification

This identification method may be achieved by the MATLAB least-mean-square identification method. This procedure enhances the modeling of controller. The plant

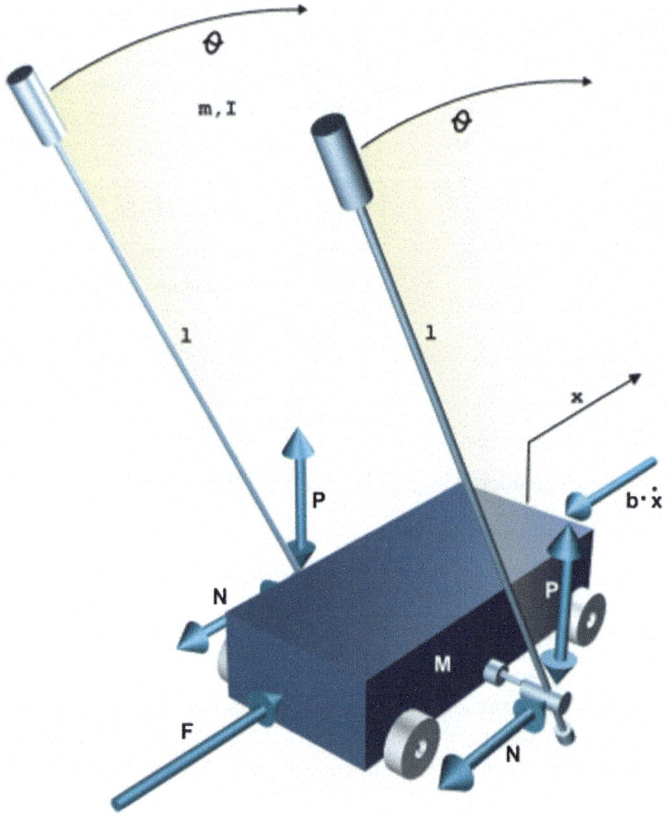

Fig. 1 Pendulum real model

is nonlinear, so to design a controller for such plants is very difficult task. The pendulum may be linearized in an equilibrium point of $\theta = 0°$. The controller should be able to balance on the upright position of the pole. The upright equilibrium point may be denoted from Eq. (5). After considering the linearization using Eq. (4), the model approximation may become as Eq. (6).

$$\begin{bmatrix} M+m & m\frac{l}{2} \\ m\frac{l}{2} & m\left(\frac{l}{2}\right)^2 + J \end{bmatrix}\ddot{\theta} + \begin{bmatrix} 0 & 0 \\ 0 & mg\frac{l}{2} \end{bmatrix}\theta = Hu \tag{6}$$

$$\ddot{\theta} = -L^{-1}G\theta + L^{-1}Hu$$

where, $L = \begin{bmatrix} M+m & m\frac{l}{2} \\ m\frac{l}{2} & m\left(\frac{l}{2}\right)^2 + J \end{bmatrix}$, $G = \begin{bmatrix} 0 & 0 \\ 0 & mg\frac{l}{2} \end{bmatrix}$.

The system equation is provided in (7), and the state-space matrices are elaborated in the next step.

$$\dot{x} = Ax + Bu;$$
$$y = Cx + Du \tag{7}$$

where,

$$A = \begin{bmatrix} 0 & 0 & 1\,0 \\ 0 & 0 & 0\,1 \\ 0 & \frac{gl^2m^2}{l^2Mm+4J(M+m)} & 0\,0 \\ 0 & -\frac{2glm(M+m)}{l^2Mm+4J(M+m)} & 0\,0 \end{bmatrix}; \ B = \begin{bmatrix} 0 \\ 0 \\ \frac{4J+l^2m}{l^2Mm+4J(M+m)} \\ -\frac{2lm}{l^2Mm+4J(M+m)} \end{bmatrix};$$

$$C = \begin{bmatrix} 1\,0\,0\,0 \\ 0\,1\,0\,0 \end{bmatrix}; \ D = \begin{bmatrix} 0\,0 \\ 0\,0 \end{bmatrix}.$$

These state-space matrices are found using the linearization method used by Kafetzis and Moysis in [2].

2.3 Controller Design

After plant identification the controller may be decided to stabilize such complex and nonlinear model. For pendulum control, a linear quadratic regulator has been opted as a controller. The simulation and real-time models are shown in the next section. Now, for the cart control and crane stabilization, similar controller may be witnessed in the next sections. As an introduction to the linear quadratic regulators, the classical optimal control may be formulated to the reknown state-feedback controller, which reduces the divergence in state trajectories of a system maintaining a minimum cost effect.

The LQR controller is used to obtain the optimal performance of the system by minimizing the cost function that relates the state vector and control input vector. Using the optimal control logic of quadratic controller, state-feedback gain is determined by the solutions of ARE [4]. There is been a great challenge to design an optimal controller using LQR due to the choice of Q and R matrices. One best solution to the problem may be Bass-Gura method [5], which produces a desired result. To select Q and R matrices, a very fast response may be observed from [3] where analytical method was proposed to select the values. A simple and modular LQR was proposed in [1] where the performance objectives were translated in time domain into cost function. The main objective of the controller is to set the cart position in a desired value even the arm angle and pendulum angle also remain as the original position [6, 7] (Fig. 2).

Here, we have considered a linear time-invariant multivariable system

$$\begin{aligned} \dot{x}(t) &= Ax(t) + Bu(t), \ t \geq 0, \ x(0) = x_0 \\ y(t) &= Cx(t) + D(t), \ t \geq 0 \end{aligned} \tag{8}$$

where $A \in R^{n \times n}$, $B \in R^{n \times m}$, $C \in R^{p \times n}$, $D \in R^{p \times m}$ are system, input, output and feed-forward values, respectively, the state vector is x, the control input vector is u, and output vector is y. The conventional LQR problem is to determine a minimum cost function, which is given in Eq. (9) to find the control input u.

$$J(u) = \int_{0}^{\infty} [x^T(t)Qx(t) + u^T(t)Ru(t)]\mathrm{d}t \tag{9}$$

The equilibrium of the departure system states may be driven by weighted matrix Q and $R = R^T$ is a positive definite matrix that penalizes the control input.

$$K = R^{-1}B^T P \tag{10}$$

Generally, solving the following algebraic Riccati equation (ARE) gain matrix K of LQR controller may be determined.

$$A^T + PPA + Q - PBR^{-1}B^T P = 0 \tag{11}$$

where $P \in R^{n \times m}$ is a solution of the equation ARE. The compositions of Q and R elements have great influences on system performance. The elements of Q and R matrices depend on the number of state variable associated with the plant and the number of input variable, respectively. When the weighting matrices are diagonal matrices, then performance index of quadratic regulator is nothing but squared error of weighted integral of the states and inputs [4].

2.4 Pole-Placement Method

Let the system is given in Eq. (1) and the control vector U is designed in the following state-feedback form, $U = -KX$, which leads to the closed-loop system described in Eq. (12) [5].

Where $K = [k_1\ k_2\ k_3\ \ldots\ k_n]$

$$\begin{aligned} \dot{X} &= AX + BU \\ \dot{X} &= (A - BK)X = A_{\mathrm{CL}}X \end{aligned} \tag{12}$$

where $A_{\mathrm{CL}} \triangleq (A - BK)$.

Now, the closed-loop poles are chosen as per the control system pole-placement and in negative plane and will have a small time constant. Here, to design a system with first companion form, we have chosen Eq. (13).

$$A = \begin{bmatrix} 0 & 1 & 0 & 0 & \cdots & 0 \\ 0 & 0 & 1 & 0 & \cdots & 0 \\ \vdots & & & \ddots & & \vdots \\ 0 & 0 & 0 & 0 & \ddots & 1 \\ -a_n & -a_{n-1} & -a_{n-2} & -a_{n-3} & \cdots & -a_1 \end{bmatrix} \quad B = \begin{bmatrix} 0 \\ 0 \\ \vdots \\ 0 \\ 1 \end{bmatrix}$$

$$A_{\text{CL}} = \begin{bmatrix} 0 & 1 & 0 \; 0 \; \cdots & 0 \\ 0 & 0 & 1 \; 0 \; \cdots & 0 \\ 0 & 0 & 0 \; 1 \; \cdots & 0 \\ \vdots & \vdots & 0 \; 0 \; 1 & 0 \\ 0 & 0 & \cdots\cdots\cdots & 1 \\ (-a_n - k_1) & (-a_{n-1} - k_2) & \cdots\cdots\cdots & (-a_1 - k_n) \end{bmatrix} \tag{13}$$

Then the characteristics polynomial will lead to the closed-loop system matrix as Eq. (14).

$$A_{\text{CL}} = \begin{bmatrix} 0 & 1 & 0 & \cdots & 0 \\ 0 & 0 & 1 & \cdots & 0 \\ \vdots & & \ddots & & \\ \vdots & & & \ddots & 1 \\ 0 & 0 & 0 & \cdots & 0 \\ -\alpha_n & -\alpha_{n-1} & -\alpha_{n-2} & \cdots & -\alpha_1 \end{bmatrix} \tag{14}$$

The gain matrix K is designed in such a way that the desired characteristic polynomial becomes as Eq. (15).

$$\begin{aligned} &|sI - (A - BK)| = (s - \mu_1)(s - \mu_2)\ldots(s - \mu_n) \\ &(s - \mu_1)\ldots(s - \mu_n) = s^n + \mu_1 s^{n-1} + \mu_2 s^{n-2} + \cdots \mu_n \end{aligned} \tag{15}$$

where $\mu_1 \ldots \mu_n$ are the desired pole locations.

For controllability of the system, Eq. (16) is given, where M is the controllability matrix.

$$M = \left[B \; AB \; A^2B \; A^3B \ldots A^{n-1}B\right] \tag{16}$$

Here, the system was not given in first companion form and to convert, it a transformation matrix T have designed in such a way that $T^{-1}AT$ will be in first companion form and W as given in Eq. (15).

$$W = \begin{bmatrix} a_{n-1} & a_{n-2} & \cdots & a_1 & 1 \\ a_{n-2} & & \cdot^{\cdot^{\cdot}} & \cdot^{\cdot^{\cdot}} & 0 \\ & & \cdot^{\cdot^{\cdot}} \;\; \cdot^{\cdot^{\cdot}} & \cdots & \vdots \\ a_1 & 1 & \cdots & \cdots & \vdots \\ 1 & 0 & \cdots & \cdots & 0 \end{bmatrix} \quad (17)$$

All the above equation from (13)–(17) can be used to design a controller where the gain matrix K can be defined as Eq. (18).

$$\begin{gathered} k_1 = (\alpha_n - a_n) \\ k_2 = (\alpha_{n-1} - a_{n-1}) \\ \vdots \\ k_n = (\alpha_1 - a_1) \\ K = \left[(\alpha_n - a_n)(\alpha_{n-1} - a_{n-1}) \ldots (\alpha_1 - a_1)\right]T^{-1} \end{gathered} \quad (18)$$

Here, an augment matrix for A, B, and C is used to modify the system and thereafter, the extended state observer was applied to estimate the states of the system. As per Remark I, the augment matrix was chosen and implemented in the plant. Using the LQR technique, the Q-weighted matrix was implemented [8]. The Q-weighted matrix was designed as the diagonal of K, which is a gain matrix. The other components L and R were decided as Eq. (19).

If (A, B) is controllable for system (11), the extended state observer is designed as follows

$$\begin{aligned} \dot{\hat{x}} &= A\hat{x} + Bu + L\left(y - C\hat{x}\right) \\ \hat{y} &= C\hat{x} \end{aligned} \quad (19)$$

where $\hat{x} = \left[\hat{x}\right]^T$ are the estimates of the state variable x. Matrix L with dimension $(n) \times r$ is the observer gain to be designed.

The standard extended state observer[9]-based control law is, $u = k_x x - \hat{d}$ (where $\hat{d} = \hat{x}_{n+1}$ and k_x is the feedback control gain). Now Eq. (19) can be rewritten as

$$\dot{\hat{x}} = \left(A - Bk_{\mathrm{p}} - LC\right)\hat{x} + Ly \quad (20)$$

where k_{p} is the plant feedback control gain after estimating the plant of Eq. (11). Leading to a model where real variables are considered with five states, ρ_1 is the additional state, which is being applied for the observer gain.

$$\begin{aligned} u(t) &= -\left[k_{\mathrm{p}} C_{\mathrm{d}}\right]\begin{bmatrix} \hat{x}_x(t) \\ \hat{x}_{\mathrm{d}}(t) \end{bmatrix} \\ &\Rightarrow k = \left[k_{\mathrm{p}}\, I_2\right] \end{aligned} \quad (21)$$

where C_d and $\hat{x}_d(t)$ are, respectively, the disturbance matrixes for system matrix C and state estimation disturbance for $\hat{x}(t)$.

3 Illustrative Example

3.1 Simulation Result

An inverted pendulum for a SIMO system has been selected as follows after using the values from Table 1.

$$A = \begin{bmatrix} 0 & 0 & 1\,0 \\ 0 & 0 & 0\,1 \\ 0 & 0.71 & 0\,0 \\ 0 & -4.11 & 0\,0 \end{bmatrix}, \text{ and } B = \begin{bmatrix} 0 \\ 0 \\ 0.38 \\ -0.16 \end{bmatrix} C = \begin{bmatrix} 1\,0\,0\,0 \\ 0\,1\,0\,0 \end{bmatrix} D = [0] \quad (22)$$

As discussed in Sect. 2.4, the basic Bass–Gura approach or renowned by pole-placement method where the transformation matrix T which leads to the first companion form of the system is given below [10].

$$s^4 + a_1 s^3 + a_2 s^2 + a_3 s + a_4$$
$$a_1 = 0,\ a_2 = 40.9036,\ a_3 = 0,\ a_4 = 0.$$

Now, as per the open-loop eigen values, we have chosen the desire poles as follows: $\mu_1 = -j6.3956$; $\mu_2 = j6.3956$; $\mu_3 = 0$; and $\mu_4 = 0$. Now, using Eq. (15), we found that

$$\alpha_4 = 0;\ \alpha_3 = 0;\ \alpha_2 = -40.9037;\ \alpha_1 = 0 \quad (23)$$

From the above equations, it is prominent that closed-loop system is stable as the poles are placed on the LHP. After replacing with Eq. (17), the state-feedback gain matrix can be found out by using Eq. (18) the value of K becomes as Eq. (24).

$$K = [0.675\ 2.6301 - 29.4225 - 215.1636] \quad (24)$$

As per the previous discussions, we have chosen the value of Q as the diagonal of K which is shown in Eq. (24)

$$Q = \begin{bmatrix} 0.675 & 0 & 0 & 0 \\ 0 & 2.6301 & 0 & 0 \\ 0 & 0 & -29.4225 & 0 \\ 0 & 0 & 0 & -215.1636 \end{bmatrix} \quad (25)$$

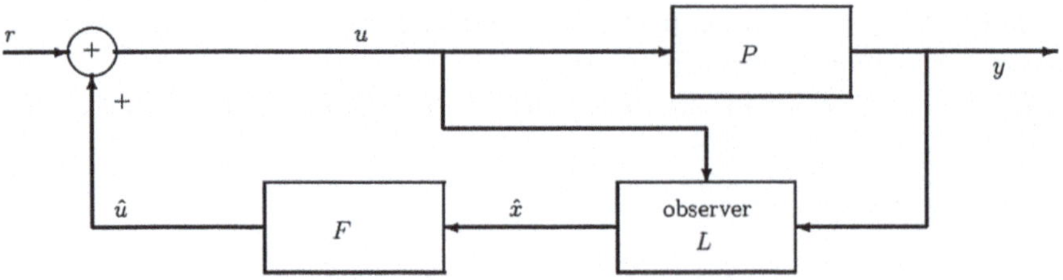

Fig. 2 Closed-loop system using output state-feedback controller

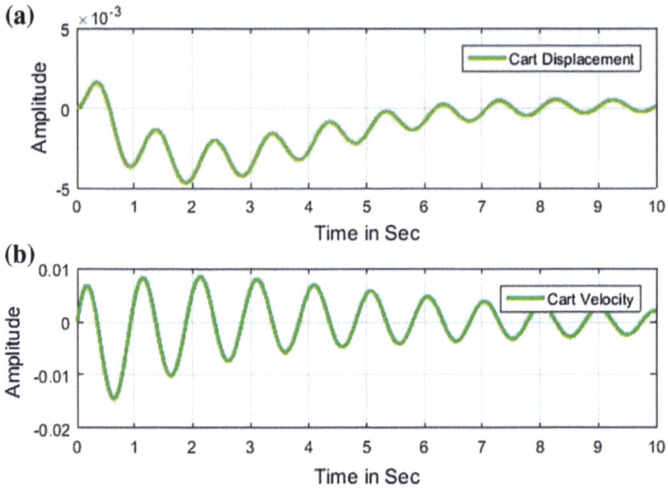

Fig. 3 Simulated cart position (**a**) and velocity (**b**) for Case 1

Now using the linear quadratic regulator method, the controller gain matrix K_1 can be found as below [11].

$$K_1 = [0.8216 - 0.2367\ 2.0883 - 0.2154] \tag{26}$$

3.2 Plant Result and Discussion

Now, various weighted matrices for Q were tested in the same model and the description is given in Table 2. The weighted matrices design is defined in Table 2. For various weighted matrices, various oscillations have been observed, where the best result was produced during Case 3. The observer poles (OP) for Case 1, 2, and 3 are found from the real plant model. The closed-loop poles (CLP) are taken from the

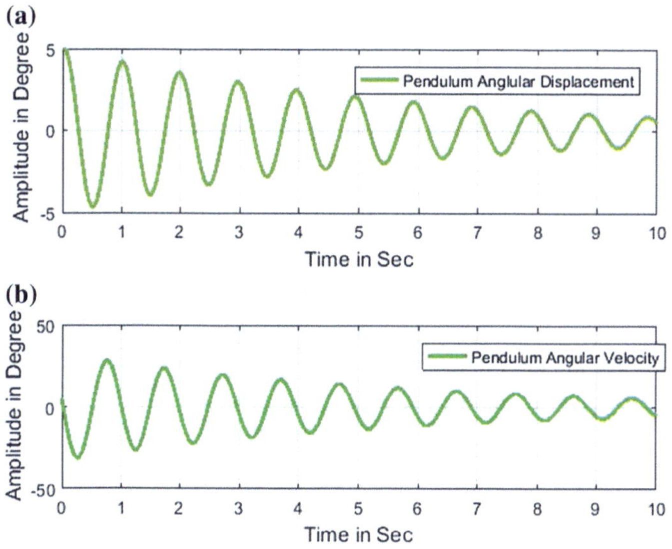

Fig. 4 Simulated pendulum angle (**a**) and angular velocity (**b**) for Case 1

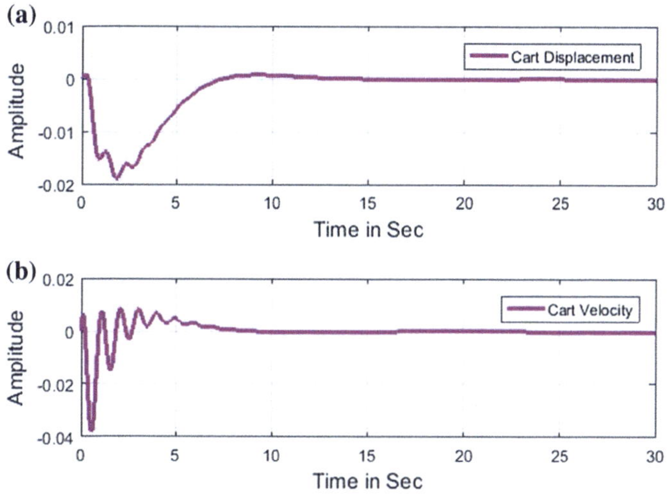

Fig. 5 Simulated cart position (**a**) and velocity (**b**) for Case 2

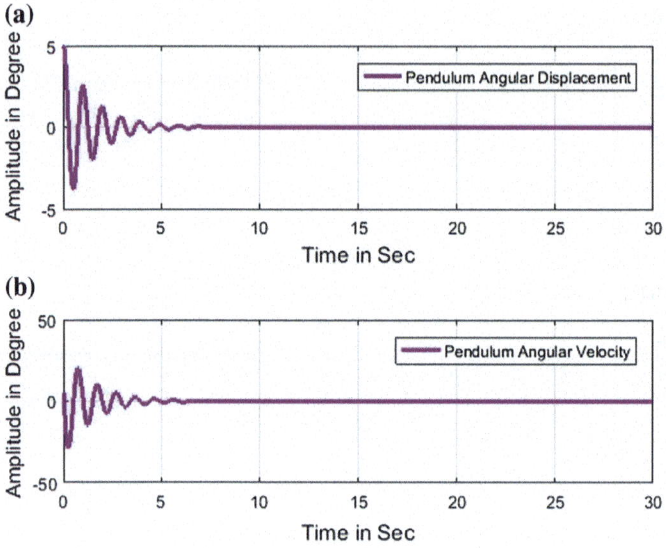

Fig. 6 Simulated pendulum angle (**a**) and angular velocity (**b**) for Case 2

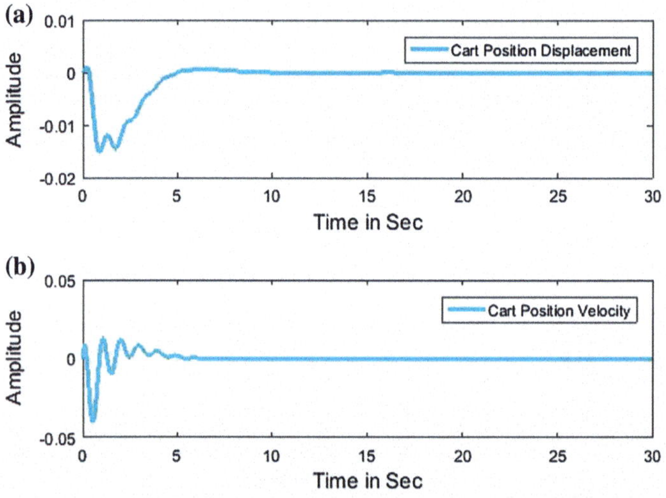

Fig. 7 Simulated cart position (**a**) and velocity (**b**) for Case 3

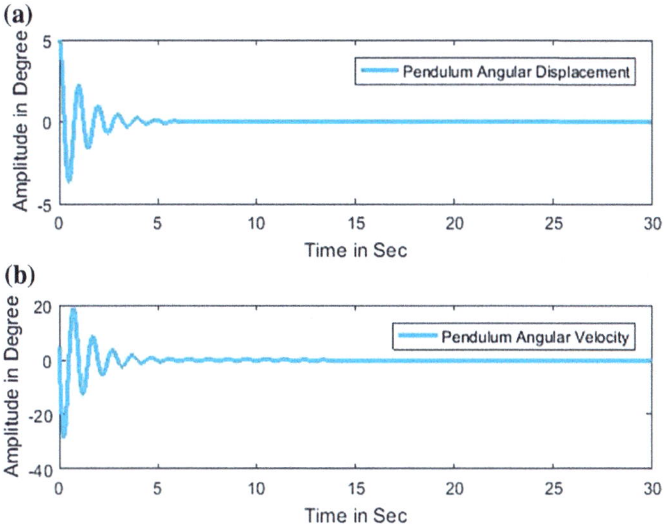

Fig. 8 Simulated pendulum angle (**a**) and angular velocity (**b**) for Case 3

Table 2 Different values of weighted matrices

Case 1 (Q)	Case 2 (Q)	Case 3 (Q)
$\begin{bmatrix} 1\,0\,0\,0 \\ 0\,2\,0\,0 \\ 0\,0\,0\,0 \\ 0\,0\,0\,0 \end{bmatrix}$	$\begin{bmatrix} 1\;\;0\;\;0\,0 \\ 0\;20\;0\,0 \\ 0\;\;0\;\;0\,0 \\ 0\;\;0\;\;0\,0 \end{bmatrix}$	$\begin{bmatrix} 5\;\;0\;\;0\,0 \\ 0\;45\;0\,0 \\ 0\;\;0\;\;0\,0 \\ 0\;\;0\;\;0\,0 \end{bmatrix}$

real plant and shown in Table 3. Due to the changes in controller gain, the stabilizing factor is changing. The observer poles are decided to be three times faster than the CLP, so as decided. For case 1, L is $\begin{bmatrix} 21.85 & 38.79 & 544.91 & -155.56 & 219.98 \\ 14.38 & 21.81 & -9.34 & 16.84 & -60.85 \end{bmatrix}$, for Case 2 is $\begin{bmatrix} 18.33 & 29.60 & 521.75 & -44.58 & 186.74 \\ 14.12 & 30.46 & 27.01 & 55.18 & -120.61 \end{bmatrix}$, and for Case 3 is $\begin{bmatrix} 24.54 & 35.84 & 558.53 & -165.07 & 224.82 \\ 14.02 & 22.09 & -8.85 & 16.59 & -59.30 \end{bmatrix}$. The simulated result of pendulum angle displacement is shown in Figs. 4, 6, 8 and 3, 5, 7 are showing cart angle control response, respectively. The different state amplitude is shown in Fig. 9, where it can be observed that four states are simultaneously damping with the control voltage supply. The simulated plant pendulum damping remains till 5.4 s whereas the real plant pendulum remains till 5.2 s. After 30 s of time, an external disturbance was introduced in the real plant, which may be rejected using an extended state observer and this may be treated as the future research work. The original plant identification

Table 3 Comparison between different weighted matrices

	CLP	OP	Gain (K)
Case 1	$\begin{bmatrix} -0.1752 + j6.398 \\ -0.1752 - j6.398 \\ -0.4355 + j0.4358 \\ -0.4355 - j0.4358 \end{bmatrix}$	$\begin{bmatrix} -0.5256 + j1.92 \\ -0.5256 - j1.92 \\ -1.307 + j3.897 \\ -1.307 - j3.897 \\ -1.5 \\ -2 \end{bmatrix}$	$\begin{bmatrix} 1 & -0.2139 & 2.303 & -0.1794 \end{bmatrix}$
Case 2	$\begin{bmatrix} -0.5522 + j6.4194 \\ -0.5522 - j6.4194 \\ -0.4326 + j0.433 \\ -0.4326 - j0.433 \end{bmatrix}$	$\begin{bmatrix} -1.66 + j19.26 \\ -1.66 - j19.26 \\ -1.298 + j1.299 \\ -1.298 - j1.299 \\ -2 \\ -3 \end{bmatrix}$	$\begin{bmatrix} 1 & -0.9669 & 2.336 & -0.6428 \end{bmatrix}$
Case 3	$\begin{bmatrix} -0.8245 + j6.4484 \\ -0.8245 - j6.4484 \\ -0.6406 + j0.6423 \\ -0.6406 - j0.6423 \end{bmatrix}$	$\begin{bmatrix} -2.47 + j19.35 \\ -2.47 - j19.35 \\ -1.922 + j1.927 \\ -1.922 - j1.927 \\ -2.5 \\ -3.5 \end{bmatrix}$	$\begin{bmatrix} 2.2361 & -2.134 & 3.5686 & -0.9323 \end{bmatrix}$

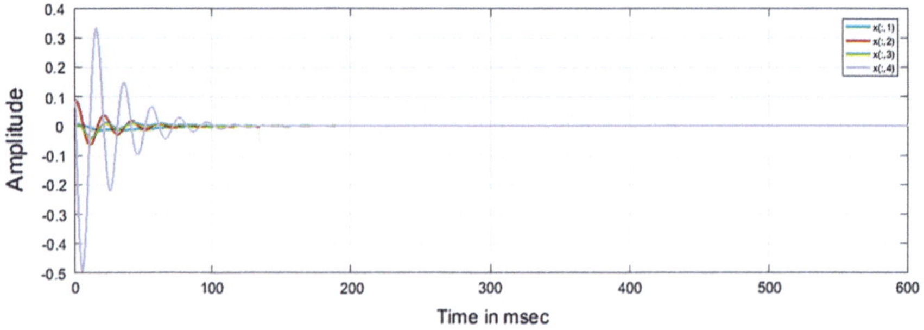

Fig. 9 Simulated states of the plant

is very tough for the selected rotational mechanical system, the real model identification waveforms are shown in Figs. 10, 11 and 12. The real model crane position and crane angle response is shown in Figs. 13 and 14 respectively where the desired and original output convergence is minimum. This response from a real plant model is very much challenging and complex to acquire still the controller is very optimal and able to sustain such complexity. An additional state of crane control is presented herewith the real plant result, which actually follows the control law of the established controller. The supply voltage u for the responses of crane is shown in Fig. 15 waveform of the voltage does not exceed ±2.5 V. From the simulated results, it has been observed that the stabilizing factor is dependent on the precise weighted matrix

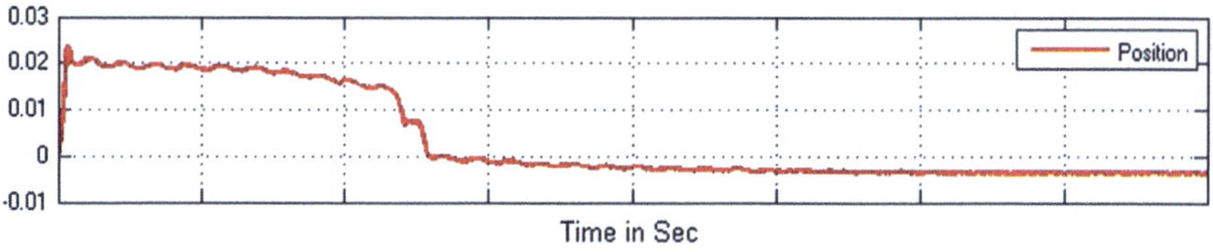

Fig. 10 Real model identification for cart position

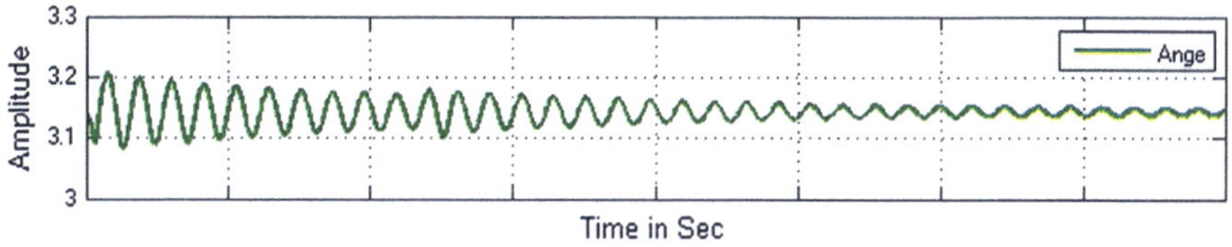

Fig. 11 Real model identification for cart angle

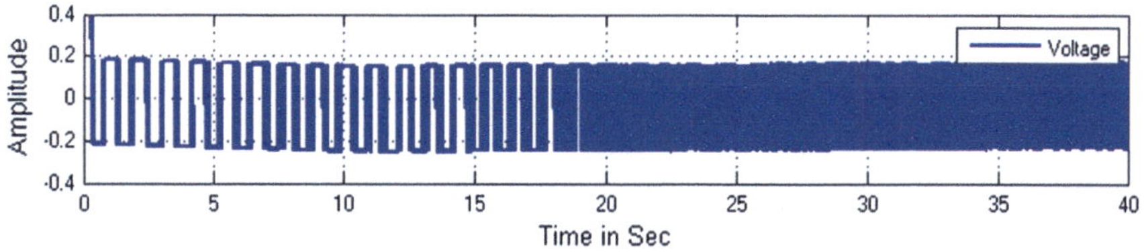

Fig. 12 Real model supply voltage for cart position and angle of motor

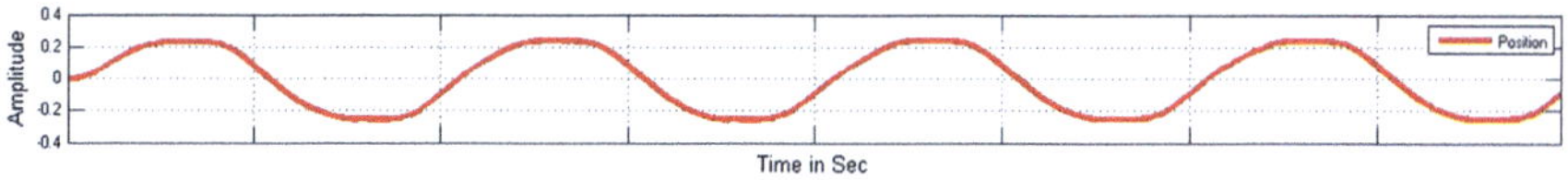

Fig. 13 Real model crane position after observer control

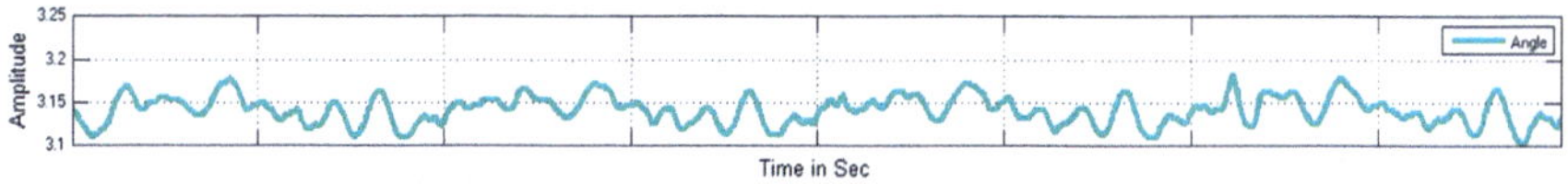

Fig. 14 Real model crane angle after observer control

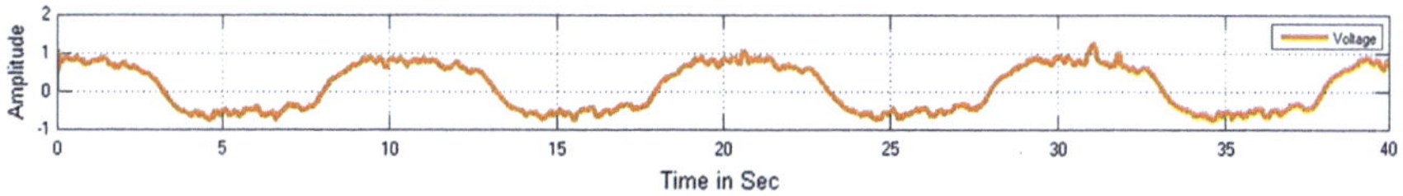

Fig. 15 Real model crane voltage supply of motor

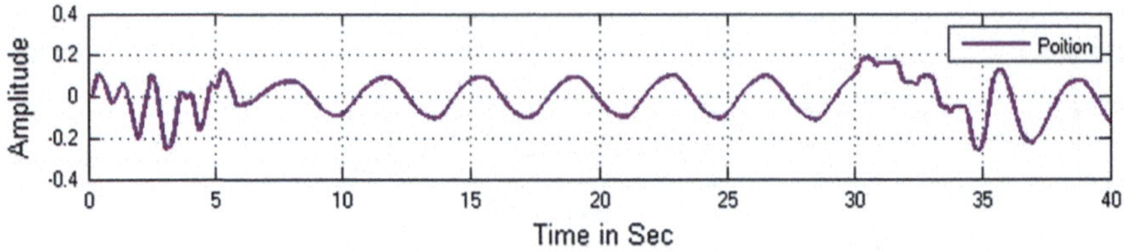

Fig. 16 Real model pendulum position after using observer

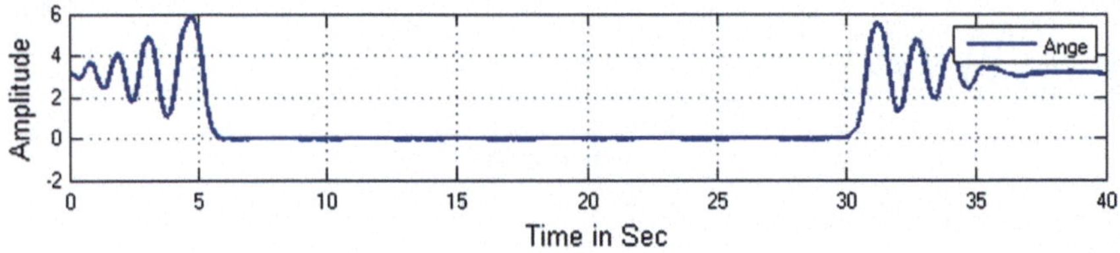

Fig. 17 Real model pendulum angle after using observer

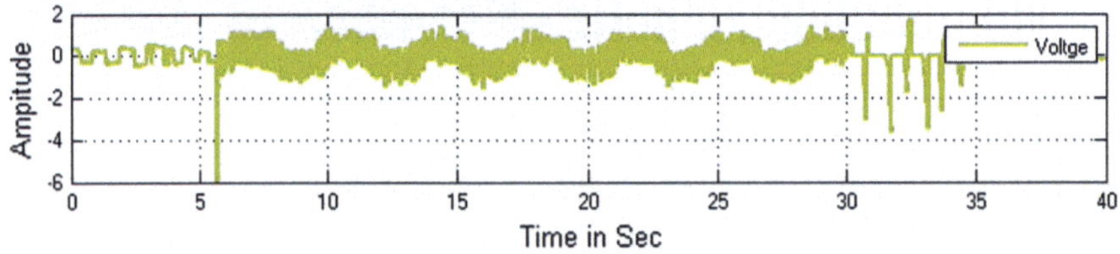

Fig. 18 Real model pendulum voltage supply to motor

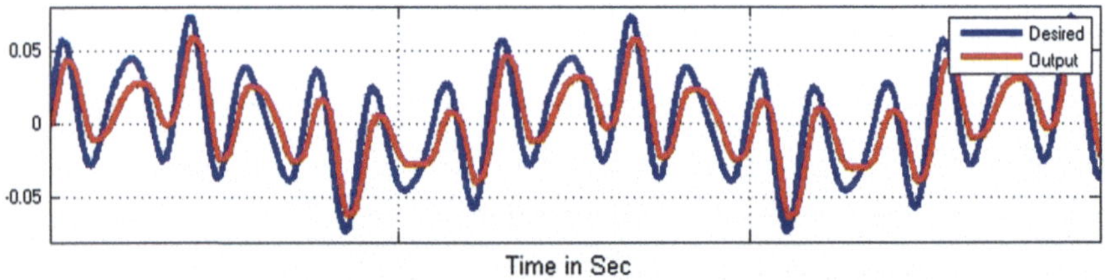

Fig. 19 Real model cart desired and original output

value, if the value is not suitable, the stability will be different. Table 3 shows that all the cases are stable because all CLPs are lying in the left hand plane of the *s*-plane (Figs. 13, 14, 17 and 19).

The pendulum position and angle is shown in Figs. 16 and 17 respectively and Fig. 18 contains the pendulum voltage. Figures 19 and 20 is describing real plant cart output and cart voltage. The cart response is following the desired output and mitigates the steady state error.

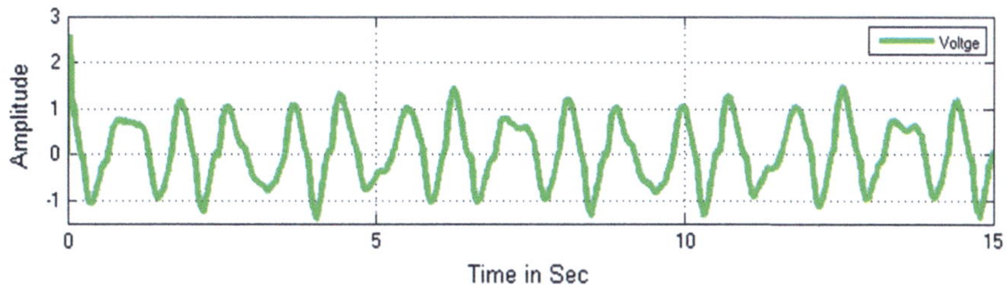

Fig. 20 Real model cart supply voltage

4 Conclusion

In this paper, an LQR controller is been established, with the help of pole-placement method. The simulated and experimental results are discussed in this paper, and the LQR controller which is applied to the real plant are discussed here. The optimal control approach is not so complex to apply for the SIMO system and the system response may be converged to stability very rapidly.

Acknowledgements The authors want to acknowledge Mr. Abaniswar Chakraborty of NIT Agartala for his help to perform the experimental validations.

References

1. Gordillo F, Acosta J, Aracil J (2003) A new swing-up law for the Furuta pendulum. Int J Control 76(8):836–844
2. Kafetzis I, Moysis L (2017) Inverted pendulum a system with innumerable applications. In: 9th international week dedicated to maths, Thessaloniki, Greece
3. Sen MA, Kalyoncu M (2016) Optimal tuning of a LQR controller for an inverted pendulum using the bees algorithm. J Autom Control Eng 4(5)
4. Vinodh Kumar E, Jerome J, Srikanth K (2014) Algebraic approach for selecting the weighting matrices of linear quadratic regulator. In: 2014 international conference on green computing communication and electrical engineering (ICGCCEE)
5. Ogata K (2015) Modern control engineering, 5th edn. Prentice Hall
6. Chatterjee S, Das SK (2018) An analytical formula for optimal tuning of the state feedback controller gains for the cart-inverted pendulum system. IFAC-PapersOnLine 51(1):668–672. ISSN 2405-8963. https://doi.org/10.1016/j.ifacol.2018.05.112
7. Harrison R (2003) Asymptotically optimal stabilizing quadratic control of an inverted pendulum. IEEE Proc Control Theory Appl 50
8. Mariano SJ, Pombo JA, Calado MRA, Ferreira LFM (2012) A procedure to specify the weighting matrices for an optimal load-frequency controller. Turk J Elect Eng Comput Sci 20(3). https://doi.org/10.3906/elk-1003-413
9. Li S, Yang J, Chen W-H, Chen X (2012) Generalized extended state observer based control for systems with mismatched uncertainties. IEEE Trans Ind Electron
10. Friedland B (1986) Control system design. McGraw Hill
11. Lungu M (2012) The estimation of an aircraft motions by using the Bass-Gura full-order observer. In: ICATE, Craiova, pp 1–5. https://doi.org/10.1109/icate.2012.6403455

Reduction of Selective Harmonics in PWM AC Chopper Using Teaching–Learning-Based Optimization

D. B. Ghorpade, A. S. Koshti and S. S. Halve

Abstract Teaching–learning-based optimization (TLBO) has been implemented to solve optimization problem in AC chopper. This optimization problem is formulated in view of getting desired voltage of fundamental frequency with selective harmonic component elimination. Pulse-width modulation (PWM) technique has been utilized for this purpose. PWM technique is utilized to set the desired voltage of fundamental frequency with elimination of selective harmonic component. Thus, TLBO is used for this purpose to obtain multiple switching angles (on and off) keeping the symmetry of switching in both half cycles of supply voltage waveform. And, based on the obtained switching angles, the operation of AC chopper is validated using PSIM software.

Keywords PWM · AC chopper · Selective harmonics elimination · TLBO

1 Introduction

Voltage control at load end is desirable. Usually, it is achieved by tap-changing transformer in power systems. With the use of power electronic converters the voltage control becomes easy but this leads to the involvement of harmonics content because of firing angle control. Thus, in AC chopper, the availability of voltage across load is controlled by adjusting the firing angle in both halves, i.e. positive half and negatives half, but this causes the output voltage waveform to possess harmonics content (multiples of fundamental frequency) along with fundamental component. Pulse-width modulation has been suggested in [1] used for AC voltage control. This pulse-width modulation technique suggests the multiple times on and off of a power

D. B. Ghorpade · A. S. Koshti (✉) · S. S. Halve (✉)
R.H. Sapat College of Engineering Management Studies and Research, Nashik, India
e-mail: atulkoshti251177@gmail.com

S. S. Halve
e-mail: shrunkhala123@gmail.com

D. B. Ghorpade
e-mail: devyanighorpade05@gmail.com

S. N. Singh et al. (eds.), *Advances in Power and Control Engineering*, Lecture Notes in Electrical Engineering 609, https://doi.org/10.1007/978-981-15-0313-9_17

Table 1 Working operation of AC chopper

Operation in	ON time	OFF time
Positive half cycle of supply voltage	D1, S1, D3 conduct	D5, S2, D7 conduct
Negative half cycle of supply voltage	D2, S1, D4 conduct	D6, S2, D8 conduct

electronic device as a switch in positive half and negative half cycle and should be symmetrical to eliminate even harmonics is view of symmetry of waveform. The voltage harmonic elimination by genetic algorithm is described in [2] for PWM AC chopper. Addoweesh and Mohamadem [3] implemented microprocessor-based controlling technique for PWM. The predominant lower order harmonics elimination is a major problem in a AC chopper, and it can be eliminated with suitably selecting the width of pulses for which the supply will be available to load for that time. Hence multiple times on and off in both cycle of supply voltage may control the harmonic content and fundamental voltage. So, selective harmonics elimination, decision of switching angles is critical and important aspect to have desired voltage of fundamental frequency. Thus, in AC chopper to get desired voltage with selective harmonic elimination is considered as an optimization problem discussed in [4–7]. This optimization problem is basically a minimization problem of control variable, i.e. switching angles on/off multiple times and can be solved using a evolutionary technique. Due to the independency of any algorithm-specific parameter required in it, TLBO is widely used for optimization work [8, 9]. Thus, TLBO technique has been implemented in this paper to get required angles at which desired fundamental voltage is obtained and with selective harmonic elimination .

2 Principle of Operation of PWM AC Chopper

Table 1 shows the operation of PWM AC chopper as given in Fig. 1, which contains two switches S1 and S2 along with diode arrangement to have the bi-directional control in the positive half and negative half cycle. Figure 2 shows the on/off operation of switch in both cycles.

3 Reduction of Harmonics in PWM AC Chopper Using TLBO

Firstly, a optimization problem in terms of desired voltage and harmonic component is formulated and expressed as

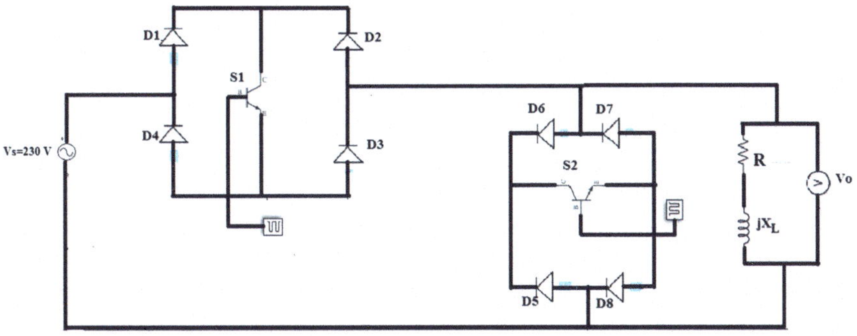

Fig. 1 PWM AC chopper with switch arrangement

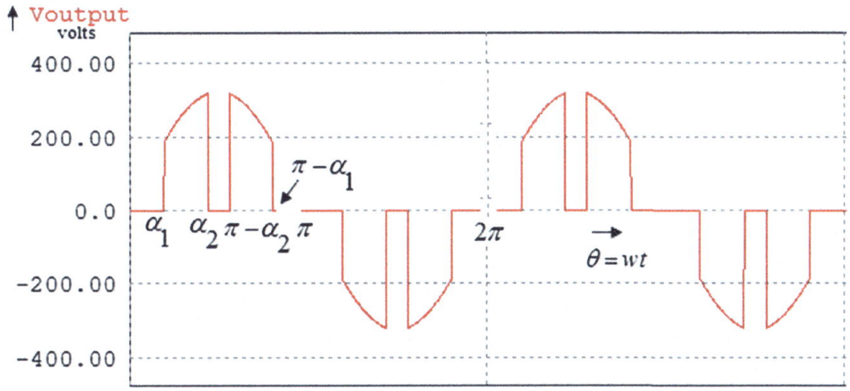

Fig. 2 Output voltage of PWM AC chopper

$$J = (Vd - B_1)^2 + (B_3)^2 + (B_5)^2 + (B_7)^2 + \cdots \quad (1)$$

where

B_1	fundamental component (max value)
B_3, B_5, B_7	odd-order harmonic components (max value)
Vd	desired voltage (max value).

In view of Fourier series as per Fig. 2, the output voltage available at load end is calculated as

$$E_0 = V_{\mathrm{dc}} + \sum_{1}^{n} B_n \sin(n\theta) + \sum_{1}^{n} A_n \cos(n\theta); \quad \text{where, } \theta = wt$$

In view of symmetry of switching operation, above equation reduces to

$$E_0 = \sum_{1}^{n} B_n \sin(n\theta), \quad n = 1, 3, 5, \ldots$$

And, value of B_n and B_1 is computed as

$$B_n = \frac{2E_m}{\pi}\left[\frac{\sin(n-1)\theta}{(n-1)} - \frac{\sin(n+1)\theta}{(n+1)}\right]_{\alpha_1, \alpha_3, \alpha_5, \ldots, \alpha_{m-1}}^{\alpha_2, \alpha_4, \alpha_6, \ldots, \alpha_m} \quad \text{where, } n \neq 1$$

where m = number of switching angles/quarter cycle and E_m is the maximum value of supply voltage.

$$B_1 = \frac{2E_m}{\pi}\left[\theta - \frac{\sin\theta}{2}\right]_{\alpha_1, \alpha_3, \alpha_5, \ldots, \alpha_{m-1}}^{\alpha_2, \alpha_4, \alpha_6, \ldots, \alpha_m}$$

TLBO is teaching–learning process in which a group of students (learners) in a class is considered as a population, control variables are the subjects offered to the students, and in this case, the switching angles are considered as control variables. This method is based on the philosophy of learning of students with the interaction between teacher and learner and amongst the learners, called as teacher phase and learner phase. The values of switching angles are considered as grades or marks in subjects. Based on the above discussion, TLBO-based algorithm for finding switching angles for PWM AC chopper is explained in the following steps:

Step-1: Population is initialized within limits given as

$$X_i^{(0)} = \left[\alpha_{1,i}, \alpha_{2,i}, \alpha_{3,i}, \ldots, \alpha_{m,i}\right]^{T}, \quad i = 1 \text{ to NP} \quad \text{where, } 0 \leq \alpha_{m,i} \leq \frac{\pi}{2}$$

Step-2: Put iteration count ($I_c = 1$).
(Teacher phase)
Step-3: X_{best} is evaluated based on best vector from $i = 1$ to NP, with respect to the minimum objective function value given by Eq. (1).
Step-4: Mean vector for switching angles is evaluated.
Step-5: Modified vector solution X_i^{modified} for teacher phase is obtained [9].
Step-6: The control variables are put within limits in case of violation of limits for this bounce-back technique is used [10].
Step-7: Compare X_i^{modified} and current generation population and retain the best one based on the minimum objective function.

(Learner phase)
Step-8: A new solution vector of all switching angles X_i^{new} is obtained. In case of violation of limits of control variables (switching angles), bounce-back technique is used to bring control variables within limits [10].
Step-9: Better solution vector is obtained with comparison of X_i^{new} and X_i^{Ic} at the end of teacher phase.
Step-10: Repeat from step-3 till maximum number of iterations is executed, say $I_c \max = 500$.

4 Results and Discussion

The algorithm presented in Sect. 3 is implemented using TLBO, and obtained results are presented in Tables 2 and 3, which illustrate the effectiveness of the method. The number of switching angles/quarter cycle (m) selected reduces particular lower order harmonics based on their optimal values. For the presented work, $m = 2$ is considered, and third-order harmonic is eliminated for desired voltage (Vd) across load. The supply voltage of 50 Hz frequency is considered with 230 (rms) or 325 V

Table 2 Obtained switching angles and voltages at $m = 2$ with p.f. = 0.80 lagging

Desired voltage (p.u.)	Obtained by TLBO				Validated by PSIM		Load
Vd	α_1(rad)	α_2(rad)	B_1(p.u.)	B_3(p.u.)	B_1(p.u.)	B_3(p.u.)	$R = 50\ \Omega$ $X_L =$ 37.48 Ω p.f. = 0.8 (lag.)
0.66	0.6108	1.3788	0.6687	0.0330	0.6670	0.0330	
0.70	0.6283	1.4137	0.6932	0.0166	0.7046	0.0165	
0.75	0.5759	1.4311	0.7474	0.0003	0.7469	0.0016	
0.80	0.5235	1.466	0.8094	0.0067	0.8087	0.0067	
0.85	0.4712	1.4835	0.8466	0.0036	0.8475	0.0024	
0.95	0.3141	1.5358	0.9425	0.0087	0.9409	0.0103	

Table 3 Obtained switching angles and voltages at $m = 2$ at p.f. = 0.90 lagging

Desired voltage (p.u.)	Obtained by TLBO				Validated by PSIM		Load
Vd	α_1(rad)	α_2(rad)	B_1(p.u.)	B_3(p.u.)	B_1(p.u.)	B_3(p.u.)	$R = 50\ \Omega$ $X = 24.38\ \Omega$ p.f. = 0.9 (lag.)
0.66	0.6108	1.3788	0.6687	0.0330	0.6672	0.0326	
0.70	0.6283	1.4137	0.6932	0.0166	0.7046	0.0164	
0.75	0.5759	1.4311	0.7474	0.0003	0.7469	0.00069	
0.80	0.5235	1.466	0.8094	0.0067	0.8086	0.00677	
0.85	0.4712	1.4835	0.8466	3.6286	0.8471	0.00231	
0.95	0.3141	1.5358	0.9425	0.0087	0.9409	0.0096	

(max). The required desired voltage across load is considered from (0.65 to 0.95 p.u.). For the desired fundamental frequency voltage with minimum third-harmonic component, the algorithm given in Sect. 3 is executed, and the optimal values obtained are presented in Tables 2 and 3. Once the optimal switching angles are obtained, then it is validated using PSIM software. The observed results are found in close agreement. It is tested at 0.8 and 0.9 p.f. loading condition. Figs. 3 and 4 show the frequency spectra of voltage harmonics for a particular case $Vd = 0.75$ p.u. at both loading conditions. Moreover, the current harmonics are shown in Figs. 5 and 6 for above-mentioned case. The obtained results show the minimum third-harmonic component for desired voltages across load. Also Figs. 7 and 8 shows the frequency spectra of voltage harmonics for $Vd = 0.85$ p.u. at 0.8 and 0.9 p.f. lagging respectively and Figs. 9 and 10 show the current harmonics spectra for the above mentioned cases.

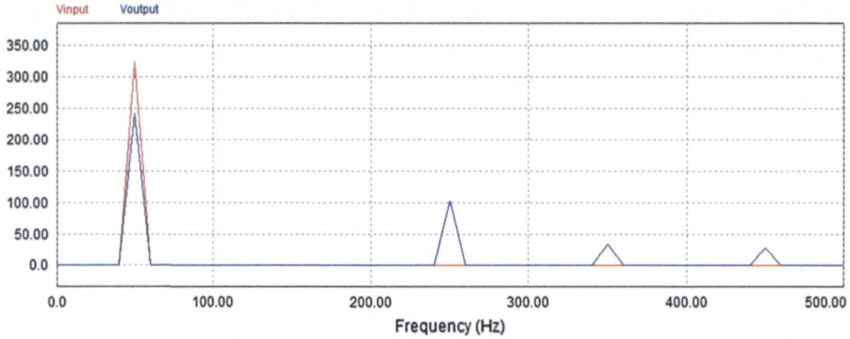

Fig. 3 Frequency spectra for $Vd = 0.75$ p.u. at 0.8 p.f. (lagging)

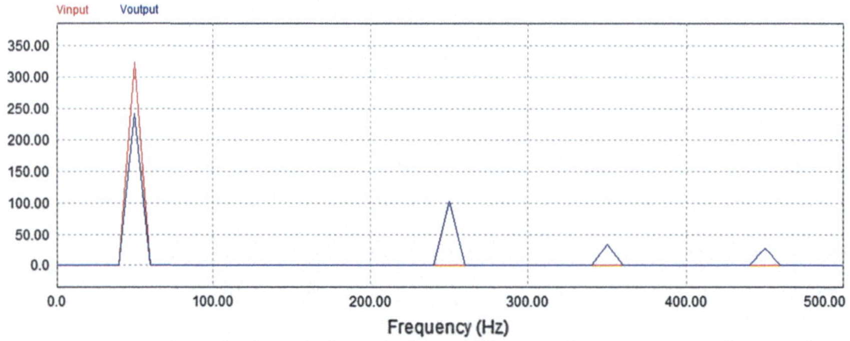

Fig. 4 Frequency spectra for $Vd = 0.75$ p.u. at 0.9 p.f. (lagging)

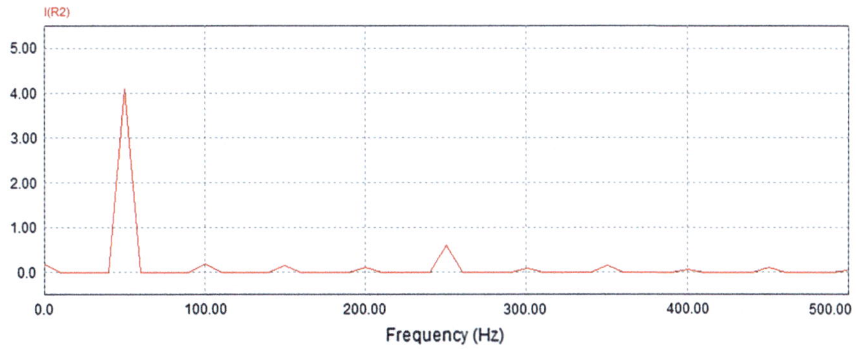

Fig. 5 Frequency spectra for current at $Vd = 0.75$ p.u. at 0.8 p.f. (lagging)

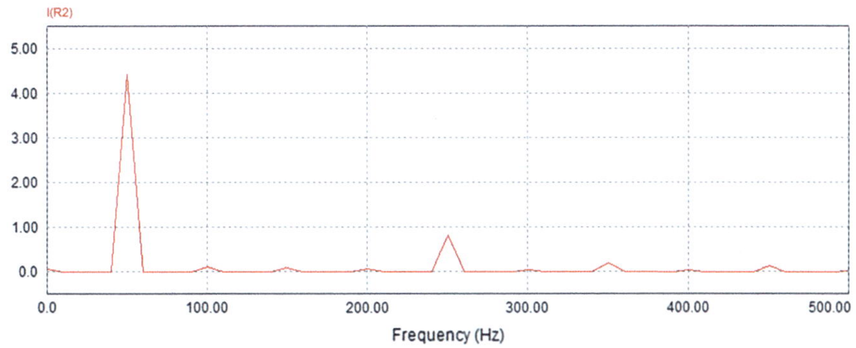

Fig. 6 Frequency spectra for current at $Vd = 0.75$ p.u. at 0.9 p.f. (lagging)

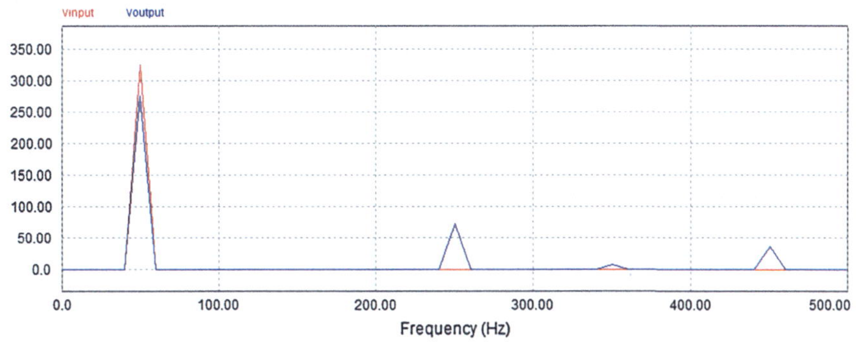

Fig. 7 Frequency spectra for $Vd = 0.85$ p.u. at 0.8 p.f. (lagging)

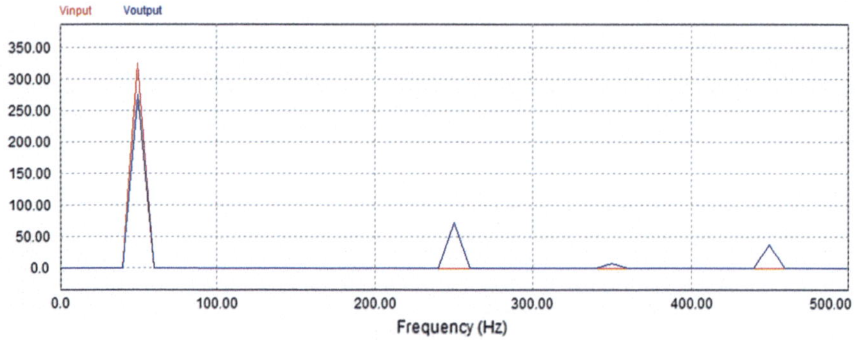

Fig. 8 Frequency spectra for $Vd = 0.85$ p.u. at 0.9 p.f. (lagging)

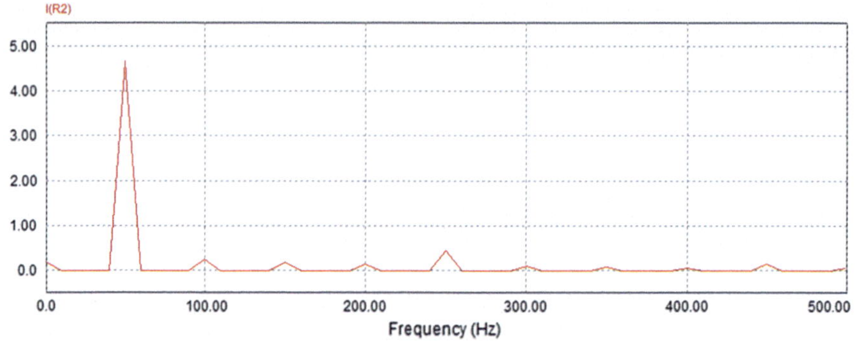

Fig. 9 Frequency spectra for current at $Vd = 0.85$ p.u. at 0.8 p.f. (lagging)

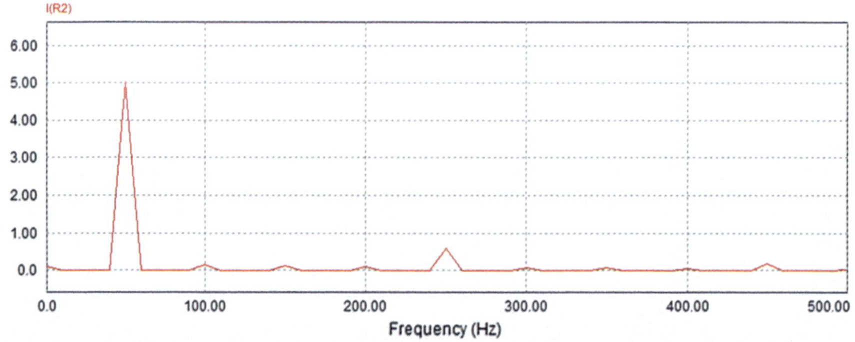

Fig. 10 Frequency spectra for current at $Vd = 0.85$ p.u. at 0.9 p.f. (lagging)

5 Conclusion

TLBO technique has been used to optimize the switching angles to minimize the selective harmonics, i.e. third order in A.C chopper to get the required voltage of fundamental frequency across load. This technique is tested for $m = 2$ switching angles/quarter cycle, and for the obtained optimum switching angles using TLBO technique the desired voltage and frequency spectra is validated by PSIM software. At obtained switching angles, the frequency spectra for voltage and current are obtained for few of the cases $Vd = 0.75$ p.u. $Vd = 0.85$ p.u. at 0.8 p.f. lagging and 0.9 p.f. which show the effectiveness of the method in view of desired fundamental frequency voltage for minimum third-order harmonic content.

References

1. Choe G-H, Wallace AK, Park M-H (1989) An improved PWM technique for AC choppers. IEEE Trans Power Electron 4(4):496–505
2. Sundareswaran K, Kumar AP (2004) Voltage harmonic elimination in PWM AC chopper using genetic algorithm. IEEE Proc Electr Power Appl 151(1):26–31
3. Khaled E, Addoweesh KE, Adel L, Mohamadeın (1990) Microprocessor based harmonic elimination in chopper type ac voltage regulators. IEEE Trans Power Electron 5(2):191–200
4. Sundareswaran K, Pavan Kumar A (2005) Performance enhancement of PWM ac chopper using random search method. J Inst Eng (India) 85(4):228–230
5. Arya LD, Koshti A, Bhatt PK (2007) Mitigation of harmonics content in PWM ac chopper using particle swarm optimization (PSO) and implementation on artificial neural network. J Inst Eng India Part EL Electr Eng Div 87:32–36
6. Al-Othman AK, Ahmed NA, AlSharidah ME, AlMekhaizim HA (2013) A hybrid real coded genetic algorithm—pattern search approach for selective harmonic elimination of PWM AC/AC voltage controller. Electr Power Energy Syst 44:123–133
7. Al-Othman AK, Ahmed NA, Al-Kandari M, Ebraheem HK (2007) Selective harmonic elimination of PWM AC/AC voltage regulator using hybrid RGA-PS approach. Int J Electr Comput Syst Eng, 227–233
8. Rao RV, Savsani VJ, Vakharia DP (2012) Teaching learning based optimization: an optimization method for continuous non-linear large scale problems. Inf Sci 183(1):1–15
9. Arya LD, Koshti A (2014) Anticipatory load shedding for line overload alleviation using teaching learning based optimization (TLBO). Electr Power Energy Syst 63:862–877
10. Arya LD, Koshti A, Choube SC (2012) Distributed generation planning using differential evolution accounting voltage stability consideration. Int J Electr Power Energy Syst 42(1):196–207

Estimating Capacitor Health Connected in Solar Power System Using Wavelet Transform

Shivam Srivastava and Sanjay Kumar Maurya

Abstract This paper presents an approach to estimates the health of capacitor connected in solar power applications using wavelet transform. In the approach, voltage and current waveform are analysed using wavelet transform. The characteristics of the voltage and current waveform under continuous conduction mode (CCM) and discontinuous conduction mode (DCM) are modelled using multiresolution analysis by wavelet transform. A relation is developed between wavelet coefficients of voltage and current waveforms for estimating the equivalent series resistance (ESR) of a capacitor. The simulation results show small error in estimation of the capacitor ESR. Further, the approach uses waveform analysis in estimating capacitor health; hence, the method is simple over the current injection method for estimation ESR.

Keywords Aluminium electrolytic capacitor (AEC) · Equivalent series resistance (ESR) · Wavelet coefficients · Solar PV

1 Introduction

The increasing dependency on non-conventional energy sources and distributed generators needs advanced operating techniques and management of the electricity grid to improve the power-supply reliability and quality [1]. There are varieties of electronics equipment used as essential parts in power converter affect the performance of power converter. In the study, it has been found that around the 60% failure of AEC affects the power converter and hence the reliability and quality of power [2].

Aluminium electrolytic capacitors (AEC) are extensively used in DC-DC converters due to their features of high capacitance and low price along with high voltage compared to polyester capacitors [3]. AEC is also having a limited life span limited to 3000 h, which defines the system lifetime, and it is the most sensitive element of

S. Srivastava
Department of Electrical Engineering, GLA University, Mathura, UP, India
e-mail: shivamnaman@yahoo.com

S. K. Maurya (✉)
Department of Electrical Engineering, GLA University, Mathura, UP, India
e-mail: sanjay.maurya@gla.ac.in

S. N. Singh et al. (eds.), *Advances in Power and Control Engineering*, Lecture Notes in Electrical Engineering 609, https://doi.org/10.1007/978-981-15-0313-9_18

a converter. Therefore, this needs research on their unpredictable failure and fault occurrence due to capacitor health [4].

With the degradation of the capacitor, the internal temperature rises because of ripple current, and hence temperature rises, which thereby decreases the electrolyte by evaporation. This phenomenon leads to an increase in ESR of the capacitor. The increased value of ESR is the best indicator of degradation in capacitance value [5]. There is an increase in an output voltage ripple with the increase in the ESR that helps in the prediction of the health of capacitor. In most of the application connected to the DC-DC converter has a limit of 5% voltage ripple. If the voltage increases beyond this, then there is a need to replace the capacitor. If the replacement of AEC is not made at the proper time and prolong to use AEC leads to damage the semiconductor switches and complete evaporation of the electrolyte [6]. According to manufactures, the lifespan of AEC is limited up to the value of C reduced to 20% of its initial value or ESR becomes double to its initial value [7]. The value of ESR varies from 10 mΩ to 1 Ω.

The capacitance health is monitoring technique is broadly classified as

(i) Offline technique
(ii) Online technique.

The offline technique is simple of all, but it needs to demount the capacitor from the system, making a system to put on hold during this process. ESR can be estimated at the resonant frequency, which requires an additional signal generator for the capacitor test [8]. Here, the capacitance is reliable on the frequency and inductance, which is its drawback. There is another method which is using a network analyser for the estimation of ESR and capacitance and simultaneously increasing the overall cost of the system [9]. The online technique having an advantage over offline technique is that there is no need to demount the capacitor from the system, and hence, system performance is not interrupted. In [10], suggests that the ratio of ripple voltage and ripple current are directly proportional to ESR called conditioning monitoring that requires a current sensor and voltage sensor. For current-sensorless online method [11] ac component of output voltage ripple is collected at two different instant using an external trigger an isolation amplifier. By the digital implementation, small computational error in average dc voltage would translate into a significant error in estimated capacitance value.

There is an instrument called LCR metre used for the measurement of ESR and capacitance. This addition of metre leads to an increase in the cost of the system; hence, it is the most expensive mode of estimation of ESR for the health prediction of the AEC. This method needs the capacitor to cut-out from the circuit for the overall analysis leading the pause in the application. For health monitoring of AEC, there is a need for current sensors that will lead to an increase in the stray inductance along with the distortion in the capacitance-voltage [12]. There is an external injection of current signals in the control loop is suggested in [13]. Result in the occurrence of low-frequency ripple in dc-link voltage and current. ESR and capacitance are estimated by measuring these low-frequency ripple components in the dc-link [14]. In [15], ESR estimation of a dc-dc buck converter is described. The input current is used to estimate the inductor current using least mean squares algorithm (LMS). Discrete Fourier transform (DFT) is used to determine the switching frequency components. In [16], estimation of health of capacitor connected to a solar PV grid is estimated by the injection of harmonic current and then further calculating the total impedance by that least mean square (LMS) estimation technique is used for estimation of ESR and capacitor.

2 Capacitor Model and Ripple Generation

A capacitor in combination with inductor and switch forms a filter in the dc-dc converter. For modelling a dc converter and study of ripple requires an equivalent circuit of a capacitor. The equivalent circuit is given in Fig. 1. The equivalent circuit for aluminium electrolytic capacitor consists of an ideal capacitor in series with equivalent series resistance (ESR) and equivalent series inductance (ESL). Since the value of ESL is negligible and does not play a vital role in the equivalent model of AEC, and hence only ESR and C is considered.

A boost converter is shown in Fig. 2 used for solar applications works in two operating conditions; continuous conduction mode (CCM) in Fig. 3 and discontinuous conduction mode (DCM) in Fig. 4. The voltage ripple in CCM is essential in the study of filter components health status. While in DCM, both current and voltage ripple are essential. The two modes of the DC converter are explained below.

The ripple voltage consists of mainly two parts:

One generated by ripple current

$$\Delta V_O = \frac{\Delta Q}{C} \tag{1}$$

From Fig. 3 ΔQ is given by the area of the shaded region:

$$\Delta V_C = \frac{I_O \times D \times T_S}{C} \tag{2}$$

The second part of ripple voltage caused by ESR.

$$\Delta V_C = \Delta V_{\text{ESR}} = \text{ESR} \times \Delta I_L \tag{3}$$

Fig. 1 Equivalent circuit of aluminium electrolytic capacitor

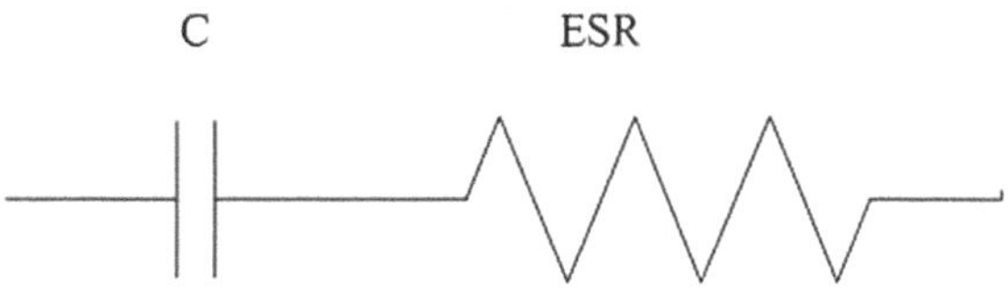

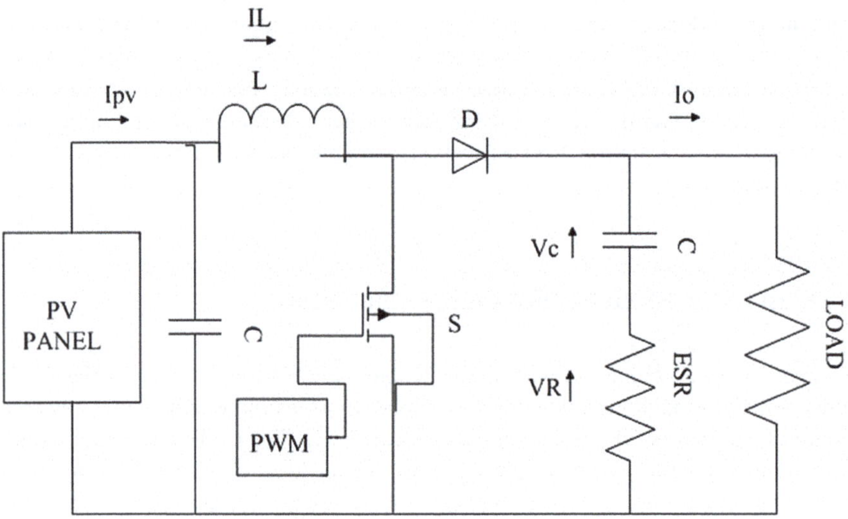

Fig. 2 Equivalent circuit of solar PV connected converter

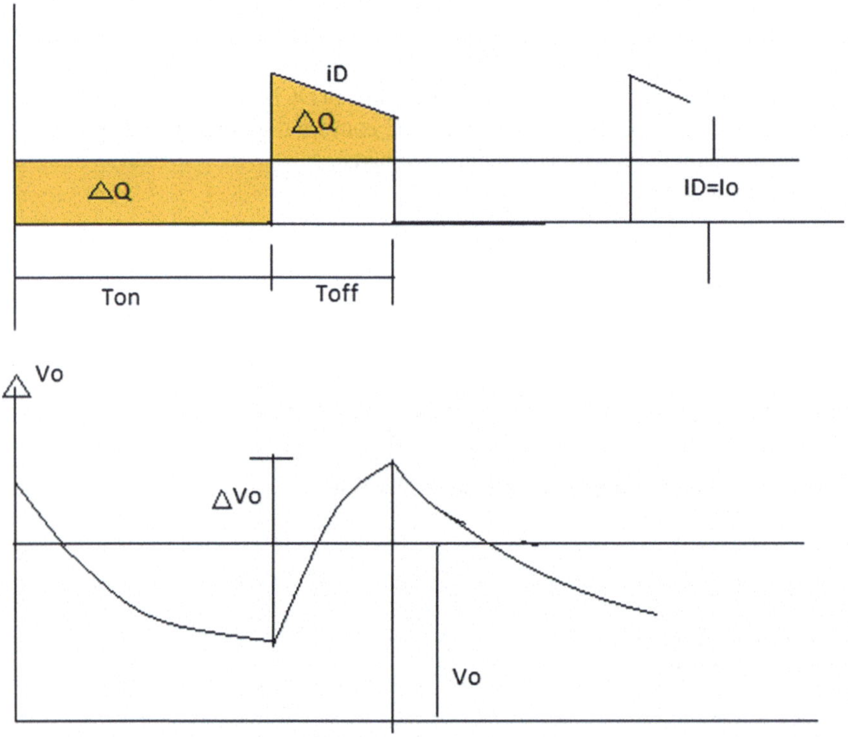

Fig. 3 Voltage ripple in boost converter

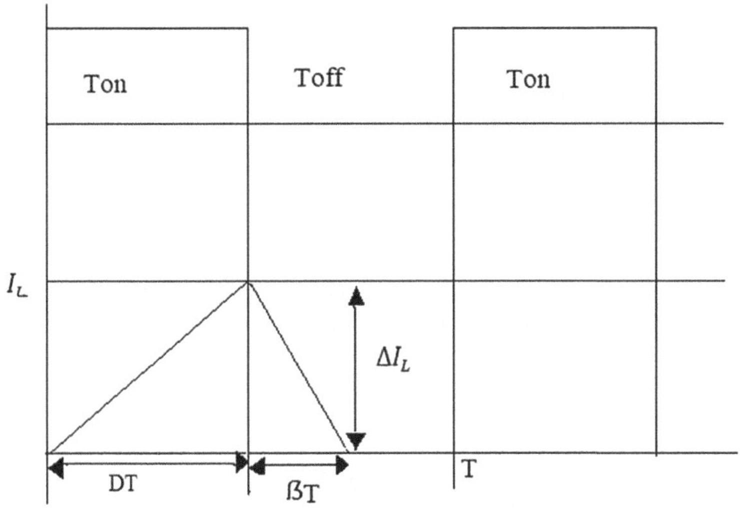

Fig. 4 Current ripple in boost converter

For DCM:
Ripple current is estimated during turn-on state as follows:

$$V_{L(\mathrm{ON})} = V_S \tag{4}$$

$$T_{\mathrm{ON}} = DT_S \tag{5}$$

$$V_{L(\mathrm{OFF})} = V_S - V_O \tag{6}$$

Form Volt-Sec Balance

$$V_S \times DT_S + (V_S - V_O)(\beta - D)T_S = 0 \tag{7}$$

Or

$$V_0 = \frac{\beta V_s}{\beta - D} \tag{8}$$

ΔI_L Ripple current in turn-on state

$$V_{L(\mathrm{ON})} = V_S = \frac{L \times \Delta I_L}{DT_S} \tag{9}$$

$$\Delta I_L = \frac{D \times T_S \times V_S}{L} \tag{10}$$

3 Proposed Method for Stand-Alone System

When a converter connected with a PV system is utilised for longer duration then the capacitor connected to the converter lead to starts to degrade, this degradation in capacitance of the capacitor can be estimated by the estimation of the ESR variation of the capacitor. Figure 5 presents a scheme for the estimation of capacitor health. From Eq. (3) and (10), it is observed that the variation in ESR value is related to the ripple generation in voltage and inductor current.

Feature of a signal in the frequency domain analysed effectively in comparison with the time domain. Fourier transform, Laplace transform, and Wavelet transform provides a tool for analysing the signal in the frequency domain. The signal with varying frequency in time is known at non-stationary signal while with the constant frequency in time is known as the stationary signal. The ripple of voltage and current from the converter is classified as non-stationary signals. A non-stationary signal can be analysed by wavelet transform. Therefore, the wavelet transform can be applied in estimation of capacitor health connected to a converter.

Case 1: CCM Mode

The variation in the ESR can be analysed by the variation in wavelet coefficients of V_{ESR} for CCM [17].

From Eq. (3)

$$\Delta I_L^{\mathrm{pk-pk}} \times \mathrm{ESR} = \left\{ V_{\mathrm{ESR}}(t)|_{t=0} - V_{\mathrm{ESR}}(t)\,|_{t=DT_S} \right\} \tag{11}$$

$$\Delta I_L^{\mathrm{pk-pk}} = \frac{D \times T_S \times V_S}{L} \tag{12}$$

From Eq. (11) and (12)

$$\frac{D \times T_S \times V_S}{L} \times \mathrm{ESR} = \left\{ V_{\mathrm{ESR}}(t)|_{t=0} - V_{\mathrm{ESR}}(t)|_{t=DT_S} \right\} \tag{13}$$

Or

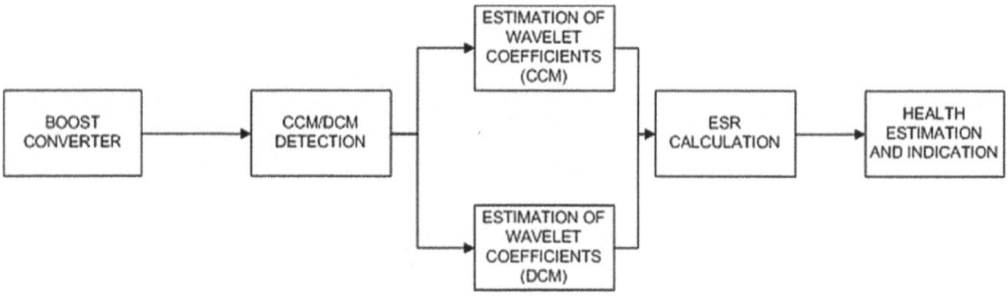

Fig. 5 Health monitoring proposed technique for AEC

$$\mathrm{ESR} = \frac{L \times \left\{ V_{\mathrm{ESR}}(t)|_{t=0} - V_{\mathrm{ESR}}(t)\big|_{t=DT_S} \right\}}{D \times T_S \times V_S} \tag{14}$$

ESR is varying from $t = 0$ to $t = DT_S$ hence from above equation

$$Y(t) = \frac{\left\{ V_{\mathrm{ESR}}(t)|_{t=0} - V_{\mathrm{ESR}}(t)\big|_{t=DT_S} \right\}}{D \times V_S} \tag{15}$$

$Y(t)$ is a term that helps in predicting the value of ESR at any instant. Here, D (duty cycle) and V_S (input voltage from solar panel) are constant. By taking a wavelet transform of Eq. (15)

$$\mathrm{CWT}_Y^{\psi}(T, S) = \Psi_Y^{\psi}(T, S) = \frac{1}{\sqrt{S}} \int_{-\infty}^{+\infty} Y(t)\Psi\left(\frac{t - T}{S}\right)\mathrm{d}t \tag{16}$$

Wavelet transform of waveform under switching condition on and off is given by Eq. (17)

$$\left[\Psi_Y^{\psi}(T, S)\right]_{(0, T_S)} = \left[\Psi_Y^{\psi}(T, S)\right]_{(0,\ DT_S)} + \left[\Psi_Y^{\psi}(T, S)\right]_{(DT_S,\ T_S)} \tag{17}$$

where $\left[\Psi_Y^{\psi}(T, S)\right]_{(0,\ DT_S)}$ wavelet coefficients of waveform from $t = 0$ to $t = DT_S$ and $\left[\Psi_Y^{\psi}(T, S)\right]_{(DT_S, T_S)}$ wavelet coefficients of waveform from $t = DT_S$ to $t = T_S$.

From experimental analysis on wavelet coefficients of V_{ESR} at higher resolution (i.e. $S > 2$) coefficients from $t = DT_S$ to $t = T_S$ is twice of coefficients from $t = 0$ to $t = DT_S$.

i.e. $\left[\Psi_Y^{\psi}(T, S)\right]_{(DT_S, T_S)} = 2\left[\Psi_Y^{\psi}(T, S)\right]_{(0,\ DT_S)}$

Hence

$$\left[\Psi_Y^{\psi}(T, S)\right]_{(0, T_S)} = 3\left[\Psi_Y^{\psi}(T, S)\right]_{(0, DT_S)} \tag{18}$$

For ESR calculation,

$$\frac{\mathrm{ESR}(t)}{\mathrm{ESR}(0)} = \frac{\left[\Psi_Y^{\psi}(T, S)\right]_{(0, T_S)}(t)}{\left[\Psi_Y^{\psi}(T, S)\right]_{(0, T_S)}(0)} \tag{19}$$

Hence, ESR can be estimated by analysing the wavelet coefficients of V_{ESR} waveform from $t = 0$ to $t = DT_S$.

$$\frac{\mathrm{ESR}(t)}{\mathrm{ESR}(0)} = \frac{\left[\Psi_Y^{\psi}(T,S)\right]_{(0,DT_S)}(t)}{\left[\Psi_Y^{\psi}(T,S)\right]_{(0,DT_S)}(0)} \tag{20}$$

Case 2: DCM Mode

The variation in the ESR can be analysed by the variation in coefficients of V_{ESR} for DCM.

$$\Delta I_L^{\mathrm{pk-pk}} \times \mathrm{ESR} = \left\{ V_{\mathrm{ESR}}(t)\Big|_0^{DT_S} + V_{\mathrm{ESR}}(t)\Big|_{DT_S}^{\beta T_S} \right\} \tag{21}$$

$$\mathrm{ESR} = Z = \frac{\left\{ V_{\mathrm{ESR}}(t)\Big|_0^{DT_S} + V_{\mathrm{ESR}}(t)\Big|_{DT_S}^{\beta T_S} \right\}}{\Delta I_L^{\mathrm{pk-pk}}} \tag{22}$$

Taking wavelet transform of Eq. (22)

$$\mathrm{CWT}_Z^{\psi}(T,S) = \Psi_Z^{\psi}(T,S) = \frac{1}{\sqrt{S}} \int_{-\infty}^{+\infty} Z(t)\Psi\left(\frac{t-T}{S}\right)\mathrm{d}t \tag{23}$$

Wavelet coefficient of waveform is given by

$$\left[\Psi_Z^{\psi}(T,S)\right]_{(0,\beta T_S)} = \frac{\left[\Psi_Z^{\psi}(T,S)\right]_{(0,DT_S)} + \left[\Psi_Z^{\psi}(T,S)\right]_{(DT_S,\beta T_S)}}{I_{L\,\mathrm{avg}}} \tag{24}$$

The wavelet coefficient of waveform after pre- and post-switching are found similar at $S > 2$. Hence

$$\begin{aligned}\frac{\mathrm{ESR}(t)}{\mathrm{ESR}(0)} &= \left(\frac{\left[\Psi_Z^{\psi}(T,S)\right]_{(0,DT_S)}(t)}{I_{L\,\mathrm{avg}}(\mathrm{t})}\right)\left(\frac{I_{L\,\mathrm{avg}}(0)}{\left[\Psi_Z^{\psi}(T,S)\right]_{(0,DT_S)}(0)}\right) \\ &= \left(\frac{\left[\Psi_Z^{\psi}(T,S)\right]_{(DT_S,\beta T_S)}(t)}{I_{L\,\mathrm{avg}}(\mathrm{t})}\right)\left(\frac{I_{L\,\mathrm{avg}}(0)}{\left[\Psi_Z^{\psi}(T,S)\right]_{(DT_S,\beta T_S)}(0)}\right)\end{aligned} \tag{25}$$

Hence, in DCM mode ESR can be calculated as follows:

$$\frac{\mathrm{ESR}(t)}{\mathrm{ESR}(0)} = \left(\frac{\left[\Psi_Z^{\psi}(T,S)\right]_{(0,DT_S)}(t)}{I_{L\mathrm{avg}}(\mathrm{t})}\right)\left(\frac{I_{L\mathrm{avg}}(0)}{\left[\Psi_Z^{\psi}(T,S)\right]_{(0,DT_S)}(0)}\right) \tag{26}$$

The colour scale range gives the idea about variation in the frequency range; high scales correspond to low frequencies while low scales correspond to high frequencies. Along with this variation in frequency this colour variation in the waveform wavelets gives the idea about the wavelet coefficients. The different colour regions show the variation in the wavelet coefficient magnitudes as shown in Figs. 6 and 7. The high intensity colour regions shows the large variations in the coefficient magnitudes while the low intensity region shows small variation in coefficient magnitudes. The coefficients of V_{ESR} vary in the same proportion as the ESR changes. Hence, by analysis of wavelet coefficients of V_{ESR} alone the health estimation of capacitor is obtained.

By this variation in the wavelet coefficient, one can estimate in the change in ESR value. When there is rapid change in the coefficient then there is a sharp change ESR value. This increasing value of wavelet coefficients led to estimate the health of capacitor. This result is can be collaborate with the data obtained from the temperature detection sensor. All AEC have their predefined temperature range.

4 Simulation and Result Discussion

A PV system is simulated in MATLAB with the converter specification of following specifications (Tables 1, 2, 3 and 4).

Capacitor = 470 μF
Load resistance = 10 Ω
Load inductance = 10 mH
Initial ESR = 56 mΩ
Switching frequency = 25 kHz

Table 1 Evaluated values of ESR in CCM mode of operation for irradiance = 1000 W/m^2

ESR (mΩ)	Calculated ESR (mΩ)	Error (%)
84	83.9	0.07
112	111.8	0.10
168	167.6	0.238
224	223.8	0.089

Table 2 Evaluated values of ESR in CCM mode of operation for irradiance = 850 W/m^2

ESR (mΩ)	Calculated ESR (mΩ)	Error (%)
84	83.832	0.2
112	111.944	0.05
168	167.664	0.2
224	223.328	0.3

Table 3 Evaluated values of ESR in DCM mode of operation for irradiance = 200 W/m^2

ESR (mΩ)	Calculated ESR (mΩ)	Error (%)
84	83.44	0.60
112	111.496	0.45
168	166.88	0.66
224	222.88	0.50

Table 4 Evaluated values of ESR in DCM mode of operation for irradiance = 250 W/m^2

ESR (mΩ)	Calculated ESR (mΩ)	Error (%)
84	83.56	0.60
112	111.44	0.50
168	166.88	0.667
224	222.88	0.50

The simulation is performed for solar irradiance of 1000 W/m^2. The converter is operated in CCM operation mode. It is seen that in this mode of operation with the regular utilisation capacitor, there is significant fall in capacitance arises with the increment in the value of ESR from 56 to 224 mΩ. By applying the proposed method, the value of ESR is calculated using Eq. (20). For CCM operation mode, the error is calculated at different solar irradiance as shown in Tables 1, 2, 3 and 4.

Similarly, the same system is allowed to operate under DCM operation mode. To operate at DCM operation solar irradiance in reduced to 200 W/m^2. By applying the proposed method, the value of ESR is calculated using Eq. (26). For DCM operation mode, the error is calculated at different solar irradiance. Figures 6 and 7 show the wavelet transform of voltage waveform under CCM and DCM mode.

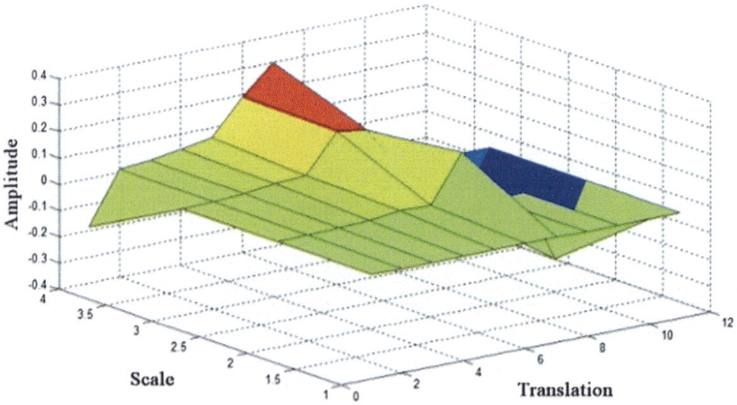

Fig. 6 Wavelet transform voltage ripple, ESR 56, CCM mode, and irradiance = 1000 W/m^2

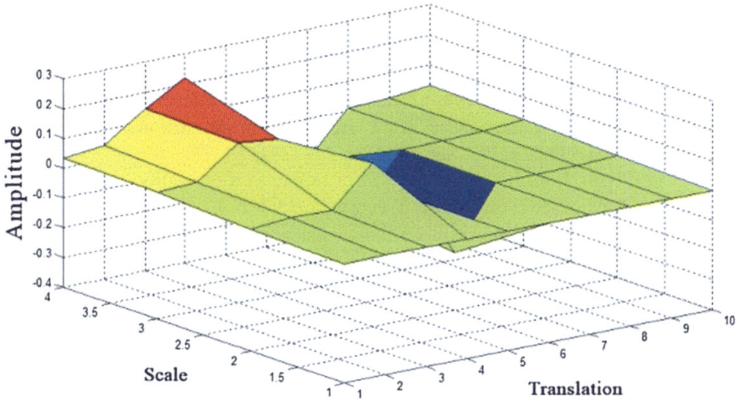

Fig. 7 Wavelet transform voltage ripple, DCM mode irradiance = 250 W/m2, and ESR = 224 mΩ

Here, the error in calculated ESR and estimated ESR is found to be less than 0.5% in CCM operation mode. Further the error in calculated ESR and estimated ESR is found to be less than 0.68% in DCM operation mode.

5 Conclusion

This paper has presented the method of health monitoring of AEC in the converter applied solar PV system by estimation wavelet coefficients of V_{ESR}. The increase in ESR changes the shape of voltage and the inductor current shape. Here, the wavelet coefficients of V_{ESR} in both CCM and DCM are obtained for estimating the ESR value and capacitor health. The proposed technique having an advantage over others as there is no need for any other type instrument or current and voltage injection for evaluation of ESR variation. Hence, it is cost effective. This scheme can also be implemented solar grid-connected system also. The error in calculated ESR and estimated ESR is found to be less than 0.5% in CCM operation mode. Further, the error in calculated ESR and estimated ESR is found to be less than 0.68% in DCM operation mode

References

1. Carrasco JM et al (2006) Power-electronic systems for the grid integration of renewable energy sources: a survey. IEEE Trans Ind Electron 53(4):1002–1016
2. Yao K, Tang W, Hu W, Lyu J (2015) A current-sensor-less online ESR and C identification method for output capacitor of buck converter. IEEE Trans Power Electron 30(12):6993–7005
3. Guan Y, Jin SZ, Wu LF, Pan W, Liu YM, Zhang J (2011) Power supply prognostics and health management of high reliability electronic systems in rugged environment. Key Eng Mater

474–476:1195–1200
4. Gasperi ML (2005) Life prediction modeling of bus capacitors in AC variable frequency drives. IEEE Trans Ind Appl 41(6):1430–1435
5. Venet P, Darnand H, Grellet G (1993) Detection of faults of filter capacitors in a converter. application to predictive maintenance, vol 2. In: The 15th international telecommunication energy conference, INTELEC '93, Sep 1993, pp 229–234
6. Wang H et al (2014) Transitioning to physics-of-failure as a reliability driver in power electronics. IEEE J Emerg Sel Topics Power Electron 2(1):97–114
7. CDE Cornell Dubilier, "Aluminum electrolytic capacitor application guide." [Online] http://www.cde.com/resources/catalogs/AEappGUIDE.pdf
8. Amaral AMR, Cardoso AJM (2005) An ESR meter for high frequencies. In: Proceedings of the 6th IEEE conference on power electronics and drive systems, Kuala Lumpur, Malaysia, 24 Nov–4 Dec 2005
9. Li Y, Figueroa D, Rodriguez J, Huang L, Liao J (1998) A new technique for high frequency characterization of capacitors. In: Proceedings of 1998 electronic components and technology conference, Seattle, WA, USA, 25–28 May, pp 1384–1390
10. Harada K, Katsuki A, Fujiwara M (1993) Use of ESR for deterioration diagnosis of electrolytic capacitors. IEEE Trans Power Electron 8(4):355–361
11. Yao Kai, Tang W, Hu W, Lyu J (2014) A current-sensorless online ESR and C identification method for output capacitor of buck converter. IEEE Trans Power Electron, Early Access Articles
12. Agarwal N, Arya A, Ahmad MW, Anand S (2016) Lifetime monitoring of electrolytic capacitor to maximize earnings from grid-feeding PV system. IEEE Trans Ind Electron 63(11):7049–7058
13. Nguyen TH, Lee DC (2015) Deterioration monitoring of DC-link capacitors in AC machine drives y current injection. IEEE Trans Power Electron 30(3):1126–1130
14. Abo-Khalil AG, Lee DC (2008) DC-link capacitance estimation in AC/DC/AC PWM converters using voltage injection. IEEE Trans Ind Appl 44(5):1631–1637
15. Pu X, Nguyen TH, Lee DC, Lee SG (2009) Capacitance estimation of DC-link capacitors for single-phase PWM converters. In: Proceedings IEEE ınternational power electronics and motion control conference, Wuhan, China, May 2009, pp 1656–1661
16. Agarwal N, Arya A, Ahmad MW, Anand S (2018) Quasi-online technique for health monitoring of capacitor in single phase solar inverter. IEEE Trans Power Electron 33(6):5283–5291
17. Ahmad Md, Agarwal N, Anand S (2016) Online monitoring technique for aluminium electrolytic capacitor in solar PV based DC system. IEEE Trans Ind Electron 63(11):7059–7066

Design of Equidistant Hexagonal Coil System for Demagnetization of Naval Vehicles

Ratan Singh, Sonal Jain, Vikas Singh and Faruk Kazi

Abstract In various applications of defence and research such as demagnetization of naval vehicles and satellites, removal of residual magnetic field plays an essential role. For such use cases, conventional way of demagnetization is the use of Helmholtz coils and Merritt coils. These coil systems produce uniform magnetic field, but the volume of the uniform region is relatively smaller. Therefore, for the applications which require patch of uniform magnetic field for longer longitudinal length of the region such as deperming process of Submarine, Aircraft Carrier; these coil systems become infeasible to implement practically due to their enormous size. Another impediment in implementation of above-mentioned coil system is that when size of coil system becomes larger, the requirement current also increases for the same degree of results. This increasing amount of current poses issues of more complexities in electrical control circuits, heat dissipation and cost. Another disadvantage with such system is that a slight shift from the optimized position causes a larger change in uniformity, and thus, such systems are not robust with shift in location. This paper proposes a novel method for the design of a higher-order equidistant coil system which overcomes these flaws and presents a technique to compute the ampere-turns requirement of each coil for generation of higher-degree homogeneous magnetic field. The system so designed has more feasible design parameters as compared to the conventional systems for high uniformity applications. The proposed system is sturdier and fault tolerant against any deviation or error from the ideal design parameters. The theoretical results are matched against the finite element simulation software Opera which are in close agreement.

Keywords Hexagonal coils of higher orders · Equidistant coil · Field homogeneity · Deperming · Opera

---

R. Singh (✉) · V. Singh · F. Kazi
Department of Electrical Engineering, Veermata Jijabai Technological Institute, Mumbai, Maharashtra, India
e-mail: ratansingh648@gmail.com

S. Jain
Department of Information Technology, A.P. Shah Institute of Technology, Thane, Maharashtra, India
e-mail: sonaljain21691@gmail.com

S. N. Singh et al. (eds.), *Advances in Power and Control Engineering*, Lecture Notes in Electrical Engineering 609, https://doi.org/10.1007/978-981-15-0313-9_19

1 Introduction

Naval vehicles consist of ferromagnetic material as the major composition in their structure. Due to the very high susceptibility of ferromagnetic metals, they tend to get magnetized over the period of time due to continuous exposure to the earth's magnetic field. This induced magnetic field in the naval vehicle's body is called the magnetic signature of the vehicles. This magnetic signature makes vehicles prone to magnetic mines. Due to the continuous exposure of the ship into earth's magnetic field, it causes failure of degaussing coils which are the first line of defence against the magnetic mines. In order to shield ships and submarines against magnetic mines, it becomes essential to wipe out the residual magnetic field by subjecting the vessel to the alternating magnetic field. Therefore, it becomes necessary to subject the naval vessels for the wiping under the deperming coils periodically. A number of laboratory experiments are done to analyse the effect of transient and static magnetic field on various systems, living or non-living using field generating electric coils. Homogenous magnetic field is of great importance in the area of magnetic resonance imaging (MRI) [1–3], magnetometers, and eddy current probes, magnetic traps, magnetic hypersonic flights [4], optical pumping and lightbeat experiment, double resonance [5, 6] etc. In light of previous theoretical studies suggesting perfect uniformity in the axial direction, various coil systems namely Helmholtz and Merritt are used for this purpose at different frequencies and for different volume size. Merritt coil is used for large volume while Helmholtz coil is used for small laboratory-based experiments.

However, these coil systems produce uniform patch of magnetic field for a relatively smaller region as compared to the space occupied [7]. In case of deperming of submarines, aircraft carriers where the longitudinal length of object is more than 10 times of the transverse length, the implementation of above-mentioned coil systems pose two major impediments. Firstly, the size of coil system becomes tremendously large allowing only a very small portion of the system being useful. Due to the tremendous geometrical size, it leads to very high rate of power consumption which in turn leads to the need of expensive peripheral circuits for generation and control of current of order 10^3 A depending on design of coil. Moreover, it becomes difficult to maintain the shape of circular coils if their radius increases more than 25 cm [8].

The amount of heat generated because of such high current is tremendous and it requires special materials and coolants for the coil systems which results in additional incurring cost of maintenance and durability. Secondly, attempts of optimizing the coil systems comprising coils more than four are restricted due to the rigorous mathematical computation and optimization required for the design of the system. Figure 1 depicts the audacity of the coils required for the demagnetization of a submarine in its 100% of the region occupied which does not offer a high degree of uniformity and hence poor demagnetization effect.

In such cases, equidistant coil systems perform better as compared to the optimized coil system. They provide a very large region of uniform magnetic field as

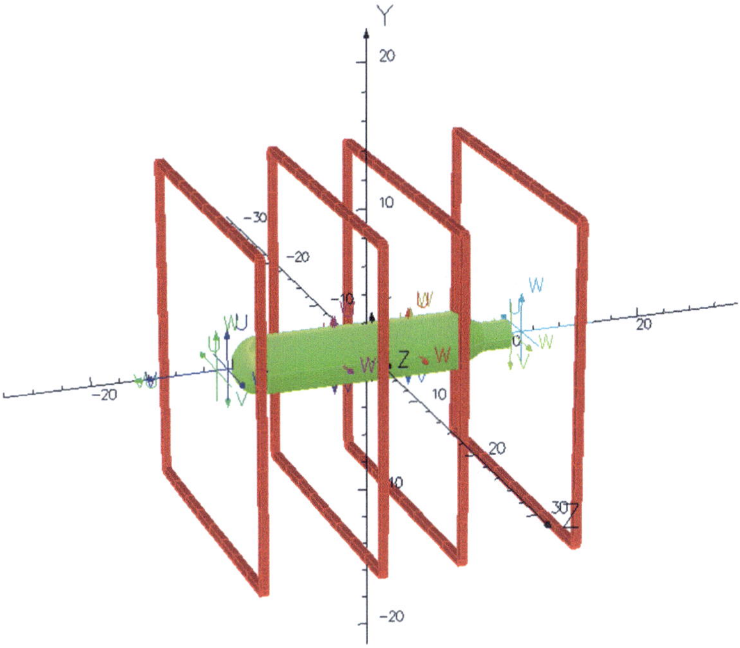

Fig. 1 Meritt coils used in deperming of submarine

compared to the optimized coil systems on similar coil dimensions with minor overshoot in the amplitude of the magnetic field. Another advantage of using equidistant coil systems is that they provide a wide range of design which makes the physical implementation more practical and robust against the designing errors. The paper is divided into the following five sections—Sect. 2 comprises the theoretical discussion of the measurement of uniformity of the magnetic field produced by a coil system. Section 3 elaborates the proposed methodology employed in design of the equidistant hexagonal coil system of higher order in detail. Section 4 discusses the theoretical results and their verification using the finite element analysis software Opera, followed by conclusions in Sect. 5.

2 Homogeneity of Magnetic Field of a Coil System

The magnetic flux density produced by the hexagonal coil can be calculated using the Biot–Savart's law [9]. Using the Biot–Savart's law, the magnetic flux density (magnetic field) at a point situated along the axis of coil system at a distance 'z' from the centre of hexagonal coil of side length 'd' carrying current 'I' is given as

$$B(z) = \frac{2.598\,\mu I d^2}{2\pi\left(0.7499d^2 + z^2\right)\left(d^2 + z^2\right)^{1/2}} \tag{1}$$

For an equidistant hexagonal coil system, net magnetic field at any point is computed by superposition of magnetic field produced by each of the coils [9]. In order to optimize the uniformity of the magnetic field generated by the coil system, there are multiple optimizing techniques available like Taylor's series expansion, Legendre's polynomial, Maclaurin series expansion [10] and Chebyshev polynomials. However for the coil systems of higher order, the optimization function of the magnetic field becomes highly nonlinear, non-convex and complex to solve which poses an impediment of expensive computational resources.

There are multiple ways to estimate the uniformity of magnetic field. In this paper, we are following the given formula to compute the deviation of magnetic field of the system,

$$\eta = 1 - \frac{B_{\max} - B_{\min}}{B_{\text{avg}}} \tag{2}$$

Here, $B_{\max}$ is maximum magnetic flux density, $B_{\min}$ is minimum magnetic flux density, B_{avg} is average magnetic flux density, and η is uniformity of the field [11].

3 Proposed Methodology

In this paper, we have demonstrated the methodology for designing the coil systems comprising seven, eight and nine hexagonal coils. With similar methodology, coil systems of higher orders can also be designed.

3.1 Seven Equidistant Hexagonal Coil Systems

Let us assume a coil system comprising seven hexagonal coils each with a side length 'd' separated by a common equal distance of $0.5d$. Let the location of coils be at 0, $\pm 0.5d$, $\pm d$, $\pm 1.5d$ for the pair of coils starting from the centre as shown in Fig. 2. In this case, we have chosen the constant separation as $0.5d$, and this can be chosen any other value as well without losing the sense of generality. Let N_1, N_2, N_3 and N_4 be the corresponding ampere-turns in the coil and 'I' be the common current through them. According to the superposition theorem [9], the magnetic field at any point in the coil system can be expressed as

$$\vec{B}(z) = N_1\overrightarrow{B_1}(z) + N_2\overrightarrow{B_2}(z) + N_3\overrightarrow{B_3}(z) + N_4\overrightarrow{B_4}(z) \tag{3}$$

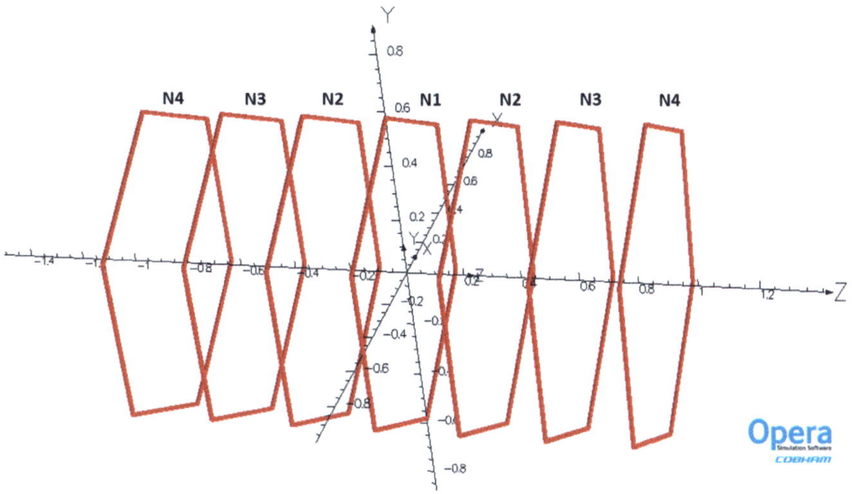

Fig. 2 Seven equidistant hexagonal coil systems

Here, $\overrightarrow{B_1}(z)$, $\overrightarrow{B_2}(z)$, $\overrightarrow{B_3}(z)$ and $\overrightarrow{B_4}(z)$ represent the total magnetic field at a point situated distance 'z' from the origin created by the pair of coils placed symmetrically about the origin. In rest of the paper, we will represent the same in matrix notation as

$$\vec{B}(z) = \left[\overrightarrow{B_1}(z) \ \overrightarrow{B_2}(z) \ \overrightarrow{B_3}(z) \ \overrightarrow{B_4}(z) \right] \begin{bmatrix} N_1 \\ N_2 \\ N_3 \\ N_4 \end{bmatrix} \tag{4}$$

To evaluate the turns in each of the coils, it is important to choose equipotential points distributed across the length of coil system. Equipotential points are those points in the space within the coil system which have magnetic field of equal strength [12]. Number of equipotential points should be equal to the number of pairs of coils placed symmetrically around the origin. It has been observed that if equipotential points are chosen close to each other, the uniformity in that region is very high. However, this choice causes increase the load over the ampere-turns of the coils. So there is a trade-off between the ampere-turn of the coils and uniformity. In this case, we are choosing equipotential points distributed equidistantly within the 50% of the space occupied by the coil system, i.e. $\pm 0.75d$, $\pm 0.5d$, $\pm 0.25d$ and 0. On evaluating the above equations at equipotential points and solving for N_1, N_2, N_3 and N_4, we get

$$\frac{N_2}{N_1} = 1.07808, \ \frac{N_3}{N_1} = 0.85678 \text{ and } \frac{N_4}{N_1} = 2.04046 \tag{5}$$

This design results in a uniformity of 99.9569% in the central half of the coil system. On substituting the ampere-turns so computed into the equation of magnetic field, the central magnetic field becomes

$$B(z) = \frac{2.3742 \times 10^{-6}\, I}{d} \tag{6}$$

3.2 Eight Equidistant Hexagonal Coil Systems

In this coil system, eight hexagonal coils of side length 'd' are kept along the same axis with 'a' as the common length of separation. Let the location of coils be at $\pm 0.25d$, $\pm 0.75d$, $\pm 1.25d$ and $\pm 1.75d$ for the pair of coils starting from the centre as shown in Fig. 3. Let N_1, N_2, N_3 and N_4 be the corresponding ampere-turn of the pair of coils and 'I' be the current in central coil. The magnetic field generated can be computed using the (4) as number of pair of coils in both of the system is four. As discussed in the previous section, we are selecting ± 0.875, $\pm 0.58333d$, $\pm 0.29167d$ and 0 as the equipotential points.

In our case, we have selected the constant difference of separation as $0.5d$. On evaluating magnetic field on equipotential points and solving for the ampere-turns of the coil system results in

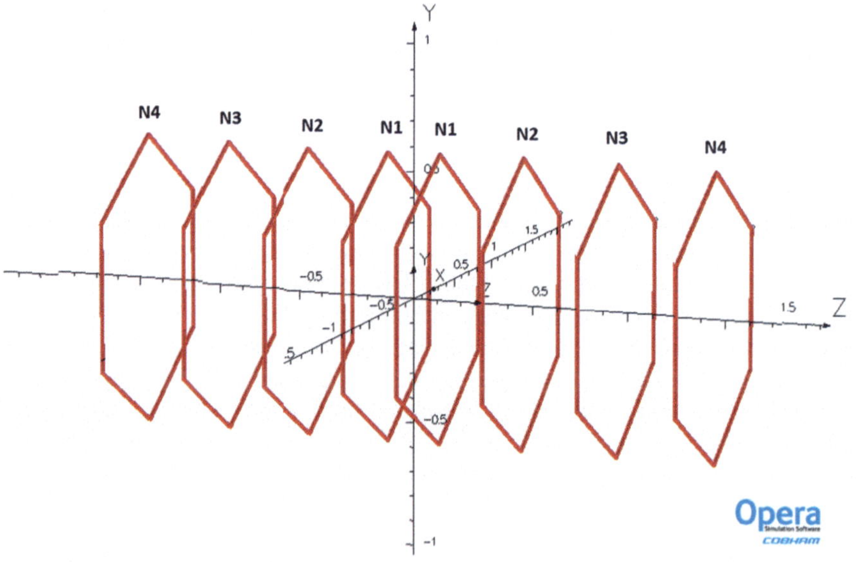

Fig. 3 Eight equidistant hexagonal coil systems

$$\frac{N_2}{N_1} = 0.96983, \ \frac{N_3}{N_1} = 1.18190 \text{ and } \frac{N_4}{N_1} = 1.38164 \tag{7}$$

For this choice of design parameter, we are getting a uniformity of 99.9569% over the half of region occupied by the coil system. On back-substituting the values of N_1, N_2, N_3 and N_4 in (4), central magnetic field for the coil system becomes

$$B(z) = \frac{2.4469 \times 10^{-6}\, I}{d} \tag{8}$$

3.3 Nine Equidistant Hexagonal Coil Systems

Consider nine hexagonal coils each with a side length 'd' placed symmetrically around origin separated by a constant distance $0.5d$. Let the location of coils be at 0, $\pm 0.5d$, $\pm d$, $\pm 1.5d$ and $\pm 2d$ for the pair of coils starting from the centre as shown in Fig. 4. Let N_1, N_2, N_3, N_4 and N_5 be the number of turns in pair coils starting from centre to end. Let 'I' be the common constant current through each of the coil. The magnetic field of the coil system can be computed using

$$\vec{B}(z) = \left[\overrightarrow{B_1}(z)\ \overrightarrow{B_2}(z)\ \overrightarrow{B_3}(z)\ \overrightarrow{B_4}(z)\ \overrightarrow{B_5}(z)\right] \begin{bmatrix} N_1 \\ N_2 \\ N_3 \\ N_4 \\ N_5 \end{bmatrix} \tag{9}$$

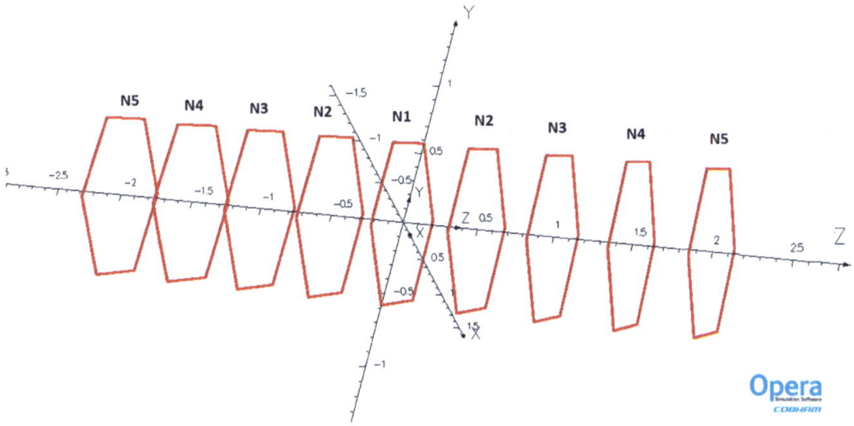

Fig. 4 Nine equidistant hexagonal coil systems

As discussed previously, we need to select five equipotential points distributed across the length of central half of the coil system. In this case, the selected equipotential points are situated at $\pm d$, $\pm 0.75d$, $\pm 0.5d$, $\pm 0.25d$ and 0.

On evaluating the (5) at the above-mentioned equipotential points and solving for N_1, N_2, N_3, N_4 and N_5, we get

$$\frac{N_2}{N_1} = 1.08812, \quad \frac{N_3}{N_1} = 0.80012, \quad \frac{N_4}{N_1} = 2.85648$$
$$\text{and } \frac{N_5}{N_1} = -1.79081 \tag{10}$$

On substituting the above-computed value of ampere-turns and positioning of the coil system equation for maximum magnetic field at centre becomes

$$B(z) = \frac{2.4141 \times 10^{-6}\, I}{R} \tag{11}$$

The uniformity for the system designed for the above parameters is 99.923% for the central half patch of the volume occupied by the system.

4 Finite Element Method-Based Simulation on Opera

The aforementioned theoretical results are verified by computing the magnetic distribution using a FEM-based software Opera. The inputs of a finite element simulation like geometry of the mesh, material and the meshing technique are handled using Opera. With the application of appropriate mesh generator, finite element simulation proves an essential method for solving complex problems [9, 13]. FEM can handle boundary condition with precise results.

4.1 Uniformity of Seven Equidistant Hexagonal Coil Systems

The simulation for seven equidistant hexagonal coil systems with side length 1 m was performed with 1 A current in central coil. The magnetic field of strength 2.4469 μT was generated at the centre. The homogeneity over this region 99.9569% as depicted in Fig. 5. This produces a uniform region of length 350% of that of Merritt coil of the same dimensions.

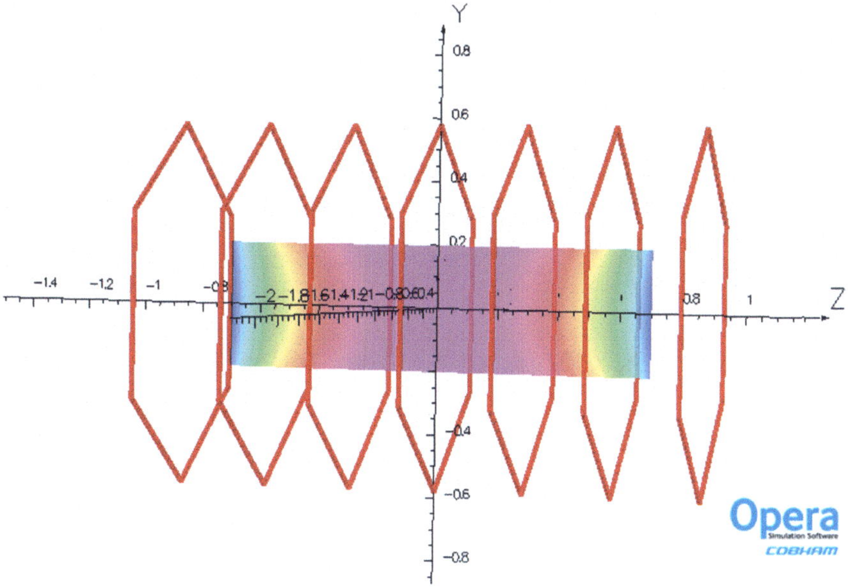

Fig. 5 FEM simulation of seven equidistant hexagonal coil systems

4.2 Uniformity of Eight Equidistant Hexagonal Coil Systems

The simulation for eight equidistant hexagonal coil systems with side length 1 m was performed with 1 A current in the central coil. The magnetic field at the centre was observed as 2.3742 μT. The homogeneity over this region 99.8% is depicted in Fig. 6 which is in close agreement with analytical results. The uniform region is approximately 400% of that of in Merritt coil of same dimensions.

4.3 Uniformity of Nine Equidistant Hexagonal Coil Systems

The simulation for a Nine Equidistant Hexagonal coil system with side length 1 m and 1 A as current in central coil was performed using OPERA. The central magnetic field was 2.4141 μT. The homogeneity over this region 99.865% is depicted in Fig. 7. The length of uniform region is 450% times of the uniform region in Merritt coil of same dimensions.

Table 1 contains the summary of the design parameters versus the percentage uniformity of the magnetic field produced by the various coil system designed using the above methodology.

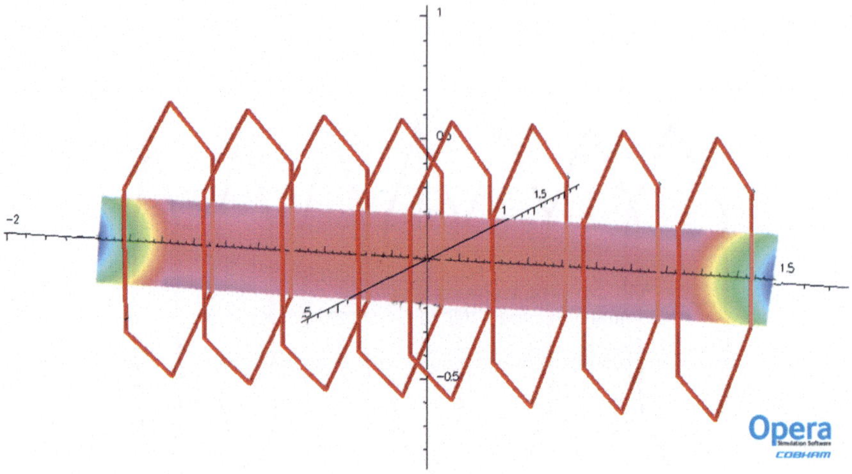

Fig. 6 FEM simulation eight equidistant hexagonal coil systems

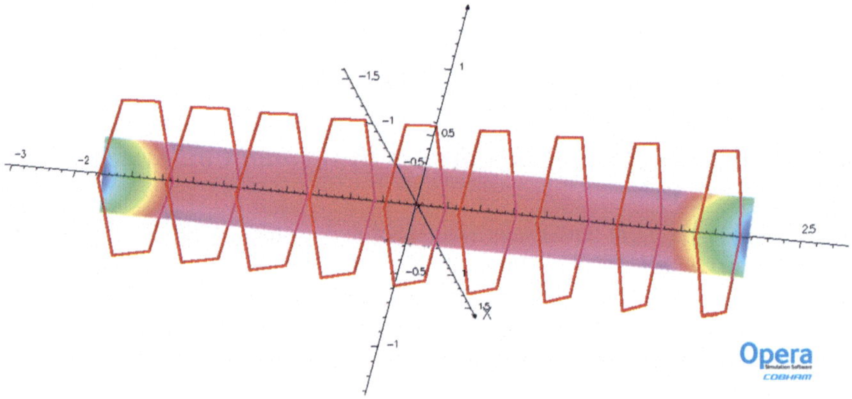

Fig. 7 FEM simulation of nine equidistant hexagonal coil systems

Table 1 Summary of hexagonal equidistant coil systems

Number of coils	Positioning of coils	Ampere-turn ratio	% Unit
7	0, $\pm 0.5d$, $\pm d$, $\pm 1.5d$	1: 1.08: 0.86: 2.04	99.957
8	$\pm 0.25d$, $\pm 0.75d$, $\pm 1.25d$, $\pm 1.75d$	1: 0.97: 1.18: 1.38	99.923
9	0, $\pm 0.5d$, $\pm d$, $\pm 1.5d$, $\pm 2d$	1: 1.08: 0.8: 2.85: −1.79	99.868

5 Conclusion

The analysis of seven, eight and nine equidistant hexagonal coil systems produces magnetic fields with higher uniformity and comparatively involves less intensive computations in design using Maclaurin series. Degree of uniformity over the half of the space occupied by the coil system is 99% which offers high uniformity in comparatively larger region of space occupied. In comparison to that, a Helmholtz coil of unit radius produces a magnetic field of uniformity of 99% in only 0.25 m long patch which is only 25% of the total space occupied by the coil system. Although the Merritt coil of unit dimensions produce a magnetic field of the same degree of uniformity in a region of 0.5 m, it is not feasible to use it for the applications where longitudinal length is several times greater than transverse length.

In above-discussed cases, the magnetic field of uniformity 99% exists in region more than thrice the length of Helmholtz and Merritt coil system for coils of equal dimensions. A triaxial magnetic field can be simulated using the three sets of coil in transverse direction.

Acknowledgements The authors would like to acknowledge the support of Centre of Excellence in Complex and Nonlinear Dynamic Systems (CoE-CNDS), VJTI, Mumbai, India, under TEQIP-II (subcomponent 1.2.1).

References

1. Pittman ME, Waidelich DL (1964) Three and four coil systems for homogeneous magnetic fields. IEEE Trans Aerosp 2(1):36–45
2. Di Barba P, Dughiero F, Sieni E (2010) Magnetic field synthesis in the design of inductors for magnetic fluid hyperthermia. IEEE Trans Magn 46(8):2931–2934
3. Bhatt V, Rautela RS, Sharma P, Tiwari DC, Khushu S (2010) Design and development of helmholtz coil for hyperpolarized MRI. Timarpur, Delhi (India) 110054
4. Brooks N, Baldwin T (2002) Methodology for universal synthesis of magnetic designs based on field specifications. In: Proceedings of the thirty-fourth southeastern symposium on system theory (Cat. No. 02EX540), pp 113–117
5. Cvetkovic D, Cosic I (2007) Modelling and design of extremely low frequency uniform magnetic field exposure apparatus for in vivo bioelectromagnetic studies. In: 2007 29th annual international conference of the IEEE engineering in medicine and biology society, Lyon, pp 1675–1678
6. Firester AH (1966) Design of square helmholtz coil systems. Rev Sci Instrum 37(9):1264–1265
7. Magdaleno Adame S, Olivares Galvan JC, Campero Littlewood E, Escarela Perez R, Blanco Brisset E (2010) Coil systems to generate uniform magnetic field volumes. In: Excerpt from the proceedings of the COMSOL conference, pp 1–13
8. Merritt R, Purcell C, Stroink G (1983) Uniform magnetic field produced by three, four, and five square coils. Rev Sci Instrum 54(7):879–882
9. Griffiths D (2014) Introduction to electrodynamics. Pearson, Harlow
10. Azipurua MA (2012) A semi-analytical method for the design of coil-systems for homogeneous magnetostatic field generation. Prog Electromagn Res B 37:171–189
11. Herceg D, Juhas A, Milutinov M (2009) A design of a square coil system for a biomagnetic experiment. 22(3):285–292

12. Rubens SM (1945) Cube-surface coil for producing a uniform mag-netic field. Rev Sci Instrum 16(9)
13. Michaelides A, Aird G, Riley C, Simkin J (2004) Electromagnetic simulation: successful applications and future challenges. Vector Fields Ltd 24 Bankside, Kidlington, Oxford OX5 1JE United Kingdom
14. Kirschvink JL (1992) Uniform magnetic fields and double wrapped coil systems: improved techniques for the design of bioelectromagnetic ex-periments. Bioelectromagnetics 13(5):401–411

Design Aspects of the Future IoT-Based On-Road Charging of Electric Vehicles

Victor George, B. Deepak, Shubham Kumar, Shobhit Kumar, Sindhu Reddy and S. Ashok

Abstract Electric vehicles (EVs) are gaining popularity from all parts of the world even though the public charging stations are not getting the required attention. As battery size, range of the vehicle, and charging time of batteries are the main concerns in EV domain, an alternate solution to these issues could be dynamic on-road charging. Although more investments in dynamic charging infrastructure would occur at the installation stage, much more cost-saving can be accomplished from the reduced battery size. Various design aspects include different technologies used for wireless power transfer through a prototype model are studied. An enhanced communication scheme for the EV charging station is proposed for the effective coordination between the dynamic charging scheme and the charging stations. Simulation studies are also carried out to test for the maximum efficiency of transferred power and variation of state of charge of the battery.

Keywords Electric vehicle · IoT · On-road charging · Wireless power transfer · State of charge

1 Introduction

The electrification for transportation has been carrying out for many years to reduce the demand on fossil fuels and switch to renewable resources. Until now, the electric vehicles (EVs) are not so attractive to consumers even with many government incentive programs. Lack of sufficient EV charging station may lead to interruptions in the journey and transportation. A better alternative could be dynamic wireless power transfer. Wireless power transfer (WPT) using magnetic resonance is the technology which could set human free from the annoying wires. In fact, the WPT adopts the same basic theory which has already been developed for at least 30 years with the

V. George (✉) · B. Deepak · S. Kumar · S. Kumar · S. Reddy
Department of Electrical and Electronics Engineering, M S Ramiah Institute of Technology, Bangalore, India
e-mail: victorgeorge@msrit.edu

S. Ashok
Department of Electrical Engineering, NIT Calicut, Kerala, India

S. N. Singh et al. (eds.), *Advances in Power and Control Engineering*, Lecture Notes in Electrical Engineering 609, https://doi.org/10.1007/978-981-15-0313-9_20

term inductive power transfer. It can eliminate the use of the wires and batteries, thus increasing the mobility, convenience, and safety of an electronic device for all users. WPT technology is developing rapidly in recent years. For a dynamic WPT system, the battery capacity of EVs with wireless charging could be reduced to 20% or less compared to EVs with conductive charging [1]. Dynamic charging system is more beneficial for large-scale EV systems, and the battery costs could be significantly reduced when the power track cost drops. There are different charging levels and charging modes for different types of vehicles depending on many factors [2]. Plug-in charging is one of the methods of charging EVs which is widely used. Public charging may be time-consuming as the queue at charging station can be huge during peak hours. Battery swapping is another technology used to reduce the waiting time at the charging stations. Swapping station is employed to change the discharged battery with a charged one and requires standardization of battery. Different techniques for wireless power transfer are discussed in [3]. Magnetic resonance coupling is best suited for this work. The performance of a wireless charging station is discussed in [4]. Oak Ridge National Laboratory (ORNL) demonstrated a 120-KW wireless charging system with 97% efficiency [5]. Even with all the advantages offered by dynamic charging of vehicle over static charging, it is still less popular globally. Smart communication technologies should be incorporated in dynamic charging scheme in order to make it more viable. Hence, the various technical aspects involved in dynamic charging of EVs and the data communication possibilities through IoT are analyzed. The various design aspects of the proposed system are discussed in the following sections.

2 Hardware Design of the Proposed System

2.1 High-Frequency Inverter

As only AC power can be transmitted from primary to secondary coil, the DC power obtained from the DC source has to be converted into AC power. For this job, an inverter is employed which converts DC to AC. A full-bridge inverter is preferred as it has higher efficiency. A full-bridge inverter can be rigged up using four switches. Since the frequency of AC considered is very high, a device with high switching frequency has to be selected and thus MOSFETs are used for this purpose.

Solar energy can be used as a DC source. Inverter efficiency depends on the switching losses and the conduction losses of the device and also the reverse recovery losses of the body diodes. The MOSFETs are switched on/off by the pulse width modulation (PWM) technique in which a digital signal is cycling off and on at a fast enough rate, and with a certain duty cycle. The minimum voltage when the gate capacitor is charged and the device can just about conduct is the threshold voltage (V_{TH}). For operating a MOSFET as a switch, a voltage sufficiently larger than V_{TH} should be applied between the gate and source/emitter terminals. Dead-band refers to

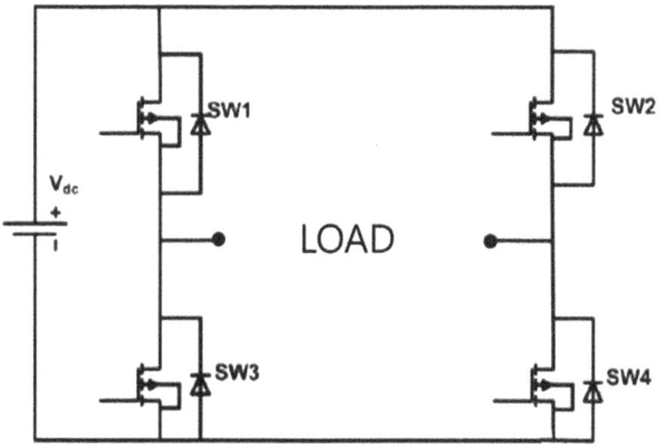

Fig. 1 H-bridge circuit

the discontinuities introduced into the reference waveform intentionally. This dead-band PWM technique increases the output voltage of the inverter and decreases the effective switching frequency and hence the switching losses are reduced as compared to the simple PWM technique. So, triggering pulses has to be given with proper dead-band and amplitude. Figure 1 shows the basic diagram of an H-bridge inverter circuit. The proper switching combination produces the positive and negative half cycles of the signal. A driver circuit is used to achieve the desired amplitude and proper dead-band is given using DSP module.

2.2 *Driver Circuit*

Driving a large power MOSFET directly might overheat and damage the control due to possible current overdraw in the circuit. The driver circuit is used to set the amplitude and frequency of the triggering pulses, thereby setting the frequency of the AC output. A gate driver IC IR2110 with high driving capabilities enables fast switching with rise and fall times of few microseconds [6]. DSP module C2000 32 is used to generate PWM signals that can provide a better dead-band for proper triggering of MOSFET. When a gate signal of some minimum amplitude is given to the MOSFET, it acts like a closed switch and the connection between drain and source is established. The triggering pulses given to the MOSFETs are given through the driver circuitry shown in Fig. 1. The design shown in Fig. 2a is for higher frequencies can be changed according to the application requirements. Pulses from pin numbers 1 and 7 are fed as gate signals to the connected MOSFETs. Pins 10 and 12 of U1 are fed with PWM pulses from a processor. This is needed in the hardware setup only. In simulation, two signal generators with required amplitude can be used [7]. The desired frequency for the test circuit can be given by making the corresponding time in the code of DSP module. DSP module is required to set the precise dead-band between the pulses. Figure 2b shows the screenshot of the code written to generate PWM pulses.

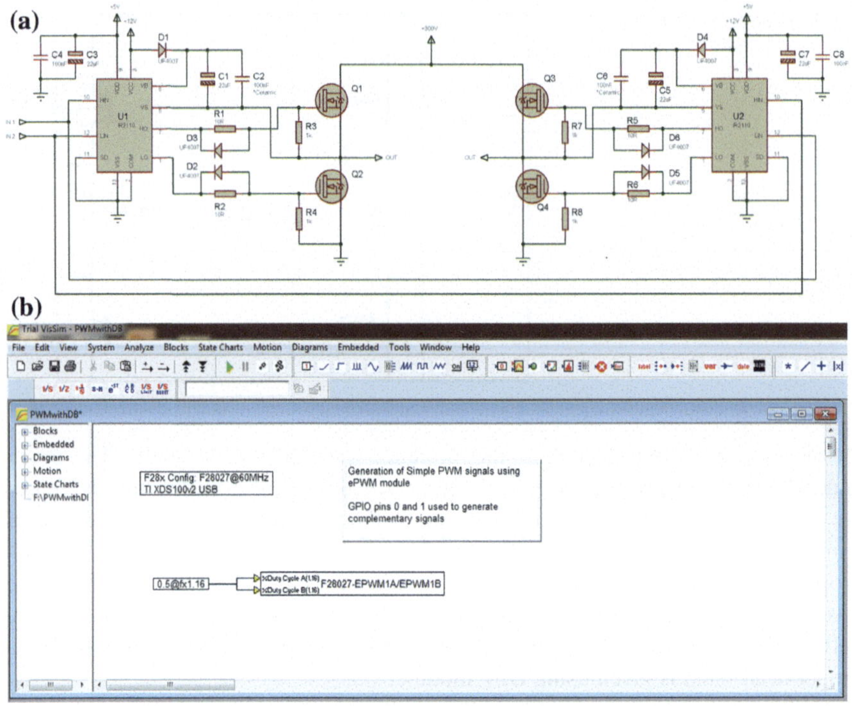

Fig. 2 **a** Driver circuit for full-bridge inverter (*Source* EEEIC, p. 2), **b** Screenshot of DSP code for generating PWM pulses

2.3 Transmitter and Receiver Coil Design

Skin effect is the major concern during the high-frequency operation of the wireless power transmission. Hence a special type of coil, called Litz wire, is used for the coil design which minimizes the skin effect and proximity effect. The transmitter coil is the one that is placed on the road and which is energized by the power obtained from the DC source. Its shape is chosen as rectangular spiral so that it can cover more portion of the road and gives enough time for the vehicle to cross over it and get charged during this time. The design calculations for the inductance of the transmitter coil [8] are illustrated below in Eq. (1).

$$L = \frac{\mu_0 . n^2}{\pi} . \left(b . \ln \frac{a}{r} + a . \ln \frac{b}{r} \right) \tag{1}$$

The transmitter coil designed has the following dimensions;

$a = 1.6$ m (length of the coil), $b = 0.3$ m (width of the coil), $n_1 = 3$ (number of turns of the coil), $r = 0.0003$ m (radius of the wire used). Substituting these values in Eq. (1), we get $L_1 = 33\ \mu\text{H}$. Three different transmitter coils are designed

Table 1 Different transmitter coil designs

Transmitter name	Length (a) in meters	Width (b) in meters	Number of turns n	Inductance (L) (μH)
$T1$	1.6	0.3	3	33
$T2$	0.4	0.1	5	30.4
$T3$	0.8	0.15	4	39

Table 2 Different receiver coil designs

Receiver	Diameter d in m	No. of turns n	Inductance L (μH)
$R1$	0.4	6	27
$R2$	0.2	6	26.6

with different dimensions in order to experiment the change in flux linkage with the receiver coil.

The receiver coil is the one that is placed at the bottom of the vehicle. It is energized by the power transferred from the transmitter coil. The shape of the receiver coil is chosen as circular spiral so that it can create more linkage between transmitter coils. The diameter of the receiver coil should be greater than the width of the transmitter coil. The design calculations for the receiver coil are illustrated below. The formula to calculate inductance of the circular spiral is given in Eq. (2).

$$L = \mu_0 \cdot R \cdot n^2 \cdot \left(\ln \frac{8R}{r} - 2 \right) \tag{2}$$

The dimensions of the receiver coil are as follows;

$R = 0.15$ m (radius of the coil), $n^2 = 6$ (number of turns of the coil), $r = 0.0003$ m (radius of the wire used), substituting these values in Eq. (2), we get $L_2 = 27\ \mu$H. The other receiver coil is designed with different dimensions in order to experiment the change in flux linkage with the transmitter coils. Tables 1 and 2 represent the different configurations of transmitter coil and receiver coil, respectively. Each of the transmitter coils is tested with each of the receiver coil to calculate the efficiency of power transfer. Figure 3 shows one of the receiver coils and the transmitter coils designed.

2.4 Compensation Network

To increase the power transfer efficiency, it is very important to operate the network at resonance. The compensation network is a circuit of capacitor filters, designed to operate the system in resonance [9]. The design steps are shown below.

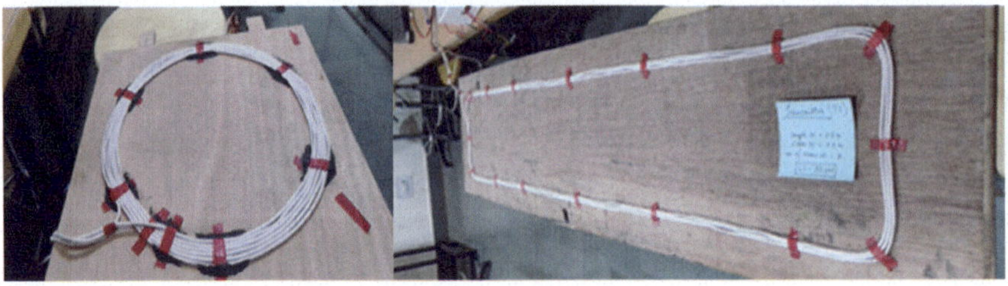

Fig. 3 Receiver coil ($R1$) and transmitter coil ($T1$)

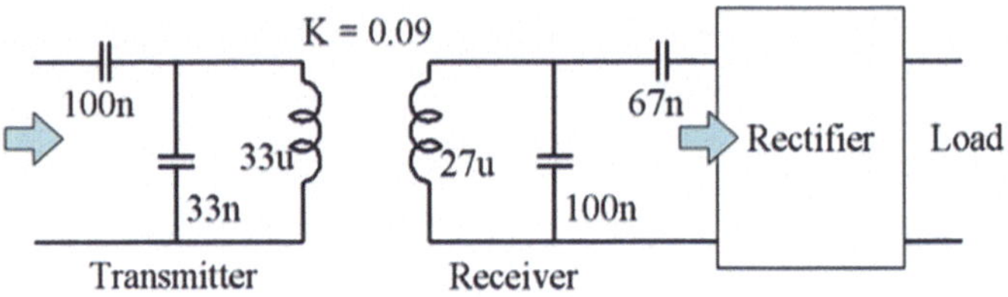

Fig. 4 Compensation network

The condition for resonance is $X_L = X_c$. The value of L_1 and operating frequency f is known, hence Eq. (3) can be used to determine the value of capacitors to be used.

$$C = \left(4\,\Pi^2 f^2 L\right)^{-1} \quad (3)$$

Primary-side capacitance $C = 126$ nF at 78 kHz frequency. So, $C_1 = 100$ nF, $C_1{}'$ $= 33$ nF. Similarly, secondary-side capacitance $C = 154$ nF at 78 kHz frequency. So, $C_2 = 67$ nF, $C_2{}' = 100$ nF. AC power output is observed after the secondary compensation network. The designed compensation network is shown in Fig. 4.

3 Simulation and Experiment Results

The rate of power transfer and the distance between the two transmitter coils is obtained through the simulation models done in Matlab/Simulink software. Figure 5 shows the functional block diagram of the Simulink model implemented to test the rate of power transfer for different distances between the two transmitter coils. The distance between the two coils $h = 0.1$. All the simulation models are obtained at a value $k = 0.09$ assuming the distance between the coils is 10 cm. The battery parameters are chosen by considering an electric vehicle model which is available in the market which is shown in Figs. 6 and 7. The battery ratings of Mitsubishi i-MIEV are replicated in the designed model. The model was tested for both 110 and 220 V supply. Figure 8 shows the implemented hardware circuit of the model. The coupling between the transmitter and receiver is greatly affected by the distance

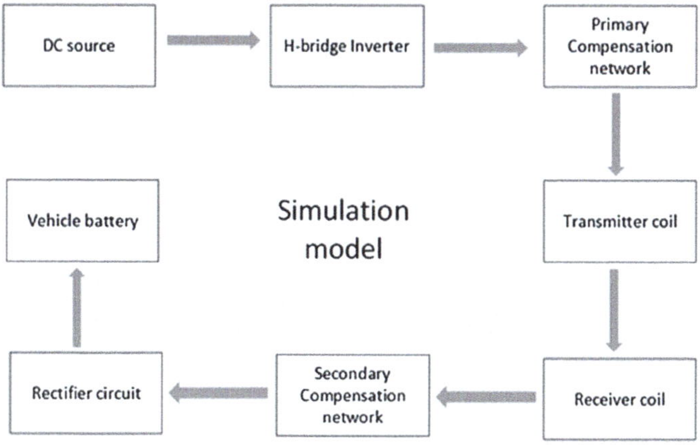

Fig. 5 Functional block diagram of the Simulink model used

Fig. 6 Vehicle battery parameters

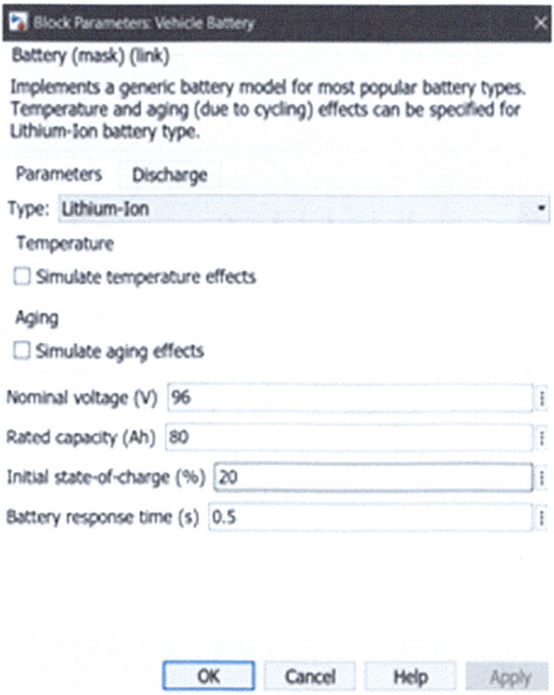

Fig. 7 Vehicle battery discharge parameters

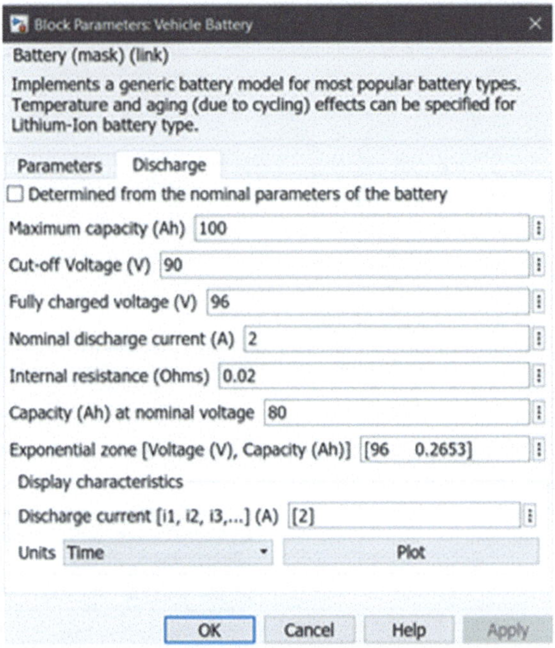

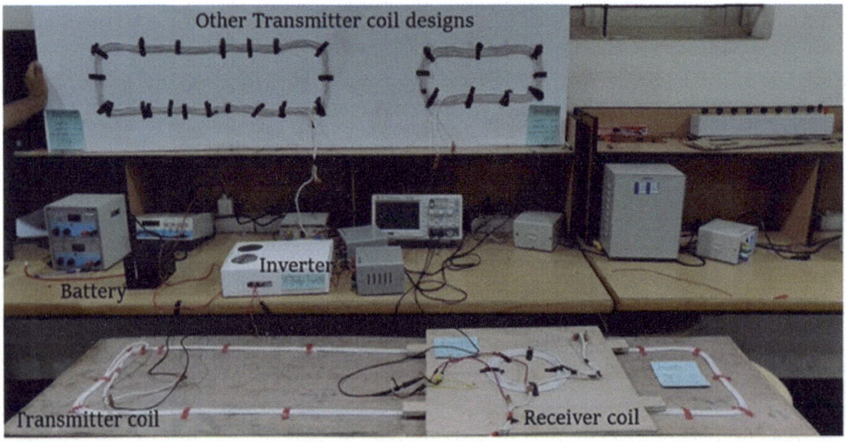

Fig. 8 Experimental setup to test wireless power transfer

Fig. 9 Variation of SoC of the 110 V battery and the output receiver current

Table 3 Experimental results of $T1$ and $R1$

Case	Frequency (kHz)	Distance b/w the coils (cm)	Supply voltage (V)	Receiver voltage (V)
A	78	10	12	8
B	20	10	12	3.2

between them. As distance between the two coils increases, the coupling coefficient value decreases. Equation (4) gives the formula to calculate coupling coefficient for different distances between the two coils [8].

$$k = \frac{2R}{a} . \left[\frac{R}{\sqrt{b.b + 4.h.h}} \right]^3 . \frac{n2}{n1} \quad (4)$$

where R, b, a, n_1, n_2 have the same meanings as in Eqs. (1) and (3).

In order to find the rate of power transfer, the change in state of charge (SoC) of the battery is observed. Figures 6 and 7 show the parameters of vehicle battery and its discharge. The efficiency of power transfer is calculated from the given input power and power obtained at the load is found to be more than 87%. The simulation results for the variation of SoC of the 110 V battery and the output receiver current obtained is shown in Fig. 9. But in hardware studies with the designed transmitter and receiver coils, the obtained output current was negligible. Hence, the design of the coils plays an important role in deciding the amount of power transfer along with the other designed circuit parameters.

Table 3 shows the experimental result obtained for transmitter ($T1$) and receiver ($R1$) at different frequencies. The corresponding output waveform at the inverter and the receiver for both the cases are shown in Fig. 10.

4 IoT-Based Infrastructure for Data Communication

The purpose of data communication is to continuously monitor the state of charge (SoC) of the electric vehicle battery to the cloud so that a control mechanism can

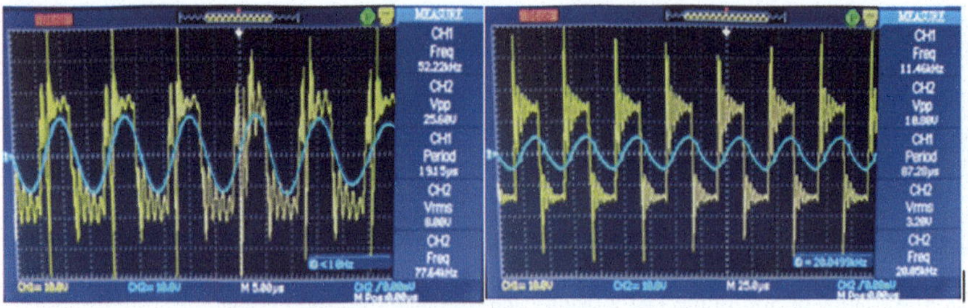

Fig. 10 Output waveform at inverter and receiver for case A and case B

monitor their charging level at any point of time and direct them real time either to charging stations or to battery swapping stations according to the choice of the user. Raspberry Pi 3 is used for the communication purpose which comes with a 1.2 GHz 64-bit quad-core processor, on-board Wi-fi, Bluetooth, and USB boot capabilities. LTC4150 C counter is used to measure the SoC of electric vehicle battery [10]. SMS API used instead of GSM module to send SMS to the user regarding the information about his location and link to the website where information regarding charging stations is given. SMS API saves serial port of Raspberry Pi which can be used for GPS module or any other component. SMS API is easy to use and reliable and saves memory space of the controller. It runs on code and does not require SIM card or any form of cellular service. An Application Programming Interface (API) Fast2SMS API is used to send bulk SMS to the users. The user ID is specified in the program which will appear as the name when SMS is received by the user. The user will be alerted by a special ID that his battery level has fallen below specified level and the same data will be sending to the cloud for further analysis.

Digital Ocean cloud services are used to update data from the electric vehicle to the website. GPS module is connected to the electric vehicle which continuously updates the location of the electric vehicle and receives the coordinates in the form of latitude and longitude. The values received by the GPS and the SoC of the Battery as read by the LTC 4150 are pushed to the cloud using Raspberry Pi. The cloud communication of the data is shown in Fig. 11. For the analysis of the acquired data, an integrated development environment (IDE) of python called "spyder" is used.

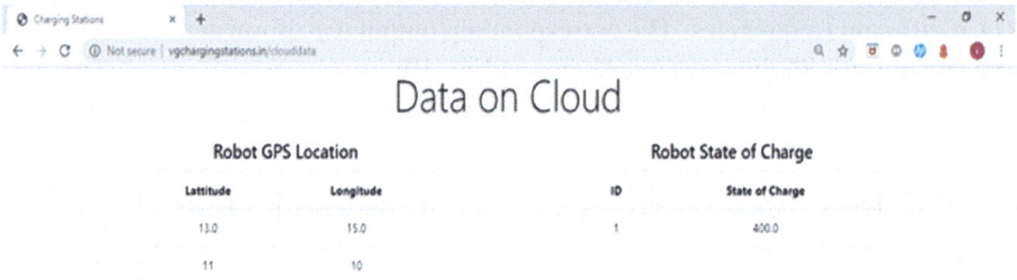

Fig. 11 Data communication to the cloud developed

Python is emerging as one of the best platforms for data analytics. It can handle huge sets of data with lesser lines of code. Spyder, in particular, has many libraries such as numpy and pandas that support importing and analysis of data. The data imported from the cloud is changed to data frames of different parameters that can be fed to the algorithm for processing. The simulation study carried out in analyzing various design aspects of the dynamic wireless power transfer scheme shows the maximum power transfer rate as 87%.

All the designed transmitter and receiver coils are tested for the input and output parameters. In all the cases, the actual current value was found to be negligible. The coil alignment and design are very crucial in developing the power transfer between transmitter and receiver. The various studies carried out show the design challenges involved in the on-road dynamic charging of electric vehicles. The inclusion of a database in the cloud can be used to make the model more interactive and efficient in predicting the short-term load at a particular charging station as well as analyzing the driving characteristics of the users. More research in this area will make this concept more economical and practical in near future.

References

1. Tavakoli R, Pantic Z (2018) Analysis, design, and demonstration of a 25-kW dynamic wireless charging system for roadway electric vehicles. IEEE J Emerging Sel Top Power Electron 6:1378–1393
2. Throngnumchai K, Kai T, Minagawa Y (2011) A study on receiver circuit topology of a cordless battery charger for electric vehicles. IEEE Energy Convers Congr Exposition, Phoenix, AZ, 843–850
3. Jiwariyavej V, Imura T, Hori Y (2015) Coupling coefficients estimation of wireless power transfer system via magnetic resonance coupling using information from either side of the system. IEEE J Emerging Sel Top Power Electron 3:191–200
4. Jeong S, Jang YJ, Kum D (2015) Economic analysis of the dynamic charging electric vehicle. IEEE Trans Power Electron 30:6368–6377
5. Oak Ridge National Laboratory. https://www.ornl.gov/news/ornl
6. Mohamed AAS, Allen D, Youssef T, Mohammed O (2016) Optimal design of high frequency H-bridge inverter for wireless power transfer systems in EV applications. In: IEEE 16th international conference on environment and electrical engineering (EEEIC), Florence, pp 1–6
7. https://www.tes.com/lessons/xki0zIClZxifsw/p5-and-p6-revision
8. Throngnumchai K, Hanamura A, Naruse Y, Takeda K (2013) Design and evaluation of a wireless power transfer system with road embedded transmitter coils for dynamic charging of electric vehicles. In: World electric vehicle symposium and exhibition (EVS27), Barcelona, pp 1–10
9. http://tahmidmc.blogspot.com/2013/01/using-high-low-side-driver-ir2110, 1 (2013)
10. Ng KS, Moo CS, Chen YP, Hsieh YC (2009) Enhanced Coulomb counting for estimating state-of-charge and state-of-health of lithium-ion batteries. Appl Energy 86(9):1506–1511

Author Index

S. N. Singh et al. (eds.), *Advances in Power and Control Engineering*, Lecture Notes in Electrical Engineering 609, https://doi.org/10.1007/978-981-15-0313-9

The manufacturer's authorised representative in the EU is Springer Nature Customer Service Centre GmbH, Europaplatz 3, 69115 Heidelberg, Germany. If you have any concerns regarding our products, please contact ProductSafety@springernature.com

Printed and bound by CPI Group (UK) Ltd, Croydon, CR0 4YY
30/09/2026
02225452-0009